THE
TELESCOPE
IN THE
ICE

THE

TELESCOPE

IN THE

ICE

Inventing a New Astronomy
at the South Pole

Mark Bowen

ST. MARTIN'S PRESS 🏠 NEW YORK

Excerpts from "Open Letter to the Group of Radioactive Persons at the Conference of the District Society in Tübingen" originally appeared in "On the Earlier and More Recent History of the Neutrino," *Writings on Physics and Philosophy* (1994), page 198, Wolfgang Pauli © Springer-Verlag Berlin Heidelberg, and are used with permission of Springer.

www.stmartins.com

Designed by Meryl Sussman Levavi

Library of Congress Cataloging-in-Publication Data

Names: Bowen, Mark (Mark Stander), author.
Title: The telescope in the ice : inventing a new astronomy at the South Pole / Mark Bowen.
Description: New York : St. Martins Press, [2017] | Includes bibliographical references and index.
Identifiers: LCCN 2017026874| ISBN 9781137280084 (hardcover) | ISBN 9781466878983 (ebook)
Subjects: LCSH: Neutrinos. | IceCube South Pole Neutrino Observatory.
Classification: LCC QC793.5.N42 B69 2017 | DDC 522/.686—dc23
LC record available at https://lccn.loc.gov/2017026874

Our books may be purchased in bulk for promotional, educational, or business use. Please contact your local bookseller or the Macmillan Corporate and Premium Sales Department at 1-800-221-7945, extension 5442, or by email at MacmillanSpecialMarkets@macmillan.com.

First Edition: November 2017

10 9 8 7 6 5 4 3 2 1

In Memory of
Bruce Koci and Per Olof Hulth

I readily believe that there are more invisible than visible things in the universe. But who shall describe for us their families, their ranks, relationships, distinguishing features and functions? What do they do? Where do they live? The human mind has always circled about knowledge of these things, but never attained it.

—THOMAS BURNET (1692)
(From the Latin motto to Coleridge's
"The Rime of the Ancient Mariner")

Contents

Part IV
The Real Thing

Illustrations

Introduction: Making Mistakes

The universe can't exist the way it is without the neutrinos, but they seem to be in their own separate universe, and we're trying to actually make contact with that otherworldly universe of neutrinos. And as a physicist, even though I understand it mathematically and I understand it intellectually, it still hits me in the gut that there is something here around surrounding me, almost like some kind of spirit or god that I can't touch, but I can measure it. I can make a measurement. It's like measuring the spirit world or something like that.

—PETER GORHAM

IN NOVEMBER 2013, THE INTERNATIONAL COLLABORATION THAT OPERates the IceCube Neutrino Observatory announced that they had detected high-energy neutrinos coming from outer space. This heralded the birth of a new form of astronomy, based not on the usual cosmic messenger, light, but on perhaps the strangest of the known elementary particles, the neutrino. It was also the culmination of a quest that had first fired the imagination of a small group of visionaries more than fifty years earlier and seen many heroic attempts and failures along the way.

Part of the reason this journey has taken so long is that it takes an unusual telescope to see an unusual particle. IceCube is unlike any other telescope you've ever seen or heard of, and in fact no one ever will see it, because it's buried more than a mile deep in the ice at the geographic South Pole. The collaborators couldn't even see it while they were building it. Francis Halzen, the Belgian theorist at the University of Wisconsin who dreamed up the idea, says it was like building a telescope in a darkroom.

This instrument doesn't employ lenses and mirrors in the fashion of the usual telescope. Presently, and it may grow, it consists of eighty-six kilometer-long "strings" of unadorned light detectors, housed in spherical glass pressure vessels about the size of basketballs. These "strings of pearls" have been lowered into eighty-six two-and-a-half-kilometer-deep holes that were drilled in the ice by a gargantuan hot water drill and allowed to freeze in place. Thus, the topmost pearls are one and a half kilometers—or about a mile—down. The holes have been drilled in a hexagonal grid pattern that covers a square kilometer on the surface of the ice. Hence, the more than five thousand detectors in this unique device monitor about one cubic kilometer, or a billion tons, of what the scientists were thrilled to discover to be very clear, deep Antarctic ice. It is the clearest natural substance known, clearer even than diamond.

Scientific American once called this telescope the "weirdest" of the seven wonders of modern astronomy. And perhaps the weirdest thing about it is that it doesn't look *up* at the southern sky from its location on the bottom of the world; it looks *down* into the ice. IceCube is designed to look through the planet at the northern sky. Since the neutrino is the only known particle that can pass all the way through a planet without being absorbed or deflected off course, any particle that reaches this particular cube of ice from the northern direction must be a neutrino. The instrument uses the Earth as a shield to block other types of particle, which might create a false signal.

The reason the neutrino can pass so easily through a planet is that it doesn't like to show its face. It is sometimes known as the ghost particle. It may be the most plentiful particle in the universe—several hundred billion will pass through your eyeballs by the time you finish reading this sentence—but it is rarely seen, and it won't hurt your eyes, because it barely interacts with any kind of matter. This makes it very hard to detect. As Nobel laureate and amateur stand-up comedian Leon Lederman once said, "A particle that reacted with nothing whatever could never be detected. It would be a fiction. The neutrino is barely a fact." Your average neutrino will pass unscathed—and therefore undetected—through a slab of lead one light year, or six trillion miles, thick. Thus it has no problem passing through the Earth, which is considerably less dense than lead and less than paper-thin in comparison to a light year, and most will pass right on through IceCube as well. Every once in a while, however, one will react with the ice in or around the detector or the bedrock below it and produce a charged particle, which will speed along in the same direction as its parent neutrino, dragging a cone of pale blue light along with it. IceCube's light sensors pick up this light, and by watching the way it passes through the three-dimensional grid of detectors, the scientists can determine the direction of the charged particle and the direction of its parent neutrino, in turn. This makes IceCube a telescope.

As it happens, the reticence that makes the little particle so hard to detect has the beneficial side effect of making it a wonderful complement to light when it comes to astronomy. Since the neutrino can pass through extremely dense media that are opaque to all wavelengths of light, it can carry information from regions of space that are inaccessible to the usual telescope, such as the interiors of stars—including the exploding ones known as supernovae—or regions of our galaxy that are obscured by interstellar dust—the black hole at our galaxy's core, for example.

One motivation for inventing this new astronomy is to see into the inner workings of the most violent events in the universe: supernovae, active galactic nuclei, supernova remnants, gamma ray bursters, colliding galaxies, and other strange beasts, some not yet imagined. The scientific possibilities also extend to cosmology and the detection of the mysterious and so far unseen cold dark matter, which constitutes most of the mass of the universe. They reach into pure particle physics as well, since all the violent creatures just named are basically huge particle accelerators, operating by the same basic principles as the manmade variety here on Earth—including the multi-billion-dollar Large Hadron Collider, which produced evidence for the Higgs boson in 2012—but on a vastly larger scale.

The neutrino itself has become a focus of particle physics in recent years, since in 1998 it produced the first and still only chink in the armor of the standard model of particle physics. This is the theoretical framework that describes the building blocks of matter, the elementary particles, and how they interact with each other through three of the four fundamental forces: the weak nuclear force, the strong nuclear force, and the electromagnetic force. The standard model, which was constructed in the 1970s, is turning out to be so successful that it's beginning to feel like a straightjacket. With the discovery of the Higgs, which was the last standard-model particle remaining to be detected, it's looking as though there's not much left to discover, and physicists don't like things to be tied so neatly with a bow. They're always looking for something new, and the surprising behavior of the neutrino suggests unknown phenomena yet to be explored. For the heart of physics, and indeed all science, is pure exploration.

This brings us to the main reason for building this unusual instrument. The newborn field of neutrino astronomy has opened a new window on the universe, and rarely in the history of astronomy has such a new window *not* led to a discovery that was unimaginable beforehand. Galileo is the classic example.

The first optical telescopes were developed in Flanders for merchants getting a jump on the market by taking advance inventories of the goods on ships as they approached across the English Channel. Galileo used his superior understanding of optics and math to craft a better one, which he presented to the Venetian Doge as a tool of war. A few months later, he trained another at the Moon on a clear night when Jupiter, the second brightest object in the sky, happened to be floating just above it and to the right. Thus he discovered the four "Medicean stars," now known as moons, heretically orbiting the planet, and got himself in trouble.

In 1965, Arno Penzias and Robert Wilson, two physicists at Bell Telephone Laboratories, made an unexpected discovery while they were designing ground-based radio antennae for communications satellites. In order to test a horn-shaped antenna they had designed to be ultra-quiet, they aimed it at what they thought was empty space and were surprised to find that it always picked up a small amount of "noise" no matter how painstaking their design. The noise turned out to be a real signal: the Cosmic Microwave Background Radiation, the afterglow of the Big Bang, the explosion out of nothingness that brought this universe into being about fourteen billion years ago. This discovery transformed the Big Bang and cosmology in general from objects of ridicule into subjects for precise study. It also illustrates another aspect of discovery: the scientist's mind needs to be *prepared* in order to interpret what he or she measures or "sees"—or even to be able to see it. Theories of the Big Bang and its microwave afterglow had been gestating for decades by the time Penzias and Wilson made their measurement. They won the Nobel Prize not simply for *finding* a signal but for interpreting it with the knowledge or "eyes" of their day. This leapfrogging between theory and experiment is what propels science forward. Sometimes theory is out in front, sometimes the accumulated weight of unexplained experimental evidence prompts new theories or even paradigm shifts. And these developments can take decades, as we shall see in this book.

After the Nuclear Test Ban Treaty was signed in 1963, the U.S. Department of Defense began sending satellites into orbit in order to verify that the Soviets weren't violating the treaty by testing bombs in space, underwater, or on the Moon. The idea was to sense the gamma rays (invisible light at shorter wavelengths than X-rays) given off by the blasts. The satellites never detected any of those, but they did detect numerous "treaty violations" in deep space: brief and astoundingly intense bursts of gamma rays in the distant cosmos. The scientific community became aware of this discovery a few years later when the data was declassified, and the enigmatic sources of the bursts were given the noncommittal name "gamma ray bursters." Over the course of one to twenty seconds or so, GRBs give off about as much light as that given off by all the other stars and galaxies in the known universe. Theory holds that they should give off neutrinos, too, so they are of great interest to IceCube.

The astrophysicist Kenneth Lang observes that "our celestial science seems to be primarily instrument-driven, guided by unanticipated discoveries with unique telescopes and novel detection equipment. . . . [W]e can be certain that the observed universe is just a modest fraction of what remains to be discovered."

The dream of making a big discovery is *one* of the things that drives the working scientist; however, media coverage and prestigious awards like the Nobel give the layperson a distorted sense of its importance, I think. The clichéd emphasis on supposedly earthshattering results is especially pernicious in physics, where it would seem that a discovery that "changes our view of the universe" takes place every few months. Some such over-the-top phrase is almost inevitably employed in any newspaper or magazine article describing even the most insignificant result, and the physicists who have now taken to trumpeting their findings in press conferences before publishing them in the peer-reviewed literature share a good part of the blame. In reality, discoveries on the order of the theory of relativity or Darwinian evolution are exceedingly rare.

But there is some truth in the noise of science journalism, nevertheless. The fact is that scientists tend to enjoy their work more than most, for the main reason that what really gets them out of bed in the morning is the thrill of the chase. They come to some minor realization, solve some esoteric technical problem, or shine a light into some new dark corner of the territory they've been exploring almost every day. More than half the time they're wrong, but at least they're on track. And finding out they were wrong—going from confusion to clarity—can be just as thrilling as finding they were right.

Francis Halzen tells me that the late John Bahcall, a respected neutrino theorist at the Institute for Advanced Study in Princeton, used to say that "physicists have two deep dark secrets that they hide from non-scientists under lock and key. The first is that physics does not progress logically; it's a series of mishaps. . . . And the second is that they're having so much fun that they'd do it even if they weren't getting paid."

My goal is to show you the truth of Bahcall's secrets by taking you inside an experiment that has provided more than enough of the kind of riches that physicists live for. I've had a front-row seat for about twenty years.

✦ ✦ ✦

I was introduced to IceCube in 1997 by Bruce Koci, the master driller. It happened in a roundabout way.

One sunny day in June of that year, I received a call at my home near Boston from an editor at *Natural History* magazine. She asked if I might be free to write an article about a paleoclimatologist named Lonnie Thompson, who retrieves ice cores from high mountain glaciers in order to study past climates and climate change.

A week later I flew to La Paz, Bolivia, and a week after that I reached base camp below the highest mountain in the country, Nevado Sajama, a dormant 21,500-foot volcano capped by a round snow dome—the perfect shape, I would soon learn, for ice core drilling. Lonnie and his team had been working on the summit for about two weeks at that point, and within the next few days they were planning to use a hot-air balloon to fly the first ice core segments down to a waiting freezer truck on the plain at the base of the mountain. Figuring it was my journalistic responsibility to get to the top in time to see this, I climbed a bit too quickly given the altitude and panted into the drilling camp in golden westerly light on the afternoon before the scheduled liftoff, carrying only a daypack and no sleeping bag.

It was immediately apparent that I had stumbled upon an extraordinary world—above and beyond the unusual work, the brutal cold, and the breathtaking views all around under a spanking azure sky. The drilling team was blasé about the surroundings. They were there to work, not moon about the beauty of the place, and their dedication was palpable. They knew each other well. They had drilled together for decades in similar locations around the world and had lived on glaciers like this for years all told. Their conversation didn't stray beyond the tasks at hand. There was a meditative silence on that mountaintop as they used their elegant solar-powered drill to mine ice core segments one meter at a time, log them into a lab notebook, pack them in insulated boxes, and bury them in a snow pit to await the descent to lower altitudes and the ultimate destination of Lonnie's walk-in freezer at Ohio State University, half a world away.

Bruce didn't stand out at first. For one thing, that wasn't his way. But I will never forget my first interaction with him, which I described in my 2005 book, *Thin Ice*:

> Soon the sun sank too low to provide power. The pyramidal shadow of the mountain stretched across the desert to meet the horizon. The sky turned purple then gray. We shoe-horned ourselves into the snow cave, a rectangular hole fifteen feet long and five feet wide—too low for standing—which served as dining room, living room, and lounge. We sat along opposite walls on benches cut in the snow, knee to knee, thigh to thigh, backs and seats against cold white surfaces, sipping tea and soup in near silence.
>
> After dinner I decided to point out that I had no gear. Bruce Koci [pronounced "ko'-see"] then forced himself slowly to his feet and climbed into the twilight. A few minutes later he called from above, holding four of the six-inch foam pads they use to insulate the ice and two of the fluffiest down sleeping bags I have ever seen—much warmer than my own back at High Camp. He helped me lay out the pads in a tent and pile the bags on top.
>
> "There," he said. "Wrap yourself in these like a fox in his tail."

This was the first of many acts of generosity, toward others and myself, that I saw Bruce perform in the years that I knew him.

At that point he was approaching the end of his twenty-year tenure as Lonnie's lead driller. The two had co-invented the solar-powered drilling technology in the early 1980s and had retrieved the world's first high-altitude ice cores together on Peru's Quelccaya Ice Cap, not far to the north of Sajama, in 1983. Bruce was gentle, quiet, and humble, clearly one-of-a-kind, and had a deep spiritual connection to the natural world. Once the expedition was over and my article was complete, I felt a desire to keep in touch, both with him and with Lonnie.

It wasn't easy at first, as they spent only two weeks at home—Bruce in Alaska, Lonnie in Ohio—between the expedition to Bolivia and another three-month effort in Tibet. Finally, in mid-November, Bruce sent me an e-mail:

> Mark;
>
> Just got back from Tibet, actually about 3 weeks ago and finally got brave enough to attack the many hundreds of messages. Sajama was quite successful as was Tibet with several cores to bedrock. . . . I am off to the South Pole soon to drill 2400 meter deep holes looking for neutrinos. This project is really interesting and at the cutting edge of high energy astrophysics so we get to make a lot of mistakes. . . .
>
> Peace;
>
> BK

I asked him about the physics project. It was named AMANDA, he told me, the Antarctic Muon And Neutrino Detector Array. "The contrast of the two projects is pretty interesting in that with Lonnie we go in with light equipment and bring out heavy ice core. With AMANDA we have a huge drill (200,000 lb.) and the data doesn't weigh anything."

He put me in touch with Francis Halzen and his colleague Bob Morse at the University of Wisconsin, the lead institution in the AMANDA collaboration. They were an affable and complementary pair: Francis the theorist and Bob the experimentalist. Bob was lead "man on the ground" in Antarctica and principal investigator, or PI, on the grant from the National Science Foundation that provided most of the funding for the project. Oddly, Francis, who had dreamed the whole thing up, was designated co-PI, meaning he didn't hold the level of responsibility that Bob did—on paper at least. But it would quickly become apparent that there is almost always more to Francis than meets the eye. Regardless of the hierarchy, he was in fact intellectual leader and *spiritus rector* of the project, and his was the head on the chopping block if the project happened to fail.

Be that as it may, both were relaxed, friendly, and quite open. They invited me to attend a collaboration meeting that would be held in conjunction with an open workshop at the University of California, Irvine, in the spring of 1998.

The collaboration was relatively small at that point in time. Wisconsin, Cal Berkeley, and Irvine had been the founding institutions in 1990; a Swedish contingent from Stockholm and Uppsala Universities had joined in 1992; and a group from a small high-energy physics institute in the former East Germany had joined two years after that. They held private collaboration meetings about three times a year, combining them with workshops every once in a while.

The purpose of this workshop was to gather together theorists with ideas about astrophysical sources of neutrinos, other neutrino astronomers, scientists and engineers with relevant technologies to bring to the table, experts in South Pole logistics, and representatives of the funding agencies to "initiate the conceptual design of IceCube—a kilometer-scale neutrino facility in Antarctica." The idea was to "extend the AMANDA technique to kilometer-scale dimensions." This, as it happened, was the first meeting ever held specifically about IceCube.

The reason they needed to extend AMANDA was that the sensitivity and angular resolving power of a neutrino telescope is directly related to its size. Owing to the neutrino's aloofness, it is necessary to monitor as large a volume of ice as possible, because the larger the volume the more likely it is that a neutrino will deign to interact inside it, die, and give birth to a visible child. Theory held that the minimum size needed for so-called discovery potential, the ability to observe the exotic cosmic accelerators that are expected to emit neutrinos, was about one cubic kilometer. AMANDA was the proof of concept for IceCube, the test as to whether it would be at all possible to see neutrinos in deep Antarctic ice. It was also the test bed for the technology that would be used in the larger instrument and the opportunity to explore the two-mile-thick East Antarctic Ice Sheet—no small task. There was dizzying technology involved, especially in Bruce Koci's drill.

At the same time, AMANDA was not exactly small. The collaboration had been working on it for about eight years. The array comprised a cylinder 120 meters in diameter, 500 meters high (a bit taller than the Eiffel Tower), and more than a mile deep, monitoring about six million tons of ice.

I soon discovered that not everyone in the AMANDA collaboration was as accommodating as Bob and Francis. The night before the spring meeting, its organizer, Steven Barwick, a professor at Irvine, held a party at his home to welcome us to town. As I thanked him on my way out, Steve informed me that a few members of the collaboration were uncomfortable with my attending the meeting and that I would not be allowed in. I was welcome to attend the open workshop, but I'd have to cool my heels for a few days until it started. I rationalized this disappointment with the thought that the most interesting conversations usually take place on the periphery of such gatherings anyway, especially after a couple of sips of alcohol.

Indeed, after the party, in the bar of our hotel, I joined Francis Halzen for a nightcap. He was a compact, youthful-looking man, who came across, quite simply, as one of the happiest people I had ever met.

Francis is not given to long pronouncements. He tends to speak in aphorisms, accompanied by a twinkle of the eye and the look of someone who's about to let you in on a joke. One of his oldest colleagues and friends, Tom Gaisser from the University of Delaware, notes the "oracular, sibyl-like" quality of the comments Francis makes during group phone calls: they can be taken in several ways, and they usually break the logjam. He is also usually two or three steps ahead of most everyone else in the room. Francis has a deep voice, speaks English with a rich Flemish accent, and will often precede a point he is about to make by saying, "I don't have to tell you this, because from what I just said it's obvious," when it isn't obvious to me at all. I liked him right away.

Many physicists seem to be perpetually high on mental activity, the bouncing around of neurons and the linking of synapses in the brain. This exhilarating and pleasurable sensation is the general buzz at these meetings, and Francis seems to thrive on it more than most. In those days, his science talk tended to be the high point of a collaboration meeting. It was generally about some physics insight he'd recently experienced, and as he spoke he was subject to surges of realization and excitement in which the ideas splashed across his mind so fast that his mouth couldn't keep up. He'd sputter like a motor burning too much oxygen with its fuel, and his words would come out in bursts.

As we sipped our drinks that evening, he came out with one of his standard, three-steps-ahead-of-the-game remarks. He assured me that even though AMANDA had not yet detected a single bona fide neutrino, the best part of the story was already over, and that had been the discovery that the instrument would actually work. This certainty had arisen in *his* mind, at least, two years earlier, when the collaboration had discovered that the deep ice under the pole is remarkably clear. To a theorist like himself, the rest was detail; at that point it was obvious the instrument would work. It had taken two years to bring the rest of the physics community along—not to mention certain recalcitrant members of the collaboration—but the workshop that would be held in a few days was a sign that the wider community, including the all-important funding agencies, was preparing to endorse the construction of IceCube.

It was an historic moment, he told me. The dream of building a neutrino telescope had been around for forty years, since the late 1950s. There had been many attempts in the intervening decades, and two or three other feasibility studies still limped along, but this was the first to show enough promise to merit a move toward the real thing.

Francis mentioned DUMAND once or twice, and I would hear this name many times in the course of the week. The Deep Underwater Muon And Neutrino Detector was the first pioneering—and by this time, unfortunately, notorious—attempt at realizing the dream. Its DNA is still found everywhere in neutrino astronomy. Several DUMAND veterans worked on AMANDA, and at least one still works on IceCube. This swashbuckling project, first funded in 1980, was supposed to have been located three miles deep on the floor of the Pacific Ocean, nineteen miles off the coast of the big island of Hawaii. It used water as the detection medium, rather than ice. After a long series of mishaps, DUMAND had been canceled just two years earlier, in 1996. There is a half-decent chance that it detected one up-going neutrino in the sixteen years of its existence. The other competing efforts were also water-based.

I was surprised to hear Francis say that it had proven easier to build one of these unlikely gadgets at the frigid South Pole than in the warmest tropical waters, for the simple reason that ice allows you to walk on your experiment. The central challenge for the water-based instruments was and remains ocean engineering. As one of the DUMAND pioneers wrote twelve years into the project, by which time they had still not placed even a preliminary design on the ocean floor, "Sailors have long known what we learned painfully, that the sea is an unforgiving medium."

By contrast, even though absurdly cold weather and six months of darkness prevented them from working more than four months a year at the pole, the AMANDA collaboration placed their first working detectors in the ice on their second try, two years into the project, and constructed their current working prototype within five years.

As our conversation ended, Francis suggested characteristically that I ignore Steve Barwick's advice and show up anyway the next morning. I did and Steve promptly kicked me out.

I took advantage of my free time over the next few days to go running on Huntington Beach and catch up on my sleep. I hung out with the Amandroids, as they called themselves, during breaks and meals, and most profitably for after-dinner drinks. I spent as much time as I could with Bruce Koci, who didn't look a whole lot different in squeaky-clean southern

California than he had on the mountaintop in Bolivia. He showed up every morning in a pair of skin-tight blue jeans and beaten-up running shoes, wearing a bulky, un-tucked chamois shirt and carrying a venerable day-pack that had seen many days at altitude. Never in the years that I knew Bruce did I see clear evidence that he had combed his hair.

For environmental reasons, he insisted on walking back and forth to the conference venue from his hotel. It rained on one or two mornings, and on those days he protected his rumpled shirt with a formerly blue Gore-Tex jacket that had been bleached nearly white by months if not years of sun, wind, and rain. One morning, he was stopped by the campus police, who suspected he was a homeless person seeking shelter in a campus building.

Bruce's main reason for coming to these meetings was to keep himself current on the science in order to psyche himself up for the rigors of drill-ing. He was one of those rare engineers who realized that it didn't matter if he built his machines up to spec if they couldn't do what the scientists needed them to, especially since, much of the time, neither they nor he knew what the specs needed to be. He once referred to his relationship with Lonnie Thompson as "one of the longest-standing friendly relationships between science and engineering ever."

He represented one side of an interesting contrast that I noticed partic-ularly at the workshop that followed the collaboration meeting. On one hand were the theorists, the inspiring John Bahcall among them, present-ing heady ideas about the cosmic accelerators IceCube might eventually observe. While Bahcall's feet seemed firmly on the ground, the other theo-rists had a febrile, untethered quality. The models they proposed seemed like examples of imagination gone wild, as if they were throwing darts at a wall, hoping that a decade or so later, when IceCube might finally produce results, one of those darts might nail an experimental finding. The lucky winner could then claim that she'd made an accurate prediction, ignoring the fact that she might have made several others that were disproven at the same time.

On the other hand were the experimentalists, who had a gritty, perse-vering quality like Bruce. It took me some time to realize that the AMANDA collaborators had to have been exhausted just then, since their frantic, an-nual, four-month Antarctic field season had ended only about a month ear-lier. It had been a successful season, but there had been plenty of frustration and interpersonal strife mixed in. They were building the world's largest particle physics detector, after all, dealing with the innumerable details involved in getting such an enormous and infernal gadget to work in one of the most inhospitable places on the planet. Unlike the theorists, they would not be bouncing to the next interesting problem in a week or two. IceCube had not even been designed yet. It wouldn't be completed for another twelve years.

On the last morning of the collaboration meeting, as the day's proceedings were to about begin, I joined the Amandroids for a self-serve breakfast of bagels, juice, and coffee in the hall outside the meeting rooms. Steve Barwick approached and said, "Mark, the collaboration has decided to give you a chance to make your case. We'd like you to give a presentation."

"When?" I asked.

"Right now."

First, I apologized for not having a set of overhead transparencies (this was before the days of PowerPoint). That brought a laugh. Then I explained that I thought I understood their needs and concerns: I would not leak preliminary results or pass negative rumors, and I believed I had a vested interest in their success. A professor from Irvine took issue with that one, referring to the book *Nobel Dreams* (subtitle: *Power, Deceit and the Ultimate Experiment*), which paints an unflattering portrait of Nobel Laureate Carlo Rubbia, a friend, or at least a colleague, of many in the room. I explained that it would be a long time before anything would appear in print (little did I know . . .). They asked me to leave the room for a while and then invited me back and welcomed me into the collaboration.

I've had open access ever since—for several years, in fact, I had total access: I even attended the closed meetings of the principal investigators. And at the end of 1999 I worked with Bruce Koci, drilling ice at the South Pole.

Those who were there generally agree that AMANDA was more fun than IceCube. It takes a looser mindset and a more risk-taking attitude to get a project off the ground than to engineer it to perfection (although that phase of the project was inspiring as well). The exploration is more wide-ranging. In some sense your job is to make mistakes. There was more derring-do on the ice during AMANDA—no one had heard of safety protocols or standard operating procedures—and Amundsen-Scott South Pole station, like so many other places, was more of a frontier outpost in those days. It is fair to say that AMANDA is the heart of this story. Francis Halzen says that AMANDA was "how neutrino astronomy was born."

It turned out that I had also met this close-knit tribe at a propitious time. For Francis wasn't quite right when he told me the best part was already over. The next six months were probably the most exciting in the history of this project, even including IceCube's groundbreaking discovery fifteen years later.

Part I

The
Birth and Youth
of the Neutrino

1. This Crazy Child

Physicists are more interesting than physics.

—ROBERT MILLIKAN

THE PARTICLE NOW KNOWN AS THE NEUTRINO FIRST APPEARED IN THE mind of the Viennese physicist Wolfgang Pauli sometime near the end of 1930.

This was about the midpoint of what was arguably the most exciting eight-year period in the history of science, as great thinkers such as Niels Bohr, Werner Heisenberg, Erwin Schrödinger, Max Born, Paul Dirac, Albert Einstein (protesting all the way), and Pauli himself addressed the bewildering puzzles of the atom and crafted the modern theory of quantum mechanics. In 1930, the quest was shifting from atomic structure to the next smallest length scale, the nucleus.

Pauli was thirty years old. He had been born in 1900—the same year, coincidentally, in which Max Planck discovered a granularity in the energy carried by a certain type of radiation, his so-called quantum of action, and opened the Pandora's Box of the atomic world (although Pauli, as we shall see, did not necessarily believe in coincidence).

Recognized as a prodigy in math and physics as a child, Pauli made his first impression on the wider world just after graduating from high school, when he wrote three papers on Einstein's mathematically sophisticated general theory of relativity, which the great man had completed only three years earlier. Pauli then went to the University of Munich to study under Arnold Sommerfeld, who was perhaps the leading authority on the "old" Bohr-Sommerfeld quantum theory, which Bohr had introduced in 1913. Pauli earned his doctorate, summa cum laude, in the minimum time allowed, three years, with a thesis on molecular hydrogen that was probably the most ambitious application of the Bohr-Sommerfeld theory ever attempted—and somehow found time to publish a magisterial, 237-page treatise on relativity at about same time. The treatise evoked the following praise from Einstein himself:

> No one studying this mature, grandly conceived work would believe that the author is a man of twenty-one. One wonders what to admire most, the psychological understanding for the development of the ideas, the sureness of mathematical deduction, the profound physical insight, the capacity for lucid, systematic presentation, the knowledge of the literature, the complete treatment of the subject matter, or the sureness of critical appraisal.

In 1924, the young man proposed what is now known as the Pauli exclusion principle, which explains how the electrons in an atom sort themselves into orbitals that line up with the rows in the periodic table of the elements. Among several of his contributions that might have been worthy of a Nobel, this was the one that won one, a couple of decades after the fact, in 1945. Pauli, who was Jewish by heritage, was in Princeton, New Jersey, at the time, having accepted an appointment at the Institute for Advanced Study in order to escape Nazi persecution. Einstein, who had nominated him for the prize, was in Princeton as well, since the institute had been founded, essentially, in order to give him a place to work outside Germany. Since Pauli didn't hold a valid passport, he couldn't travel to Stockholm for the award ceremony, so a splendid banquet was thrown for him in Princeton instead. Partway through, to everyone's surprise, Einstein rose and proposed an impromptu toast designating Pauli as his intellectual and spiritual heir. "I will never forget the speech," the younger man wrote in a letter to Max Born, ten years later. "It was like the abdication of a king, installing me as sort of 'son of choice,' as successor."

Born was not alone in considering Pauli a genius comparable to Einstein, and possibly even a greater scientist, though not as great a man. Pauli had no interest in the practical applications of science. He didn't read newspapers. When he was offered a permanent position at the Institute for Advanced Study, for example, he turned it down and returned to the position at the Eidgenossische Technische Hochschule (ETH) in Zürich that he had held before the war, in the view that nuclear weapons research was casting a pall on American physics.

He was oblivious to the cutthroat competition of science; his passion was for clarity. Nor did he care for recognition; he usually found it irritating, although he did go out of his way to give credit to others. He made many important discoveries that are now attached to other people's names, independently and frequently before they did, but rarely mentioned it. And he was unconstrained by the petty "publish or perish" mentality of most academics. His published output was relatively slim. But in his lifelong search for clear understanding, he wrote thousands of letters to the many giants who walked the world of physics in his day—and to experts in other fields, especially philosophy, psychology, and art history. His collected scientific correspondence comes to more than seven thousand pages. Often his most important ideas were first mooted in these letters, and sometimes he never went on to publish them in the scientific literature. The letters were copied, passed around, and studied, like teachings from on high.

Although Pauli is not generally associated quite as closely with the invention of quantum theory as are Bohr, Heisenberg, and Schrödinger, the letters demonstrate that he was crucially involved. No one understood the big picture as well as he did, and his colleagues respected him not only for this, but also for his caution and deep grounding in the classical traditions of physics. No less than Niels Bohr once referred to him as "the very conscience of the community of theoretical physicists" and "a solid rock in a turbulent sea" during the revolutionary years from 1925 to 1933 when quantum theory was born.

Charles Enz, his biographer and last assistant, writes that Pauli "was the critical analytical mind behind" the development of the theory: "Both Bohr and Heisenberg considered him the supreme judge." They used him as the sounding board for their latest speculations—and braced themselves for biting comments when they did. Pauli was the first person with whom Heisenberg shared his famous uncertainty principle, in a fourteen-page letter that ended with, "I know very well this is still unclear in many points, but I have to write you in order to make it somewhat clearer. Now I await your merciless criticism." He and Pauli had met in Munich, studying together under Sommerfeld, and remained friends for life. Their correspondence during the crucial period from 1925 to 1927, when Heisenberg invented quantum mechanics, indicates that much of his work was inspired by Pauli's insights and suggestions.

Though his caustic wit was legendary, Pauli could be charming and funny as well, and the huge circle of friends he amassed in his short life demonstrates that they could see through his barbs to the kind and generous heart beneath. "We always benefitted by Pauli's comments even if disagreement could temporarily prevail," wrote Bohr in gentle memoriam after the younger man died. "If he felt he had to change his views, he admitted it most gracefully, and accordingly it was a great comfort when new developments met with his approval."

Pauli's "merciless criticism" was invariably directed at vague or shoddy thinking. His friend and fellow physicist Paul Ehrenfest nicknamed him "Scourge of God." His most famous remark, tendered after reading a paper that he found especially wide of the mark, was, "Not only is it not right, it isn't even wrong!"

Professionally then, Wolfgang Pauli was thriving in 1930, the year the mysterious new particle began flitting about in his mind. He was world-renowned in physics circles and held a tenured professorship at the ETH. But his emotional life was on a different tack altogether.

He had moved to Zürich from Hamburg, where he had held various academic positions and made, as usual, a number of lifelong friends. His five years in that city may have been his most productive—he proposed the exclusion principle while he was there—but they were also the years when he began to lose his grip.

The town was famous for its night life, and he frequently availed himself of it with his friends. ("After the second bottle of wine or champagne," he wrote in a letter, "I usually become good company (which I never am when sober) and can, under these circumstances, make a very good impression on my surroundings, especially if they are feminine."). Unbeknownst to his companions, he often kept the party going after they went home, in the notorious red-light district of the city—which was named Sankt Pauli, believe it or not, another of Ehrenfest's nicknames for him. He smoked and drank in the seedy bars, got involved in brawls, picked up women—the staid professor by day, the desperate libertine by night. Dr. Jekyll and Mr. Hyde.

When he moved to Zürich, he was more circumspect. He participated in the intellectual ferment and elegant night life of the city, rubbing shoulders in the salons with the likes of James Joyce, Thomas Mann, and the artists Max Ernst and Hermann Haller. But he continued to satisfy his darker urges with periodic visits to Hamburg and Berlin.

Pauli had been raised believing he was Christian; he was baptized Catholic. As a child he was not told that his father, an internationally renowned physician and professor of chemistry, had changed his name and converted from Judaism to Catholicism as a young man, in order to advance his career in the anti-Semitic world of Austrian academia. His mother was Catholic and part Jewish, and oddly, both parents converted to Protestantism when Pauli was about eleven. His mother was especially devout. It is unclear when he learned of his Jewish heritage—it was probably sometime in his teens or early twenties—but he sensed the lack of clarity and was upset by it all through his formative years.

During his final year in Hamburg, his father, who was an inveterate womanizer, left his mother for a young sculptor about his son's age, and in November 1927, Pauli's mother committed suicide. He received the offer from the ETH that very month.

A year and a half after his mother's death, he left the Catholic Church and acknowledged his Jewish heritage. Six months after that, he married a German cabaret dancer, in an affair that was a disaster from the start: even before the marriage she announced that she was in love with another man, and she eventually left Pauli for that other man. Their union lasted less than a year, and the divorce was finalized on the twenty-sixth of November, 1930.

So, this thirty-year-old man had a lot on his mind in early December, as he pondered the problems of the nucleus.

The first pertained to a common radioactive process known as beta decay, in which the nucleus of one element transmutes spontaneously into the nucleus of a different element and gives off an electron, which is also known as a beta ray. The conundrum was that beta decay seemed to be violating one of the most hallowed laws of physics, the principle of energy conservation.

In practice, this principle is something like balancing your checkbook: Any physical system that is involved in any sort of "transaction," that is, changes in any way, must contain the same amount of energy after the transaction as it did beforehand. The energy may end up in different places, but it all has to go somewhere.

A specific example of beta decay would be the radioactive decay of carbon-14 into nitrogen-14. This is the process behind carbon dating, the method for determining the age of formerly living things, such as old pieces of wood or bone, which is widely used in archeology and geology. Carbon is the sixth element in the periodic table, which means that its nucleus contains six positively charged protons, and the "14" denotes atomic weight. This means, according to present knowledge, that along with its six protons, carbon-14 has eight neutrally charged neutrons in its nucleus, which adds up to a total of fourteen "nucleons."

A significant contributor to the confusion in 1930 was the fact that the neutron had yet to be discovered. The primitive theory of the time held that the nucleus was made up of positively charged protons and negatively charged electrons, so that the carbon-14 nucleus, to stay with this example, would have comprised fourteen protons and eight electrons. It was known to have an electrical charge of six, so the electrons would have offset just the right amount of positive charge contributed by the protons.

The carbon-14 nucleus beta decays into nitrogen-14, an isotope of the seventh element in the periodic table, which, according to the old way of thinking, would have comprised fourteen protons—the same as in carbon-14—and seven electrons, one less than before. This took care of the change in electric charge, since the nitrogen nucleus has a charge of seven; and it all seemed to add up, since an electron is caught speeding away after the decay. Until you looked at energy.

In 1905, Einstein had demonstrated the equivalence of energy (E) and mass (m) with his famous equation $E=mc^2$. (The letter c denotes the speed of light, which is a constant.) So, from an energy standpoint, before the decay we have simply the mass-energy of the carbon-14 nucleus, and afterward we have the mass-energies of the nitrogen nucleus and the electron, plus the so-called kinetic energy that the electron carries by virtue of its velocity. Since the masses of the nitrogen nucleus and the electron are fixed and add up to something less than the mass of the original carbon nucleus, the nuclear model of 1930 predicted that every electron emitted in a beta decay must emerge with the same kinetic energy, or velocity: just enough to make up the mass-energy difference between the one particle that existed before the decay and the two particles that existed after.

The problem was that the electrons emerged with a *range*, or spectrum, of energies. If they had all emerged with the highest energy in the range, everything would have been fine, but this was rarely the case. (And, in fact, we now know that it never actually happens.) A small amount of energy seemed to be disappearing somehow.

This problem had been festering for more than twenty years. Lise Meitner, an Austrian experimentalist with a background in theory, and Otto Hahn, an accomplished German radiochemist, had begun investigating the beta ray spectrum in 1907, expecting to find no spectrum at all. At first, they found what they wanted to—an uncharacteristic blunder for this superb experimental team. But it didn't take long for them to identify some flaws in their methods, improve them, and in 1911 reveal the first confusing evidence that the electrons *did* emerge in a spectrum. Meitner, however, the theorist of the two, was not ready to accept her own result. She made various suggestions about problems with the new experiment or secondary processes in the nucleus that might modify an initially pure beam. Most people's doubts were put to rest in 1914, when James Chadwick, working under the great Ernest Rutherford in the Cavendish Laboratory in Cambridge, England, completed what is now considered to be the first definitive experiment proving the existence of a spectrum. But Meitner continued to dig in her heels. This led to new experiments and other scientists joining the cause, including Charles Drummond Ellis, another Briton. The quest dragged on for another thirteen years, until 1927, when Ellis and his colleague William Wooster not only ruled out secondary processes but also proved that some energy was definitely going missing, because the *average* speed of the emerging electrons was too low to make up the mass-energy difference between the one nucleus that existed before the decay and the new nucleus and the electron that existed after. One test wasn't enough to convince the entire community, however, especially Meitner. So it wasn't until she and her assistant Wilhelm Orthmann confirmed and extended Ellis and Wooster's result two years later, at the very end of 1929, that the physics community was forced to accept the fact that something fishy was going on in beta decay.

The atom had been providing so many surprises over the previous few decades that the architects of the new quantum theory, Niels Bohr in particular, were willing to question any classical truth. In a manuscript he sent to Pauli in mid-1929, Bohr suggested that the missing energy might indicate that the hallowed conservation law did not hold in the quantum realm.

This offended Pauli's deep understanding of symmetry in the physical world. (Few laypeople realize the extent to which beauty and elegance motivate the theoretical physicist, and symmetry principles not only underlie much of that beauty, they are also among the theorist's most powerful tools.) He didn't see why electric charge would be conserved in beta decay, while energy, which was basically the central theme in Einstein's highly successful theory of special relativity, would not. He responded to his mentor with characteristic candor. (Pauli had studied with Bohr at his institute in Copenhagen.) "I must say that your paper gave me very *little* satisfaction. . . . We really don't know what is the matter here. You don't know either. . . . In any case, let this [matter] rest for a good long time, and let the stars shine in peace!" Bohr never did publish the manuscript.

Pauli followed his own advice and let the matter rest, and as time went by, began to suspect that the problem of the missing energy might be related to a more recent puzzle in the existing nuclear model, having to do with spin. This is like the spinning of a top or the rotating of a planet, except that spin is an intrinsic property of elementary particles, like mass or electric charge. It's as if they're spinning all the time.

In 1924, when he had proposed the exclusion principle, Pauli had actually intuited the existence of spin before it was discovered. Bohr's old quantum model of the atom, which was state of the art at the time, held that at most two electrons could fill each of the energy levels, or orbitals, surrounding an atomic nucleus. But this was simply a rule; he gave no underlying reason for it. Pauli supplied a reason, which turned out to be a new law of physics. In its simplest form, his exclusion principle states that no two electrons can exist in the same quantum state. And since two electrons were going into each of Bohr's orbitals, Pauli deduced that the electron must have some property that had not yet been discovered. Believing it was counterproductive to use one's classical mind to visualize the goings-on in the strange world of the quantum, however, he refused to make any claims as to what this property might be. He called it a "classically non-describable two-valuedness." One year later, the Dutch physicists George Uhlenbeck and Samuel Goudsmit explained certain fine features of the emission spectrum of hydrogen by identifying this property as spin.

Particles behave differently depending on their spin, and like most things quantum-mechanical, this property comes in quantized units. Particles with half-integer spin, such as the electron, the proton, and the neutron, obey the exclusion principle. Those with integral spin, such as the photon or particle of light, do not: they *like* to get together. Spin also tied a nice ribbon around Bohr's atomic model, since the electron, being spin one-half, will only have two spin states: up and down. An up spin will pair with a down spin in each atomic orbital, and the two will cancel each other out.

But the ribbon began to unravel early in 1929, when several experiments showed that the nitrogen nucleus had a total spin of one and did not obey the exclusion principle. This didn't work with the proton-electron model of the nucleus, which called for the nitrogen nucleus to comprise fourteen protons and seven electrons—a total of twenty-one spin-one-half particles—for there is no way that an odd number of half-integral spins can be arranged to produce a total spin of one. Ten could point up, and ten could point down, for example, cancelling each other out, but the one that was left would then give a total spin of one-half.

These two conundrums gestated in Pauli's mind for about two years altogether, through the grief of his mother's death and the anguish of his marriage. Some clarity must have been born from his divorce, however, for only eight days after it was finalized he wrote a witty "Open Letter to the Group of Radioactive Persons at the Conference of the District Society in Tübingen," proposing a tentative solution to both.

He had been invited to this conference in the German city of Tübingen, but could "not appear personally," he explained, "on account of a dance which takes place in Zürich on the night of 6 to 7 December." A friend delivered the letter for him. It was addressed mainly to two experimentalists at the meeting whom he held in the highest regard, Lise Meitner and Hans Geiger, the inventor of the Geiger counter.

"Dear Radioactive Ladies and Gentlemen," it began . . .

I have . . . hit upon a desperate remedy for rescuing the "alternation law" of statistics [the exclusion principle] and the energy law. This is the possibility that there might exist in the nuclei electrically neutral particles, which I shall call neutrons, which have spin-½, obey the exclusion principle and moreover differ from light quanta in not travelling with the velocity of light. The mass of the neutrons would have to be of the same order as the electronic mass and in any case not greater than 0.01 proton masses.—The continuous β-spectrum would then be understandable on the assumption that in β-decay, along with the electron a neutron is emitted as well, in such a way that the sum of the energies of neutron and electron is constant.

To clear up any confusion with the name, let's just say that the particle he was describing is now known as the neutrino. In fact, it was an imperfect combination of the neutrino and the particle we now know as the neutron (and there are those who say he invented both). But be that as it may, Pauli got the neutrino side of the equation almost entirely correct: the missing energy in beta decay *was* being carried away by an unseen, lightweight, electrically neutral particle of spin one-half. Fifty years later, the Italian physicist Bruno Pontecorvo observed, "It is difficult to find a case where the word 'intuition' characterizes a human achievement better than in the case of the neutrino invention by Pauli."

Pauli was suggesting that energy was being shared between his new, unseen particle and the kinetic energy of the electron. Some would power the electron as it sped away from the nucleus, the rest would go to the neutrino, the total amount would always be the same, but the fraction allotted to each would change randomly from one decay to the next. This would conserve energy for each individual beta decay and explain the continuous energy spectrum as well. By suggesting that an electrically neutral particle with spin one-half could "exist in the nuclei"—one for each electron—he was also placing an even number of particles in the nucleus and proposing a solution to the nitrogen anomaly.

But his crystal ball wasn't quite as clear when it came to the constituents of the nucleus. The neutron, which we now know does "exist in the nuclei," has two properties in common with the neutrino, electrical neutrality and a half unit of spin, but it weighs about the same as the proton, and is not emitted in beta decay. The two conundrums would require two particles and several advances in theory and experiment, which would tumble in over the next several years.

Pauli was astute enough to realize that he was groping in the dark:

I admit that my remedy may perhaps appear unlikely from the start, since one probably would long ago have seen the neutrons if they existed. But "nothing venture, nothing win," and the gravity of the situation with regard to the continuous β-spectrum is illuminated by a pronouncement of my respected predecessor in office, Herr Debye [Peter Debye, who would win the Nobel Prize in Chemistry in 1936], who recently said to me in Brussels "Oh, it's best not to think about it at all—like the new taxes." One ought therefore to discuss seriously every avenue of rescue.—So, dear radioactive folk, put it to the test and judge.

By "radioactive folk," he meant primarily Meitner and Geiger. And they responded favorably to the idea, at least as far as they could. They knew of no *conflicting* experimental evidence, but they knew of none that confirmed the idea either. This would be the case for another twenty-six years.

The Chinese American experimentalist Madame Chien-Shiung Wu, whom we shall meet, once observed that "later generations, having seen the triumphant success of the neutrino hypothesis, probably will never fully appreciate the courage and insight it took [in 1930] to put forth such an outlandish idea as the existence of an elusive particle."

It seems fitting that such a strange and ghostly creation should have occurred to this man in the midst of a deep emotional crisis. For even though the neutrino was one of the first subnuclear particles identified, it has been baffling physicists ever since. Even now, going on a century since Pauli invented it, the little particle seems to be pointing the way to new physics beyond the standard model. In a letter that he wrote in 1958, two months before he died, he referred to the neutrino as "this crazy child of the crisis of my life (1930–1)—which also behaved crazily."

In keeping with his cautious nature, its creator spoke guardedly of the "Pauli neutron" for the next couple of years. He was concerned that he may have built a castle in the sky. The British astronomer Fred Hoyle once told a story he had heard from the astronomer Walter Baade, who first met Pauli in Hamburg and became another of his lifelong friends. One evening in 1930 or 1931, so the story goes, perhaps on the very day Pauli wrote his famous letter, Baade was visiting him at his home in Zürich when Pauli declared, "I've done a terrible thing today, something which no theoretical physicist should ever do. I have suggested something that can never be verified experimentally." Hoyle related that "Baade instantly made a bet, to be paid in champagne— Pauli's favourite drink—that some day this neutrino would be detected."

Pauli spent the summer of 1931 in the United States, lecturing in Chicago, Ann Arbor, and New York. In June, he presented "his" particle in public for the first time at a joint meeting of the American Association for the Advancement of Science and the American Physical Society in Pasadena, California. A *New York Times* article expressed the reception rather well: "Physicists will accept a third particle with reluctance. What with protons and electrons, they have reduced the atom to very simple terms. A third term adds complications that they dislike. Besides there is no experimental evidence of neutrons. They are like the 'average man' of the statisticians—a mere mathematical creation." Indeed, many respected physicists would view the neutrino as nothing more than a useful mathematical trick for decades to come.

Pauli's mental state continued to deteriorate. Although Prohibition was in effect in the United States, it seems that it wasn't much of an obstacle. It was relatively easy to smuggle whiskey into Ann Arbor, for instance, which is close to the Canadian border, and at a dinner party in that town Pauli got so drunk that he fell down a flight of stairs and broke his arm. He made the long train ride to Pasadena in a cast that held it "up in the air like a traffic cop signaling," and later joked that it was the only time he ever raised his hand in a "Heil Hitler" salute.

In late October, he traveled back across the Atlantic to Rome, in order to attend a conference organized by a new vital force in physics named Enrico Fermi. Samuel Goudsmit, one of the discoverers of spin, who also attended, later described this as "the first nuclear physics meeting." And Pauli's arrival, a day or two late, provided a benign example of the by then legendary "Pauli effect." Goudsmit recalled that "he entered the lecture hall the very moment that I mentioned his name! Like magic! I remarked about it and got a big laugh from the audience."

Pauli was haunted by bizarre coincidence all through his life. Things tended to break and accidents tended to happen when he was around, although an essential feature of the effect was that these accidents never inconvenienced Pauli himself in the least. One of his students, Marcus Fierz, writes that "even quite practical experimental physicists were convinced that strange effects emanated from Pauli. It was believed, e. g., that his mere presence in a laboratory produced all sorts of experimental mishaps. . . . For this reason, his friend [from the Hamburg days, fellow Nobel Laureate] Otto Stern, the famous artist of molecular beams, never let him enter his laboratory. . . . Pauli himself thoroughly believed in his effect. He once told me that he sensed the mischief already before as a disagreeable tension, and when the suspected misfortune then actually hit—another one!—, he felt strangely liberated and lightened."

Perhaps the most famous example was the time a piece of equipment collapsed for no apparent reason in Nobel Laureate James Franck's laboratory in Göttingen, Germany. Everyone thought Pauli was in Switzerland at the time, but when Franck wrote him a tongue-in-cheek letter, absolving him of the crime, Pauli revealed that he had been traveling to Copenhagen that day and his train had made a brief stop in Göttingen just at the time of the accident. Then there was the reception in his honor at which his friends tried to stage the effect by suspending a chandelier on a rope, planning to release it when Pauli entered the room. In the event, the rope got wedged in a pulley and the "accident" never happened.

Enrico Fermi was a year younger than Pauli. They had first met in Göttingen in 1923, when they had held fellowships together under Max Born. By 1930, Fermi had gathered a spirited group of young men around him, who were nicknamed the via Panisperna Boys, after the street in Rome where their institute was located. He had a practical turn of mind (quite unlike Pauli), was deeply engaged in the details—he is one of the few great theorists who also performed breakthrough experiments—and was quite open to new ideas.

He had invited Pauli to speak about his new particle at the conference, but Pauli "was still cautious and did *not* speak in public . . . only privately." ("Horribile dictu," he added, "I had to shake hands with Mussolini.") He held useful discussions with Fermi, who "at once showed a lively interest in my idea and a very positive attitude to my new neutral particles," and he engaged in the inevitable debate with Bohr, who "in complete contrast defended his idea that in beta decay energy was conserved only statistically." Bohr was famous for his dogged debates.

Back home in Zürich, Pauli reached a new low of feverish drinking, partying, and brawling, and became so quarrelsome with his colleagues that he was threatened with dismissal from the ETH.

In desperation, he took the advice of the father he despised and sought the help of the psychologist, aka alienist, Carl Jung, who later recalled their first meeting:

> And what shall we say of a hard-boiled scientific rationalist who produced mandalas in his dreams and in his waking fantasies? He had to consult an alienist, as he was about to lose his reason because he had suddenly become assailed by the most amazing dreams and visions. . . . When [he] came to consult me for the first time, he was in such a state of panic that not only he but I myself felt the wind blowing over from the lunatic asylum.

Realizing nevertheless that he was dealing with an extraordinary individual with a mind "chock-full of archaic material," Jung decided "to make an interesting experiment to get that material absolutely pure, without any influence from myself." Suspecting that Pauli had trouble relating to the opposite sex, he assigned an inexperienced young woman to be his therapist and followed the course of his therapy from the background. (The young woman also had a disturbing first impression: Pauli was so overwhelmed with emotion during their first session together that he rolled around on the floor as he poured out his stories.)

Meanwhile, science marched on. In February 1932, at about the time Pauli entered therapy, James Chadwick discovered the neutron. The Pauli neutron now needed a new name, and it was supplied by the via Panisperna Boys. According to the contemporary Italian physicist Luisa Bonolis, " 'Neutrino,' a funny and grammatically incorrect contraction of 'little neutron' (in Italian neutronino), entered the international vocabulary through Fermi."

The particle explosion had begun. In August 1932, Carl Anderson from Caltech detected the positron in a shower of cosmic rays raining down from the sky.

✦ ✦ ✦

Cosmic ray physics is an enormous and venerable field, which predates not only particle physics but even atomic and quantum physics. It also happens to provide the most comfortable home for neutrino astronomy. Up until the mid-1950s, when accelerators began to come on line, it served as the breeding ground for breakthroughs in nuclear and particle physics as well. Anderson's discovery of the positron, which would have important implications for neutrino physics, and his subsequent discovery of the muon, which is intimately related to the neutrino, are prime examples.

Cosmic ray physics was born in 1912, when the Austrian physicist Victor Hess made the first measurements, at 17,400 feet in a hydrogen balloon, that gave proof to a pervasive "ionizing radiation," constantly streaming into the atmosphere—and through it, into our bodies and our planet—from outer space. His "rays" are now known to consist mostly of protons and larger atomic nuclei—and now that we know how to detect them, neutrinos as well. It's not possible to do astronomy with a charged cosmic ray particle, as interstellar magnetic fields will bend its trajectory as it flies through space, so that its incoming direction gives no clue as to the place it was born. Since the neutrino is uncharged, on the other hand, it travels in a straight line, like light, and *can* be used for astronomy.

Perhaps the greatest cosmic ray physicist of them all, the Frenchman

Pierre Auger, once referred to the early pioneers as "mountaineers, mine workers, divers and air riders." "The different places for carrying out measurements ridicule any description," he wrote. "Thus also the amusing story I once heard a Russian physicist tell in a French lecture: 'I have measured radiation in the sea and in high mountain ranges; I have measured it on the ground of lakes and in the highest atmosphere, in rock-salt and carbon mines and in deepest caverns. Finally I have measured it "en fer" (which means "in hell").' Of course, he wanted to say 'dans le fer,' 'in iron.'"

Today's practitioners still work in remote and far-flung places. There is a huge experiment on the Tibetan Plateau, and another at 13,500 feet on the flanks of a Mexican volcano. The Pierre Auger Observatory, named for the legendary pioneer, covers an area about the size of the state of Rhode Island on the vast Pampa Amarilla, a high plain in western Argentina.

✦ ✦ ✦

Anderson's positron was the first so-called antiparticle, anti in this case to the electron. It has the same mass and spin, but carries a positive electric charge. (And antimatter isn't quite as exotic as it may sound. Abraham Pais points out that it is "as much matter as matter is matter.") The positron also answered a question that had been raised by the so-called Dirac equation, which the young Paul Dirac had essentially dreamed up during the winter of 1927–28 and which many consider the most beautiful equation in all of physics.

Dirac was at first unsure how to interpret his creation, since it seemed to make the absurd prediction that an electron could have negative energy. But the math also worked if the particle in question had a positive electric charge, in which case it would have positive energy. After scratching his head for several years, he postulated the existence of a positively charged "anti-electron" in 1931, and Anderson's positron left its first track in his cloud chamber less than a year after that. Anderson and Victor Hess shared the Nobel Prize in Physics in 1936.

In 1927, Dirac had made another theoretical contribution that would help the neutrino along, by applying the new quantum theory to the interaction of an atom with an electromagnetic field and thus proposing the first theory of "quantum electrodynamics." Now, classically, light was thought to be a wave in the electromagnetic field; however, near the beginning of the century, Einstein had pointed out that it could also sometimes behave like a particle, the photon. By demonstrating in his new theory that the photon could appear and disappear spontaneously from the void, Dirac gave respectability to the general notion that other elementary particles might do the same. And when Chadwick's neutron effectively banished the electron from the nucleus, some scientists began to suspect that the electron emitted in beta decay might be created spontaneously. And those few who took the neutrino seriously began to suspect that it might, as well.

Pauli later recalled that "a general clarification" took place at the Seventh Solvay Conference, which convened in Brussels in October 1933.

Owing to the recent burst of discoveries, there was a last-minute decision to focus this edition of these influential conferences on nuclear structure. Many important players in the beta ray saga were there, including Ernest Rutherford, Lise Meitner, James Chadwick, and Charles Drummond Ellis. Bohr and Pauli were accompanied by their fellow theorists, Schrödinger, Heisenberg, and Dirac, and two rising stars, Enrico Fermi and Rudolf Peierls, also joined in. (Peierls was studying with Pauli at the time.)

Marie Curie, the grande dame of radioactivity, was there (she would die of radiation-induced leukemia the following year), and so were her daughter and son-in-law, Irène and Frédéric Joliot-Curie, a couple that had led a star-crossed scientific career thus far. They had produced both the neutron and the positron in their laboratories before Chadwick and Anderson had found them, but had not realized in either case what they had done.

The Joliot-Curies' luck began to turn in Brussels, when they presented the first glimmerings of perhaps the most momentous discovery of the twentieth century, the first step on the road to the discovery of nuclear fission, the "splitting of the atom" that would power the first atomic bombs thirteen years later. By bombarding thin sheets of aluminum and boron with alpha particles, that is, helium nuclei consisting of two protons and two neutrons, they had produced the first artificial radioactive substances: unstable isotopes of phosphorus and carbon. In keeping with their tradition of not quite knowing what they were up to, however, the Joliot-Curies didn't actually realize they had produced these isotopes at the time of the conference—the discussions in Brussels would spur them on to this discovery less than three months later—but they did present evidence of great relevance to the neutrino: a new form of beta decay that produced a positron instead of an electron. And now that Chadwick had placed the neutron firmly in the nucleus, this made it possible to understand the two forms of beta decay as flip sides of the same coin.

Phosphorus stands two positions to the right of aluminum on the periodic table of the elements: it has two more protons in its nucleus. Thus the Joliot-Curies had succeeded in causing the aluminum nucleus to absorb both of the protons in an alpha particle. Their artificial phosphorous then decayed to silicon, which stands between aluminum and phosphorous on the table, emitting a positron as it did so. We now know that in this form of beta decay, a proton changes to a neutron, and the decaying element therefore takes one step *down* on the periodic table—in contrast to the previously known process in which a neutron changes to a proton and the element takes a step *up*. Electric charge is conserved in each case, as the creation of a positron offsets the *disappearance* of a proton in the Joliot-Curies' process, while the creation of the electron offsets the *appearance* of the proton in the original process.

Pauli's last bit of clarification was provided by Charles Drummond Ellis, who put the final nail in the coffin of Niels Bohr's alternative hypothesis for beta decay. Recall that Bohr had proposed that energy conservation might not hold for individual decays, but that it would hold on average, overall. This implied that the energy spectrum of beta-electrons would tail off at high energies, but have no upper limit. At the conference, Ellis and his student W. J. Henderson presented results demonstrating that the energy spectrum *did* have an upper limit, and right where it was expected to be, based on mass-energy arguments. This meant that the average energy of the electrons had to lie below this limit, so that energy was lost even *on average* unless at least one other particle was involved. There are those who argue that Ellis and Henderson *discovered* the neutrino with this experiment, and indeed by today's standards they did. But Bohr was incredibly stubborn about this sort of thing. He didn't give in for another three years.

Putting all these new developments together, Pauli realized that both forms of beta decay presented another conservation problem, with spin, that seemed to *require* the emission of a neutrino: Remember that every particle involved in both forms of beta decay has spin one-half. To take the Joliot-Curies' process as the example, if the proton in an unstable phosphorous nucleus changes to a neutron and ejects only a positron, an extra half unit of spin has been created: two half spins can add to one or zero, but not to the original one-half. But if a neutrino is also ejected, and its spin is one-half, then spin is conserved. "On this basis of this new situation," Pauli wrote many years later, "my earlier caution in postponing publication now seemed superfluous. . . . I gave my ideas on the neutrino (as it was now called) in the discussion" at the conference.

The little particle wasn't quite born, but one could say after nearly three years that it was finally conceived. And as it was conceived, its creator achieved psychological clarity as well.

✦ ✦ ✦

Pauli's therapy with the young woman lasted five months, "and then for three months he was doing the work all by himself," writes Jung, "continuing the observations of his unconscious with minute accuracy. He was very gifted in this respect." In mid-1933, a few months before the Solvay Conference, he entered therapy with Jung himself.

Six months after the conference, Pauli entered into a second, more successful marriage, which lasted until the end of his life. His therapy ended a few months later, but his friendship with Jung endured. At the psychologist's request, he continued to record his vivid and remarkable dreams—more than a thousand altogether. ("They contain the most marvelous series of archetypal images," Jung exclaimed.) These dreams became the basis for some of Jung's most important lectures and papers; however, the identity of the dreamer was always kept scrupulously anonymous. The truth was revealed about two decades after both men died.

They corresponded until the end of Pauli's life, frequently meeting in Jung's mansion on the shores of Lake Zürich for an evening's conversation. Pauli's psychological emergence ("individuation" in Jungian parlance) and his invention of the neutrino marked a turning point in his scientific life. He continued to make important contributions to pure physics, to be sure, but he evolved into more of a natural philosopher in the tradition of Newton, Kepler, and the ancient alchemists, who were of great interest to Jung.

Both the psychologist and the physicist viewed the traditional scientific approach to nature as being incomplete and together began searching for ways to extend it. One of these efforts was a joint investigation into what Jung once called "the no-man's-land between Physics and the Psychology of the Unconscious . . . the most fascinating yet the darkest hunting ground of our times."

Perhaps the most revolutionary feature of quantum mechanics is the famous "inseparability of the observer and the observed," since it implies that there is no such thing as an "objective" reality. (Einstein found this consequence of the theory especially abhorrent.) What one observes depends upon how one observes it, and the act of observing inevitably changes the system being observed. But the observer in quantum mechanics is still quite detached: the inseparability is entirely physical, related only to how the measuring apparatus is set up and so on. Pauli suspected that the theory did not go far enough in this respect, "that the observer in present-day physics is still too completely detached." He and Jung believed that mind and matter were mirror images of each other, "complementary aspects of the same reality . . . governed by common ordering principles"—those principles being Jung's archetypes. I have already alluded to the profound significance and beauty of symmetry in physical law. Symmetry of all kinds, and mirror symmetry in particular, was one of Pauli's obsessions.

In 1952, he and Jung published a book together, *The Interpretation of Nature and the Psyche,* comprising two monographs, one by each. Jung's contribution was the famous (or infamous) "Synchronicity: An Acausal Connecting Principle," in which he proposed a mechanism behind meaningful coincidence and certain paranormal phenomena, including telepathy. Pauli had a personal interest in acausal connections, of course, owing to his long experience with the Pauli effect, which he suspected to be a "synchronistic [manifestation] of a deep conflict between his rational and non-rational" sides. He had encouraged the hesitant psychologist to venture into this dangerous realm and made numerous contributions to the manuscript. His own monograph, "The Influence of Archetypal Ideas on the Scientific Theories of Kepler," was an exploration of the role of the unconscious in scientific discovery. One learns to be very careful in trying to summarize Pauli's extraordinarily precise and rigorous thinking, but I think it's fair to say that he believed new discoveries become possible only as the collective mind becomes capable of visualizing or conceiving them and that this co-evolution, the leapfrogging between theory and experiment that I mentioned earlier, is driven by the emergence of archetypes. His invention of the neutrino just as the particle explosion began might serve as a good example.

Once he committed his invention to print and gained his individuation, Pauli stepped back from the neutrino's upbringing. His crazy child still had one big shock in store for him, however.

2. Infancy and Youth

THE 1933 SOLVAY CONFERENCE PROVIDED A "GENERAL CLARIFICATION" for Enrico Fermi as well. Upon returning to Rome, he dove in to the creation of a quantum theory for beta decay, completing it in December, less than two months after the conference ended.

Fermi took all the new ideas and ran with them: he assumed that the electron and neutrino could appear spontaneously, that protons and neutrons could exchange identity, and, agreeing with Pauli, that all the known conservation laws held true. Taking a cue from Heisenberg, who had suggested in Brussels (not quite accurately as it turned out) that the proton and neutron might be different states of the same fundamental particle, he guessed that the electron and neutrino might also be related—a conjecture that has resonated and continued to gain meaning ever since.

Although he didn't employ a force per se, Fermi's model for beta decay turned out to be the first rudimentary theory of the weak nuclear force. In other words, you could argue that he discovered the third fundamental force of nature, after gravity and electromagnetism. (Today there are four: the strong nuclear force was discovered soon thereafter.) Some also see his model as the first example of a modern field theory, and Francis Halzen, the guiding spirit of IceCube, points out that it could also be seen as the starting point for the standard model of particle physics, which didn't come into being for another forty or fifty years.

On the ultimate test, predictive value, the theory passed with flying colors. Fermi used it to derive from first principles the curve that had bedeviled theorists and experimentalists alike for more than twenty years: the detailed shape of the electron energy spectrum in beta decay. Granted the neutrino had yet to be seen, but its intrinsic role in a theory that lined up so well with experiment made its existence hard to deny—although many continued to do so. When he submitted his paper on the theory to *Nature*, the world's preeminent general science journal, it was rejected as containing "speculations too remote from reality to be of interest to the reader." (Fifty years later, the journal's editors admitted that this was perhaps the greatest editorial blunder they had ever made.) Fermi published it in three more specialized—and less visible—physics journals instead.

Employing his theory to estimate the mass of the neutrino, Fermi demonstrated that it must be "either zero or in any case very small with respect to the mass of the electron." This statement has also turned out to be true, and, as usual with this strange particle, it only made the neutrino more elusive.

The theory also had implications for neutrino detection—and, therefore, neutrino astronomy. For it showed that beta decay could run in reverse: a free and invisible neutrino flying unheeding through space and time could pass close enough to a neutron or proton to interact with it, change it to its counterpart, and produce a free electron or positron, which could then be detected. This process happens to be the basic principle of operation for IceCube.

Disappointingly, however, only three months after Fermi published his findings, Rudolph Peierls and another superb German theorist, Hans Bethe, showed that so-called inverse beta decay didn't happen very often. They used Fermi's theory to show that neutrinos at the energies common in beta decay could traverse about a thousand light years of water, on average—sixty-three million times the distance between the Earth and the Sun—without interacting, and therefore concluded that there was "no practically possible way of observing" the particle.

This sensible remark by two fairly intelligent individuals might stand as a cautionary tale on the dangers of prediction. Bethe and Peierls could not possibly have realized that the discoveries in physics over the next ten years would change this state of affairs dramatically—not to mention the entire human experience on this planet. Furthermore, as Peierls would admit about five decades later, they weren't counting on "the ingenuity of experimentalists."

The via Panisperna Boys became the next set of midwives to the neutrino, even as it took a back seat to the neutron for the next decade or so, as Chadwick's particle drove a plethora of new and literally earth-shattering discoveries.

Pretty much the moment the neutron appeared, Fermi and several other perspicacious individuals realized that it should be able to penetrate the nucleus more easily than the positively charged alpha particle can, since, being uncharged, it won't be repelled by positively charged nuclear protons. Taking his cue from the Joliot-Curies, he and his boys set about bombarding every known scientific element with neutrons. Partway through, by complete serendipity, they discovered that "slow," that is, low-energy, neutrons penetrate the nucleus more easily than fast ones do. And over the next several years, they succeeded in producing radioactive isotopes for every known element except the two lightest, hydrogen and helium. For this series of discoveries, in 1938, Fermi was awarded the Nobel Prize in Physics.

As it happened, he had just missed discovering nuclear fission. In fact, it is widely believed that the via Panisperna Boys induced the first manmade fission reaction while they were working with the heaviest known element at the time, uranium, but didn't realize what they had done, because they misidentified the byproducts. The via Panisperna Boys weren't quite as good at chemistry as they were at physics—and this may have been a blessing in disguise. As the gifted writer and theoretical physicist Jeremy Bernstein observes, "One can only imagine what might have happened if nuclear fission had been recognized in Fascist Italy in 1934."

Fermi won his prize at a difficult time. Among many other things, Mussolini allied with Hitler that year and began passing anti-Semitic laws in Italy. Ever the pragmatist, Fermi advised his wife Laura, who was Jewish, to invest what money they had in jewels, fur coats, and so on, and took her and their daughter with him when he accepted his prize in Stockholm. They then sailed directly for the United States, where they used the prize money and the selling of their finery as the nest egg to start a new life. The Fermis never resided in Italy again.

Many of the other scientists in the beta ray saga were also forced to flee: Pauli, Peierls, Lise Meitner. Meitner, a baptized, so-called assimilated Jew, was Austrian by birth and therefore protected from the anti-Semitic laws of Nazi Germany, where she lived and worked. But this thin platform of stability collapsed in March 1938 with the Anschluss, Hitler's annexation of Austria. In a panic, she fled her home in Berlin and the Kaiser Wilhelm Institute of Chemistry, where she and Otto Hahn had been collaborating for a quarter of a century, and made a harrowing escape from Germany by train. Neils Bohr managed to find her a position in Stockholm, and in that unhappy exile she continued to collaborate with Hahn as best she could, through the mail.

This formidable pair had also been bombarding various elements with neutrons, and Hahn, who was probably the most accomplished radiochemist of his day, had taken a special interest in uranium, every isotope of which is radioactive. A few days before Christmas that fateful year, as Meitner enjoyed the holiday with friends in the west of Sweden, she received a letter from her collaborator, revealing that Hahn and his assistant, Fritz Strassmann, had discovered an isotope of barium among the breakdown products of a sample of uranium that had been bombarded with neutrons. Barium has an atomic number of fifty-six, whereas uranium's is ninety-two. This was a transmutation of an entirely different order than the small movements on the periodic table that had previously been observed: the uranium nucleus had split nearly in half.

On Christmas Eve day 1938, while walking in a snowy wood with her physicist nephew, Otto Frisch, Meitner realized "that if you really do form two such fragments they would be pushed apart with great energy." The sum of the masses of the breakdown products is so much smaller than the mass of the original uranium nucleus that an astounding amount of Einstein's mass-energy is released. Frisch later calculated "that the energy from each bursting uranium nucleus would be sufficient to make a visible grain of sand visibly jump." And since there are some 10^{21}, or a billion trillion, nuclei in a single gram of uranium, this adds up to a very powerful explosion. In mid-January, Frisch named the process *fission*, by analogy with the binary fission of bacteria.

Not only does each individual fission of a nucleus produce an extraordinary burst of energy, but also, long before Hahn, Strassmann, and Meitner made their discovery, several far-seeing scientists realized that the splitting of a nucleus might also lead to a so-called chain reaction. When one neutron splits a single nucleus of one particular isotope of uranium, the fission products undergo beta decay and produce new neutrons, which happen to have the right speed or energy to split other nuclei, whose decay products produce more neutrons, which split more nuclei, and so on . . .

On December 2, 1942, a uranium pile (the heart of a modern-day nuclear reactor) engineered by Enrico Fermi underwent the first manmade, self-sustaining nuclear chain reaction in a squash court under the stands of the abandoned football stadium at the University of Chicago. Fermi then went on to become one of the principal architects of the first atomic bombs, which brought World War II to an end about two and a half years later.

And what does all this have to do with the neutrino? Well, each beta decay in a chain reaction also produces at least one of the ghostly particles. Thus a nuclear blast or a controlled nuclear pile produces so many neutrinos that it's hardly worth attaching a number to it. "Zillions" probably gets it about right. The existence of such powerful neutrino sources laid the ground for detection.

Other developments that took place during the war years also moved things along.

The first was a theoretical contribution by one of the more enigmatic figures in the history of physics, a thin, wealthy, pessimistic individual by the name of Ettore Majorana, who was one of the via Panisperna Boys.

The boys jokingly referred to themselves as a religious order in which the infallible Fermi played the role of pope and Majorana, grand inquisitor. Not unlike Wolfgang Pauli, he skewered any and all sloppy thinking. He had no need to work for a living, he hung around the Istituto out of a sort of bored and bemused interest, and his entire scientific output—slim, but extremely influential—was produced in a span of less than ten years. To say that he was unconventional would be an understatement. Not only was he an out-of-the box thinker, in 1938 he pulled a stunt that has turned him into a cultural icon and the subject of an enduring popular mystery in Italy: he boarded a ship with his passport and a large wad of cash and disappeared. Some believe he committed suicide, some that he took refuge in a Catholic monastery, and some, more recently, that he took up a second life under an assumed name in South America.

Majorana's signal contribution to neutrino physics was to reveal another kind of mystery that also remains unanswered more than eighty years later. In a paper he published in 1937, the year before he disappeared, he presented a variation to the Dirac equation that suggested that the neutrino might be its own antiparticle. This conjecture may seem somewhat rarified, but it had down-to-earth implications for detecting the little particle, as we'll see in a moment.

The second development also occurred in 1937, and it came from the same Carl Anderson who had rocked the world by detecting the positron five years earlier. Again in a shower of cosmic rays, this time on top of Pike's Peak in Colorado, he and his student Seth Neddermeyer detected the particle that is now known as the muon. This was a surprise, as it was unclear at the time what possible purpose the particle could serve. When Nobel Laureate I. I. Rabi heard about it, he quipped famously, "Who ordered that?"

At first, the mesotron, as it was originally known, seemed to fit the bill for a particle that had been postulated two years earlier by the Japanese theorist Hideki Yukawa. This was the proposed "field particle" that would carry or transmit the strong nuclear force, which binds protons and neutrons together in the nucleus. The analog in electromagnetism would be the photon, which carries the electromagnetic force. Yukawa had predicted his particle's mass, and since the mass of Anderson's mesotron fell in the right range, most physicists assumed it must be the one. Everything was going swimmingly until three Italians who had been conducting experiments in hiding during the war demonstrated that the mesotron could not possibly be Yukawa's field particle, because it was unaffected by the strong force.

At this point, the third great figure in the neutrino's upbringing stepped to the fore, Bruno Pontecorvo, "a tall, broad-shouldered, handsome tennis champion from Pisa." He had joined Fermi's via Panisperna Boys as an undergraduate in 1931 and been working with the Joliot-Curies in Paris in 1938 when Mussolini had allied with Hitler. Being Jewish, he, like his mentor, had decided to flee with his family to the United States, an adventure that involved among other things escaping Paris on a bicycle as German troops approached the city, and riding it all the way to Toulouse.

When Pontecorvo arrived in America, his fellow via Panispernan Emilio Segrè found him a job in the oil industry in Tulsa, Oklahoma, where he used his knowledge of neutron scattering to invent several methods for prospecting for radioactive materials, including uranium. This had strategic implications, since the most difficult part of actually making a fission bomb is to produce a so-called critical mass of weapons-grade uranium or plutonium. The lion's share of the material resources in the Manhattan Project was devoted to this task.

In 1943, Pontecorvo took a position at a research laboratory at McGill University in Montreal that was an arm of Tube Alloys, the secret Anglo-Canadian nuclear program, and put his inventions to use. The Brits and Canadians being U.S. allies, of course, Tube Alloys was effectively an arm of the Manhattan Project. At McGill, Pontecorvo also played a central role in designing what was probably the most advanced nuclear reactor in the world at the time, the NRX (Nuclear Reactor X) heavy water reactor in Chalk River, Ontario.

During a breezy, anecdotal talk that he gave in Paris in 1982 on "The Infancy and Youth of Neutrino Physics," Pontecorvo recalled that it had occurred to him in the mid-1940s "that the appearance of powerful nuclear reactors made free neutrino detecting a perfectly decent occupation." In May 1945, a few months before the first manmade nuclear explosion, Trinity, took place in the New Mexican desert, he proposed the first experimental method for detecting the particle in a technical report for the Chalk River Laboratory, which remained classified for four years, probably owing to several appearances of the word *pile*.

The basic idea was to bombard a solution of some well-chosen substance with neutrinos, which would interact via inverse beta decay with a vanishingly small fraction of the dissolved nuclei and transform them into a new, radioactive substance that could be separated out of the solution and quantified. Each transmuted nucleus would represent the product of one inverse beta decay event and thus be the clear signature of the death of a free neutrino. Scanning through the known radioisotopes, Pontecorvo found chlorine to be "by far the best" target nucleus, because it would transform into a radioisotope of argon, an inert noble gas that could easily be separated out. Another advantage was that this isotope has the relatively long half-life of thirty-five days (reverting back to chlorine via positron beta decay), so there is no great hurry in carrying out the separation. It can be done periodically, and the amount produced since the previous separation measured with a Geiger counter.

In his now-legendary report, Pontecorvo alluded to three potential neutrino sources, "a powerful reactor [which he saw as the most promising], a concentrate of radioelement(s) extracted from a reactor and . . . the Sun."

In 1939, Hans Bethe, one of the pair that had previously suggested that the neutrino was effectively undetectable, had produced a general theory of energy production in stars that showed that the Sun should be an incredibly bright neutrino source. In a nutshell, all stars are powered by nuclear fusion, the binding together of small nuclei—single protons and alpha particles mostly—into larger nuclei, and virtually every step in the cycle produces neutrinos. ("The nuclear reactions that produce the neutrinos also cause the sun to shine," observed the subsequent solar neutrino theorist John Bahcall.) The vast majority of the trillion or so neutrinos that are passing through your body as you read these words were born in our nearest star, and they do so day and night, of course, since they pass through the Earth as easily as a bullet passes through fog.

It should be obvious by now that good physicists are often decades ahead of their time, and this is doubly true in neutrino physics, where progress is so hard-won and long in coming. Even in the 1940s, long before it was detected, quite a lot was known about the particle, although this knowledge was seasoned with sensible doubt, since it had given up so few experimental facts. It was believed, for example, that nuclear reactors should emit *antineutrinos*, while the Sun should emit neutrinos.

These were the early days of particle physics. Strange new creatures were cropping up almost yearly, mostly in cosmic ray instruments sitting on mountaintops. They were being classified into groups, and new rules about their behavior were beginning to emerge.

In 1945, the same year in which Pontecorvo invented his detection method, the theorists Abraham Pais and Christian Møller coined the term *lepton*, from the Greek *lep*, meaning slight or slender, to characterize the lightest two known particles, the electron and the neutrino. Aside from their relatively light weight (at the time it was suspected that the neutrino weighed nothing at all) leptons are also distinguished from nucleons by being unaffected by the strong nuclear force; they feel only the weak force.

One of the first new rules to emerge was the notion of lepton conservation. Consider the beta decay process that had led Pauli to invent the neutrino in the first place. When the unstable carbon-14 nucleus decays to nitrogen-14, a neutron changes to a proton, and a lepton, an electron, is produced. Since there were no leptons in the picture beforehand, lepton conservation dictates that the neutrino that is produced along with the electron must be an antilepton: an antineutrino. So the neutrino that Pauli invented was actually the antiparticle. And since this is the form of beta decay that takes place in nuclear reactors, they give off the antiparticle, too—in quantity.

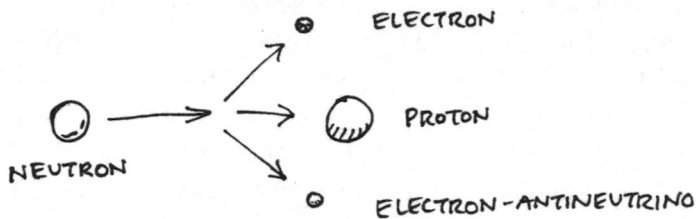

A neutron changing to a proton in the original form of beta decay. There are no leptons and there is no electric charge before the decay. After the decay, the negative charge of the electron offsets the positive charge of the proton, and the electron antineutrino offsets both the lepton number and the flavor of the electron.

Now, on the face of things, Pontecorvo's detection method should *not* be sensitive to the antiparticle. In his scheme, a stable chlorine-37 nucleus, with seventeen protons and twenty neutrons, is transformed into a radioactive argon-37 nucleus, with eighteen protons and nineteen neutrons: a neutron is changed to a proton. By electric charge conservation, the creation of this proton must be accompanied by the creation of an electron, and since the electron is matter as opposed to antimatter, the particle that initiated the reaction must have been matter as well: a neutrino. Since the Sun emits neutrinos, Pontecorvo's method *should* be sensitive to those.

But this is where Ettore Majorana's rarified conjecture comes in. If he was right and the neutrino and antineutrino are identical, Pontecorvo's method should be sensitive to both.

In 1945, a few months after the war ended, the Pontecorvo family moved to Chalk River to be near Dad's new workplace in the reactor complex. At about the same time, the three Italians who had been working in hiding revealed their surprising news about the muon, and Pontecorvo's interest was piqued. "That was indeed an intriguing particle," he recalled in Paris many years later. "I found myself caught in an antidogmatic wind and I started to put lots of questions." He set up a cosmic ray laboratory in Chalk River in collaboration with the Canadian physicist E. P. "Ted" Hincks, and over the next several years they made a series of discoveries that answered all of his questions and more.

The upshot was that the muon turned out to be the third lepton. It has the same charge and spin as the electron, it is similarly affected by the weak but not the strong force; in fact, it has so much in common with its lighter cousin that it is often described as a heavy electron. It is unstable, with a lifetime of 2.2 millionths of a second, whence it decays into an electron and two other particles. At that early stage of the game, Pontecorvo guessed correctly that these would be a neutrino and an antineutrino, and this led him to make another wise guess: that the neutrino might carry a sort of identity card that associated it with either the muon or the electron. "For people working on muons in the old times," he recalled in Paris, "the question about different types of neutrinos has always been present."

If a single muon, which is a lepton, decayed into three leptons, and one of those was an electron, then lepton conservation said that the other two must cancel each other out: they must be a lepton and an antilepton, in

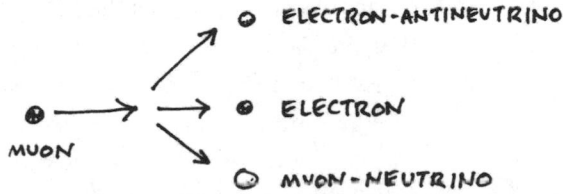

A muon decays into three particles. Before the decay, there is one lepton of muon flavor, with a negative electric charge. After the decay, the electron carries the electric charge, the muon neutrino carries the muon flavor, and the electron antineutrino offsets the flavor of the electron and the lepton number of either the electron or the muon neutrino. Thus lepton number, lepton flavor, and electric charge are conserved.

other words, a neutrino and antineutrino. But when a particle and its antiparticle find themselves in close proximity they usually annihilate and give off other particles. Since he and Hincks had found that the two uncharged products of muon decay don't annihilate, Pontecorvo deduced that some hitherto unobserved quality must make them different and that it might have something to do with the difference between the muon and the electron.

Following this train of thought: to conserve "muon-ness," which is now known as muon "flavor," the new neutrino must be a muon neutrino, and to conserve electron flavor, which was zero before the decay, the antineutrino produced in concert with the new electron must be an electron antineutrino. And we can now say precisely what it was that Wolfgang Pauli intuited back in 1930: since an electron is produced in the original form of beta decay, the accompanying neutrino must be an electron antineutrino.

To bring this all back to earth, or perhaps Antarctic ice, it so happens that flavor has important implications for neutrino astronomy. For a muon neutrino can initiate inverse beta decay in the same way that its electron counterpart can, with the important difference that it will give birth to a muon, rather than an electron. And a muon happens to be much more easy to detect as it flies through the ice than an electron is. Muon detection was the basic principle of operation for the Antarctic Muon And Neutrino Detector Array, aka AMANDA, and it is still the bread and butter for Ice-Cube. The muon is the workhorse of neutrino astronomy.

In 1947, while Pontecorvo and Hincks were still pursuing these investigations, the tracks of a new particle, the pion, were revealed in photographic emulsions exposed on mountaintops in the Pyrenees and the Bolivian Andes. It was soon realized not only that this was the field particle Yukawa had postulated but also that the pion decayed into the muon—which explained why only the latter had been found at lower altitudes.

The pion also plays a key role both in neutrino astronomy and in experimental neutrino physics, since it provides the most obvious mechanism in particle accelerators, either manmade or cosmic, for producing high-energy neutrinos. When a proton is accelerated in an electromagnetic field, either on Earth or in the cosmos, and then collides with some other particle, such as a photon or an atomic nucleus, it will give birth to a pion. If that pion is uncharged, it will decay into two gamma ray photons and back to a proton. If it's charged, it can decay by two pathways: to a muon, a muon neutrino, and a neutron, or to an electron, three neutrinos, and a neutron. So the way to make a "neutrino factory" here on Earth is to direct a manmade proton beam at a target or "beam dump" that will generate pions, and manipulate the charged pion beam and its decay products in such a way as to produce a clean beam of neutrinos. Cosmic accelerators, such as active galactic nuclei, supernova remnants, and their siblings, are presumed to accelerate protons and other nuclei in their own ways, and these particles will collide with cosmic beam dumps to produce pions and, in turn, the cosmic high-energy neutrinos that IceCube is looking for.

Bruno Pontecorvo was juggling quite a few balls during these productive years. Not only were he and his Swedish wife, Marianne, raising a family, they also moved at least four times. In 1948, after turning down several offers from top universities in the United States and Italy, he accepted a senior position at the main national laboratory for applied nuclear research in England, the Atomic Energy Research Establishment in Harwell, and the family moved back across the Atlantic.

By then, the most unwieldy of the balls he was keeping in the air was the suspicion surrounding the fact that he and Marianne were ardent communists. Like many Italian intellectuals, he had joined the communist party in 1936 with the advent of the Spanish Civil War. The couple had met in Paris when he was working with the Joliot-Curies, who were also active in the communist cause, and the Pontecorvos had worked against Nazism and Fascism during their years there. Bruno's brother, Gillo, who would achieve international renown as a filmmaker (best known for *The Battle of Algiers*), was a member of the Italian communist party, and a cousin also held a high post in the party.

The year after the Pontecorvos moved to England, the Soviet Union detonated its first atomic bomb, an exact copy of the American version, and in March 1950 the German physicist Klaus Fuchs, a socialist true believer who also worked at Harwell, was convicted of passing nuclear secrets to the Soviets. A wave of anti-communist hysteria swept the west. Over the summer, U.S. Senator Joseph McCarthy commenced his infamous campaign against the communist "fifth column." All of this brought more pressure to bear on Bruno Pontecorvo.

In September 1950, he, his wife, and their three young children took a holiday to Italy and quietly disappeared. About a month later, the British government was forced to admit that one of the top nuclear physicists in the country had most likely defected to the Soviets. It made front-page news in Britain.

People who lived during that incendiary time still remember Pontecorvo as being a spy. He was lumped in with Fuchs, Julius and Ethel Rosenberg (who were sentenced to death in the United States about six months after his defection), Kim Philby, and others as one of the top turncoats of the Cold War era. When the Rosenbergs were convicted, *Time* ran an article under the title "SPIES: Worse than Murder" that placed him in the inner circle of evil and explained how his alleged crimes (for which they offered no evidence) fit into a global communist conspiracy.

No evidence has ever been found that Pontecorvo was a spy. The contemporary science historian Simone Turchetti, who is probably the foremost expert on the "Pontecorvo Affair," believes he was innocent. Recently released documents demonstrate that British government officials were aware of his innocence but didn't disclose it because they were involved in delicate negotiations with the United States about the transfer of nuclear technology to Britain and the myth of his guilt served the agendas of powerful U.S. political factions and agencies, such as the FBI, which they had no good reason to resist. The witch hunts served many political ends, of course, and had no particular use for the truth.

It was five years before Pontecorvo made his whereabouts known. In 1955, in a press conference at the leading nuclear and high-energy physics laboratory in the Soviet Union, the Joint Institute for Nuclear Research in Dubna, near Moscow, he claimed that he had left England because he was terrified of the witch hunts and "the pressure put on him by security forces during vetting." He said "he had defected to correct the balance between East and West and that he had only ever worked on the peaceful uses of atomic energy." After 1978, when the Soviets finally allowed him to travel to the west, he campaigned for nuclear disarmament.

One can never know, of course, but Turchetti has found many documents that support Pontecorvo's innocence, including records of meetings *he* called with his superiors in order to discuss his apprehensions before he made his fateful decision. He would certainly have been an asset to the Soviet nuclear energy and weapons programs—and perhaps he was, after he defected. It seems unlikely that he was beforehand.

It was also a bad career move. For, in defecting, this world-class scientist left the vanguard of his field for a backwater. He would continue to make important theoretical contributions from behind the Iron Curtain, but the Soviet Union fell far behind the west in the development of accelerator technologies, which is where particle physics went for the rest of the century. Eleven people, all experimentalists, have received Nobel Prizes based on Pontecorvo's theoretical work in neutrino physics. (Thus it is ironic that he was basically an experimentalist.) Had it been possible for him to remain in the west—and had he lived long enough—it seems likely that he would have shared in at least one.

He died in Dubna in 1993. In accordance with his will, half of his ashes were buried there and the other half in Rome.

If Wolfgang Pauli conceived of the neutrino and Enrico Fermi gave it life, then Bruno Pontecorvo gave it personality. His guess that it came in different flavors turned out be correct: the muon neutrino was discovered in 1962. And in 1958, behind the Iron Curtain, he also intuited what may be the particle's weirdest and most ghostlike property: that it will change flavor, or "oscillate," as it speeds along.

There are now three charged leptons: the electron, the muon, and the tau particle, which was discovered in 1975. The tau is the heaviest of the three, weighing in at about 3,500 times the mass of the electron, and its neutrino was finally detected in 2001. So there are six leptons altogether, in three pairs, and each has an antiparticle.

Oscillation, which will only occur if the neutrino has mass, means that an electron neutrino will change into, say, a muon neutrino, then perhaps a tau neutrino, then back into an electron neutrino, and so on, as it races along—something like watching your dog change into a cat and then an ocelot and then back into a dog in the course of his morning walk. Pontecorvo's 1958 conjecture was proven true forty years after he made it and five years after he died, by an instrument very much in the tradition of AMANDA and IceCube. But that's getting ahead of the story.

3. From Poltergeist to Particle

Neutrinos induce courage in theoreticians and perseverance in experimenters.

—MAURICE GOLDHABER

BY THE EARLY 1950S, THE STAGE WAS SET FOR DETECTING THE NEUTRINO. The challenge was taken up by two men with very different personalities.

One was a self-effacing, painstaking, and as events would show, exceedingly patient man by the name of Ray Davis, a physical chemist at Brookhaven National Laboratory on Long Island, where the stated mission was to explore "peaceful uses of the atom." In those good old days scientists were actually encouraged to go exploring. When Davis arrived and asked the chairman of the chemistry department what he ought to do, he was told to go figure it out himself.

Off he went to the library, and a lifelong fascination with neutrinos was born. For the first few years, he worked on an indirect method for detecting the particle, and in 1951 he began working on a practical realization of Pontecorvo's direct radiochemical method.

Davis's stated goals for his first set of experiments were to try to detect neutrinos coming from the Sun and to see whether or not "reactor" antineutrinos would interact in the same way as the solar variety. Now, the radiochemical method consists of sweeping a small number of argon atoms from a large tank of target liquid sometime after any neutrino interactions have occurred, so it does not reveal the direction of the incoming neutrinos. Thus Davis's instruments could not be called telescopes. Nevertheless, by focusing on the Sun he was embarking on the very first experiment in neutrino astronomy.

He chose simple dry cleaning fluid as his target liquid, perchloroethylene, a compound that contains four chlorine atoms, and he built two detectors that were relatively large for the time, one employing 200 liters of the stuff (science-speak for a 55-gallon oil drum) and the other 3,900 liters or a thousand gallons.

Employing the High Flux Research Reactor at Brookhaven as what he hoped would be an antineutrino source, he measured a noticeable signal with the larger instrument, which he attributed not to neutrinos but to stray protons from the reactor. He then buried the tank about nineteen feet deep, a good distance from the reactor, hoping to shield it from cosmic rays, and the signal disappeared. This second step, incidentally, initiated the peculiar tradition that survives to this day of locating neutrino detectors and telescopes in mineshafts, tunnels, and the like, in unexpected and remote locations, often in the mountains.

Davis then carried on in the august tradition of cosmic ray physics by taking his smaller instrument to the top of 14,265-foot Mt. Evans in Colorado, where he measured another spurious signal that he attributed to cosmic rays. He was, as I say, a painstaking and cautious man. When he finally published his results in 1955, he made no claim to have detected neutrinos, anti- or otherwise, although he did use his null result along with the admittedly low sensitivity of his instrument to provide an upper limit for the neutrino brightness of the Sun. (Limits are an important matter in physics, as we shall see: If you don't see something, but you know how sensitive your instrument is, you can say that whatever source you're examining isn't bright enough to give you a signal, and this can tell you something about its physics.) He would later tell the story of how one of the paper's reviewers was unimpressed by his upper limit, because his experiment was too insensitive to have any bearing at all on the question of the existence of neutrinos. The reviewer had illustrated his point by observing that "one would not write a scientific paper describing an experiment in which an experimenter stood on a mountain and reached for the moon, and concluded that the moon was more than eight feet from the top of the mountain." Indeed, Davis's own results would later demonstrate that his first limit was high by a factor of about 15,000—or, in plain English, that his instrument was 15,000 times too insensitive to detect neutrinos from the Sun.

He went back to the drawing board.

The other man to take up the challenge was a physicist about four years younger than Davis, named Fred Reines. Reines had started out as a theorist. He had earned his doctorate in 1944 with a thesis on "The Liquid Drop Model of Nuclear Fission." This being a timely topic (publication was withheld until after the war), he was recruited directly from school to the Manhattan Project in Los Alamos, too late to contribute significantly to the bombs that were used in the war. He remained at Los Alamos afterward, participating in many of the bomb tests that took place in the South Pacific, and eventually rose to the level of director for the Operation Greenhouse tests on the Eniwetok Atoll, which prepared the ground for the hydrogen or fusion bomb.

In 1951, tired of such mission-oriented work and yearning for something more fundamental, he asked his boss, who was also open-minded, if he could take a sabbatical in order to "ponder." Reines later "moved to a stark empty office, staring at a blank pad for several months searching for a meaningful question worthy of a life's work. It was a very difficult time. The months passed and all I could dredge up out of the subconscious was the possible utility of a bomb for the direct detection of neutrinos." Enrico Fermi happened to be visiting Los Alamos that summer, so Reines "went down the hall, knocked timidly on the door and said, 'I'd like to talk to you a few minutes about the possibility of neutrino detection.' [Fermi] was very pleasant, and said, 'Well, tell me what's on your mind?' I said, 'First off as to the source, I think that the bomb is best.' After a moment's thought he said, 'Yes, the bomb is the best source.' So far, so good! Then I said, 'But one needs a detector which is so big. I don't know how to make such a detector.' He thought about it some and said he didn't either. Coming from the Master that was very crushing." Reines returned to his blank pad.

Several months later, he and a Los Alamos colleague named Clyde Cowan were flying east on business when their plane developed engine trouble and they were forced to land in Kansas City. Finding themselves at loose ends for several hours while the engine was being repaired, they batted around ideas about challenging problems and finally agreed to "work on the neutrino." Cowan "knew as little about the neutrino as I did," wrote

Reines, "but he was a good experimentalist with a sense of derring-do. So we shook hands and got off to working on neutrinos."

They actually did design a detector for use with a bomb. The plan was to suspend it in an underground vertical shaft, pumped free of air, about two hundred feet from the bomb tower, and release it as the bomb was detonated, so that it would be falling in a vacuum and therefore unaffected as the shock wave passed through. After landing on a bed of feathers and foam rubber, it would begin detecting the copious antineutrinos emitted by the numerous fission byproducts produced by the explosion. A few days later, when the surface radioactivity had diminished enough, the detector would be dug up and the results would be read from a recorder.

This audacious concept never "flew," however. When they presented it at a seminar at Los Alamos, a colleague suggested they replace the bomb with a reactor. They swiftly developed a new concept, and queried Fermi again, by letter. This time the master was more optimistic: "Certainly your new method should be much simpler to carry out and have the great advantage that the measurement can be repeated any number of times. . . . I do not know of any reason why it should not work."

Their method was entirely different from Pontecorvo's and had the added advantage of aiming specifically at the *anti*neutrinos that were expected to emerge from reactors.

Their detector was arranged in the manner of a club sandwich, with two "meat" or "target" layers and three "bread" or "detection" layers, one in the middle of the meat layers and the other two on the top and bottom. The target layers were tanks of cadmium chloride dissolved in water, and the detection layers were tanks of liquid scintillation material, monitored by light detectors. (Scintillation materials give off light when charged particles or gamma rays pass through them.)

Their method was based on the inverse of the Joliot-Curies' beta decay process involving the positron: an antineutrino from the reactor collides with a free proton in the water of the target layer, which changes to a neutron and ejects a positron. This results in two flashes of light. The first comes almost immediately, as the positron finds a nearby electron and they mutually annihilate, sending off two gamma ray photons in exactly opposite directions. This is why the target layers are each sandwiched by two detection layers: the annihilation event will light up the two adjacent scintillation tanks simultaneously. The newborn neutron then stumbles around in the water of the target tank for about five millionths of a second before being captured by one of the dissolved cadmium nuclei, and this gives off a second flash of gamma rays, which lights up the same two scintillation tanks. The five-microsecond delay time acts as a signature for the neutrino interaction, which helps distinguish it from the background created by the charged cosmic ray particles that will inevitably rain through the detector and stray neutrons and protons from the reactor.

Cowan and Reines ran their first set of experiments in the early spring of 1953 at a reactor in Hanford, Washington, that had been used to produce plutonium for the bomb that was dropped on Nagasaki. Although they were plagued by a high cosmic ray background, they observed an increase in signal when the reactor was on as opposed to off, and after a few months convinced themselves that the increase was real. In November, they announced the "probable" detection of the free neutrino. This was a bit of a stretch, and one guesses that Reines was the one who pushed for it. He was far more aggressive than Cowan, who was modest and devoutly Catholic, and overshadowed him all through their collaboration—as he did most of the people he worked with. Cowan later wrote that the "evidence would not yet [have stood] up in court" and reflected that only in retrospect did it appear "genuine."

It was big news, nevertheless, and it soon filtered across the Atlantic to the man who had invented the particle more than two decades earlier.

A young post-doc who was working with Pauli at the time, one William Barker, writes that when word reached Zürich, a group of friends and faithfuls joined the great man in a walk up the Üetliberg, a hill near the city, for a celebratory dinner. "On the way down, [Konrad] Bleuler and I noticed that Pauli was a little wobbly from the red wine we had at dinner. (He had graciously responded to many individual toasts.) Bleuler said to me: 'Take his left arm—I'll take his right arm, we can't afford to lose him now.'"

This celebration was a bit premature, as there were real problems with the experiment, but Pauli didn't need much of an excuse to throw a party, generally speaking.

Cowan and Reines went back to Los Alamos to improve their detector, and soon received a tip from the theorist John Archibald Wheeler that the most powerful reactor in the world was nearing completion at the Savannah River nuclear reservation in Aiken, South Carolina. In the fall of 1955, they caravanned with their families across the country to run their second set of experiments. There was a summer camp feeling in the air, since they had shared the tip with Ray Davis and he was running his second series of experiments side-by-side with theirs. Davis came up empty again, but Cowan and Reines struck gold.

The new reactor produced many more neutrinos, and their improved methods also helped tamp down the various backgrounds. By early June 1956, they had what they felt was a definitive result. To give a sense of the neutrino's bashfulness, they estimated that the reactor was sending about twelve trillion electron antineutrinos through each square centimeter of their detector every second, but they detected only about three inverse beta decays per hour.

"We were done," Cowan wrote. "The experience of knowing a fact new to mankind and knowing it for a while all alone is an unforgettable one. The neutrino existed as an objective, demonstrable fact of nature. The great laws of conservation stood firm. And our small group had had the privilege of sharing in the work that made them so."

This time they felt confident enough to notify Pauli directly. On the fourteenth of June, they sent a telegram to Zürich:

> We are happy to inform you that we have definitely detected neutrinos from fission fragments by observing inverse beta decay of protons. Observed cross section agrees well with expected six times ten to minus forty-four square centimeters.

It had to be forwarded to Pauli, who was attending a meeting at CERN, the European accelerator laboratory on the outskirts of Geneva. When he received this definitive announcement of his brainchild's birth, twenty-six years after he had conceived of it as "a desperate remedy" for the crisis in beta decay, he interrupted the proceedings to read the telegram aloud and make a few impromptu remarks. He then responded to Cowan and Reines by night letter, a less expensive form of overnight telegram, quoting a Chinese proverb: "Thanks for message. Everything comes to him who knows how to wait." But the Pauli effect may have been at work again, for the letter never arrived! A copy was finally sent to Reines thirty years later by Charles Enz, Pauli's last assistant.

There remained the small matter of the bet with Walter Baade, involving the case of champagne. At a meeting on neutrinos that took place at the Royal Society, London, in 1967, astronomer Fred Hoyle remarked that Pauli "paid up—as I happen to know since I drank some of the payment." Pauli, true to form, drank his share as well.

✦ ✦ ✦

One would think that a discovery of this magnitude would be an occasion for unreserved celebration, but the little particle had other ideas in mind. As Charles Enz put it, "Several experiments [soon] discovered a birth defect."

Cowan and Reines probably *had* detected the neutrino, but they had overreached again in publishing a hard number for the so-called cross-section. A cross-section gives a measure of the likelihood that an interaction involving a collision or some similar sort of interaction will occur. It has the dimensions of an area. One way to get a grip on the idea is to think of a window made of glass hard enough that if a kid throws a baseball at it, it will break only one in ten times. The cross-section for hitting the window will be its area, pure and simple, while the cross-section for breaking it will be a tenth of that.

Aside from publishing the number—6×10^{-44} square centimeters, as in the telegram to Pauli—they had gone a step further and claimed that this value was "within 5 percent" of the theoretical prediction for the cross-section, with an error of plus or minus about 10 percent. This put the theoretical value comfortably inside their experimental range.

Even as they were savoring the knowledge of their new discovery "all alone," however, two Chinese American theorists on the east coast, Tsung-Dao Lee from Columbia University and Chen-Ning Yang from the Institute for Advanced Study, were beginning to suspect that the neutrino (or, more precisely, the weak force) might have a surprising quality that would increase the theoretical value by a factor of two. When Reines became aware of their idea, he stuck to his guns and stoutly defended his and Cowan's analysis. Within six months, Lee and Yang's suspicion was confirmed by other experiments and the discrepancy with the Savannah River number became difficult to ignore. Reviewing their methods, Cowan and Reines then realized that they had "grossly overestimated the detection efficiency." In 1958, they carried out a third series of tests with methods improved yet again and came up with a number almost twice the original, snapping into line with the new theory.

But the damage had been done. Reines's initial defensiveness, along with what might have seemed a habit of coming up with numbers to match the theory of the day, had induced a certain mistrust, and in some cases dislike, that persisted for decades. There were even those who suspected that he and his partner hadn't actually detected the neutrino. And Reines's behavior in subsequent years did nothing to defuse the controversy. It was four decades before half the Nobel Prize in Physics was awarded for the discovery, by which time Cowan had died, so that Reines received their share of the prize alone. It might be fair to say that his aggressiveness in staking claims and his combativeness in defending them robbed his self-effacing partner of a Nobel Prize.

Reines, who died in 1998, three years after receiving the prize, was "a man of imposing physical stature" and outsized character as well. John Wheeler once described him as "talented in both theory and experiment, a bear of a man given to thinking big about nearly impossible problems as he paced up and down in his oversized shoes." He was undoubtedly one of the great experimentalists of the twentieth century, and his restless footprints wind through all the choicest glens in neutrino physics and astronomy—as we'll see. He had a great love of poetry, wrote poems himself, and was talented enough as a baritone to have faced a career choice between opera and physics in his youth. Later in life, he sang in the Shaw Chorale with the Cleveland Symphony Orchestra under the direction of the composer and the legendary conductor George Szell. John Learned, one of the cofounders of DUMAND, who is a singer himself, says that "Fred's voice was deep and rich and just much better than most mortals." But Reines had a dark side as well. He was extremely competitive, even with his own students, whom he rarely supported, and he earned numerous enemies over the years.

Learned remembers Reines telling him that he and Cowan weren't "trying to *measure* any physical parameters" in their first experiment at Savannah River, "only to show that they had found the elusive neutrino." "With a little humility and openness," Learned adds, "this could have all been avoided."

The writer-physicist Jeremy Bernstein refers to the year and a half following Lee and Yang's conjecture as the Glorious Revolution. (At the time, some referred to it with less political correctness as the Chinese Revolution.) It shook physics as it hadn't been shaken since the discovery of fission. The two theorists had realized that the weak force—and the neutrino in turn—might violate one of the most hallowed laws of physics: the law of mirror symmetry, which says that a mirror image of any physical system should behave exactly as the real thing. The technical term for this is parity.

What prompted Lee and Yang to call parity into question was the behavior of a new breed of particle that had recently turned up in cosmic ray showers, with a hitherto unobserved quality that was eventually named *strangeness*. Two of these strange beasts, the theta (θ) and the tau (τ), appeared to be identical—they had every "personal" quality in common, such as mass, spin, and lifetime—except that the theta decayed into two pions, while the tau decayed into three. "Physicists would have been happy to put the theta and the tau down as identical,"* writes Bernstein, except that this would have violated parity conservation, a law that "was not to be tampered with lightly."

Lee and Yang were energetic collaborators. When they were seized with an idea they would talk it through in Chinese together for days on end—often quite loudly, according to their colleagues. They would break periodically to go off to their separate corners and do competing calculations, and then reconvene and pick up the discussion. One day, during an exchange in a restaurant on upper Broadway in Manhattan, across the street from Columbia, they realized that one way into the theta-tau puzzle might be to examine every experiment involving the weak force that had ever been done, to see what it might have to say about parity conservation. When they then went about doing so, they found that none had actually tested the law. Only eight days after Cowan and Reines sent their triumphant telegram to Wolfgang Pauli, Lee and Yang submitted a classic paper to the venerable U.S. journal *Physical Review*, suggesting that "one way out of the [theta-tau] difficulty is to assume that parity is not strictly conserved, so that the theta and tau are two different decay modes of the same particle," which is to say, mirror images of each other. Observing that parity was conserved to a high degree of accuracy in experiments involving the strong and electromagnetic forces, they posed a challenge to the experimental community by suggesting some ways to test the parity of the weak force. The vast majority of physicists expected parity to survive.

The first experiment to yield an answer was based on straightforward beta decay. It was led by Chien-Shiung Wu, a close friend of Lee and Yang's who

---

* The tau particle we are speaking of here is not the tau lepton. This tau and the theta were eventually recognized to be different states of the same particle, which was named the kaon.

also worked at Columbia and had also been raised in pre-communist China. Wu was a highly cultured, elegant, and attractive woman, internationally renowned for her precise and cautious experimental work.

She settled upon the idea of using a magnetic field to line up, or polarize, the spins of radioactive cobalt-60 nuclei, which are basically tiny magnets, and observing the direction of the beta-electrons that they would emit. If the electrons sprayed out randomly in all directions, parity would be conserved; otherwise, it would not. It was necessary to cool the cobalt nearly to absolute zero in order to prevent the jostling that occurs at higher temperatures from knocking the nuclei out of alignment, so she collaborated with a team from the National Bureau of Standards in Washington on that aspect.

After six painstaking months setting up her apparatus, Wu got her answer within minutes of starting the experiment: the beta-electrons emerged preferentially in the direction opposite the magnetic field. Parity was in its death throes.

She was in constant communication with Lee and Yang, of course. On the third of January 1957, a Thursday, she called Lee to inform him of her results. Lee had initiated a tradition in the Columbia physics department of gathering for lunch on Friday at one of the many excellent Chinese restaurants near the Columbia campus. The next day, he shared the news with his colleagues . . . and set the mind of a fellow professor, Leon Lederman, to thinking. In what may be the shortest time between concept and result for a finding of such magnitude, Lederman and his colleague Richard Garwin designed a new method for testing the parity of the weak force by late that evening; had it up and running on the cyclotron at Columbia's Nevis Laboratory, a few miles up the Hudson River, by two a.m.; and produced a definitive result, confirming Wu's, by six a.m. the following Tuesday.

A group at the University of Chicago that had been running a similar experiment for a few months came up with a second confirming result within days of Garwin and Lederman. Parity wasn't just dead; it was annihilated.

On the fifteenth of January, less than two weeks after Wu's call to Lee, Columbia University took the unusual step of holding a press conference to announce a physics result (thus stealing the limelight from the Chicago group, who weren't even mentioned). In a story that appeared on the front page of the *New York Times*, I. I. Rabi, the chairman of Columbia's physics department, was quoted as saying, "In a certain sense, a rather complete theoretical structure has been shattered at the base and we are not sure how the pieces will be put together." Such is the damage that the neutrino tends to leave in its wake.

Lee and Yang were awarded the Nobel Prize later that year. Many believe Chien-Shiung Wu should have been honored as well.

Also in that eventful year, an elegant experiment at Brookhaven demonstrated that the neutrino is left-handed: it spins counterclockwise as it zips through space. And this brings us back to Ettore Majorana.

One might conclude from Ray Davis's null result with "reactor" neutrinos and Cowan and Reines's positive one that the neutrino and its antiparticle are different—or, to put it another way, that the neutrino *has* an antiparticle. It turns out, however, that this is not necessarily true. Theory says that the chlorine-to-argon reaction that is the basis for Davis's method is not sensitive to the difference between the particle and the antiparticle, but to the particle's handedness: only a left-handed particle can initiate it. And eighty years after Majorana suggested that the neutrino might be its own antiparticle, the evidence still is not in. If he was right, then Cowan and Reines weren't detecting antineutrinos; they were detecting right-handed neutrinos.

As physics has evolved in the intervening decades, Majorana's conjecture has taken on added meaning. Physicists now realize that it has implications not only for neutrino physics, but also for some of the most compelling unanswered questions in particle physics and cosmology.

Louis Pasteur once said that "in the field of observation, chance only favors the prepared mind." Ray Davis and Fred Reines had prepared their minds equally well, but chance happened to favor Reines. Considering what was known about the neutrino at the time, this was really just a matter of luck.

Davis, being a true and in fact visionary scientist, did not skip a beat. Once he proved to himself that his method could not detect "reactor" neutrinos (it took him several more years) he turned his attention to the Sun.

✦ ✦ ✦

From his perch in Zürich, Wolfgang Pauli attended to the Glorious Revolution with great interest. Parity was one of his major obsessions, after all; he admitted to having a "mirror complex."

Four years before the Revolution began, even though, as he later wrote to Carl Jung, "there was not actually anything going on in the world of physics to justify focusing on that particular subject," he had begun his own investigation into a deeper form of symmetry, which included not only parity (P); but also time reversal (T), running time backward; and charge conjugation (C), which is the act of converting every particle in a system into its antiparticle. (It is possible to run time backward on the atomic scale in the laboratory.) If a system remains unchanged by the act of mirroring all three of these properties at the same time, a transformation that Pauli called "strong reflection," it would be said to have CPT symmetry or to be CPT invariant.

He began investigating whether the fundamental equations of quantum mechanics and relativity obeyed this symmetry in 1952, and finally proved that they did near the end of 1954. To this day, no experiment has disproven Pauli's CPT theorem. It is seen as his third great contribution to physics, after the exclusion principle and the invention of the neutrino. With parity's downfall, and especially since "his" neutrino was so intimately involved, Pauli and his new theorem were "on everybody's lips." "To many physicists," according to T. D. Lee, "CPT was a fixed point around which all else turned."

Remember that Pauli shared with Jung the notion of a mirroring between psychology and physics. "Now, 'mirroring' is an archetype," he told an interviewer in 1957, after the news of parity violation had broken. "This has something to do with physics. Physics relies on a relation of mirror symmetry between mind and nature. . . . At that time [while he was working on the CPT theorem] I had vivid, almost parapsychological dreams about mirroring, while I worked mathematically during the day. . . . I would call that, for instance, a kind of synchronicity, since there are unconscious motives when one is engaged in something."

Not long after he finished writing his definitive paper on the theorem, he experienced what he called "a very impressive dream":

> I am in a room with the "dark woman," and experiments are being carried out in which "reflections" appear. The other people in the room regard the reflections as "real objects," whereas the Dark Woman and I know that they are just "mirror images." This becomes a sort of secret between us. This secret fills us with *apprehension*.
>
> Afterward, the Dark Woman and I walk alone down a steep mountainside . . .

In describing this dream to Jung in a long letter in 1957, Pauli reminds his friend of a "Chinese woman" who recurred in his dreams, whom he saw "as a special aspect—maybe a parapsychological one—of the 'Dark Woman.'" In an earlier dream, he points out, the Chinese woman had "had a child, but 'the people' refused to acknowledge it." He believes the "other people" in both dreams represent his own "conventional objections to certain ideas—and my fear of them." The secret he shares with the Dark Woman, which fills him with apprehension, is the fact that there is "no symmetry of [objects] and reflections in this dream, since the whole point is about distinguishing between the two." In other words, that parity is broken.

This impressive dream occurred about a year and a half before Lee and Yang began to question the law of parity conservation and about two years before a Chinese woman proved that it wasn't true. Pauli had met Madame Wu in Berkeley in 1941 and been "very impressed by her (both as an experimental physicist and as an intelligent and beautiful Chinese young lady)." He had not gone to the trouble of examining the fundamental nature of parity alone in his CPT investigations, since at the time he was sure it was conserved. He continued to hold that belief a year and a half later when the Glorious Revolution began to unfold, and when Lee and Yang put out the call for experiments to test parity, he firmly expected it to survive.

When the results came in six months later, it took them several days, evidently, to cross the Atlantic. For, on the seventeenth of January, 1957, the very day that the *New York Times* announced parity's downfall above the fold on its front page, Pauli nearly placed another of those bets that he was bound to lose: "I do *not* believe that the Lord is a weak left-hander," he wrote to his student Victor Weisskopf, "and I am ready to bet a very large sum that the experiments will give symmetric results."

His day of reckoning came four days later, beginning with his morning mail. There he found a copy of the *Times* article, sent by another former student, along with two theoretical papers by Lee and Yang that explored the implications of parity violation. In the afternoon, he received a letter on the specifics of all three experiments from Valentine Telegdi, the leader of the Chicago group. By "coincidence," Pauli was scheduled to give a talk on the history of the neutrino at the Zürich Society of Natural Sciences that evening. Witnesses say it was an excellent talk; he was quite excited. At the end, he broke the news of parity violation and offered some "reflections" on the problem and its importance.

Despite his apparent bravado, the death of parity came as a shock to a man for whom symmetry held near-mystical significance. On the afternoon of his day of reckoning, he took a moment to write Madame Wu. "I congratulate you (to the contrary of myself). This particle neutrino, of the existence of which I am not innocent, still persecutes me." And in his 1957 letter to Jung, he described being "very upset" after receiving the news, and behaving "irrationally for quite a while."

"Now after the first shock is over, I begin to collect myself," he wrote to Weisskopf, six days after his reckoning. "It is good that I did not make a bet. It would have resulted in a heavy loss of money (which I cannot afford); I did make a fool of myself, however (which I think I can afford to do). . . . I am shocked not so much by the fact that the Lord prefers the left hand as by the fact that He still appears to be left-right symmetric when he expresses himself strongly. In short, the actual problem now seems to be the question: Why are strong interactions right-and-left symmetric?"

There is nothing in the equations of physics that explains why the weak force should violate parity. The strong force does not, and neither do the electromagnetic force or gravity.

As time went by, Pauli was calmed by the knowledge that strong reflection, CPT symmetry, still held, and his dreams reflected this gradual calming.

Wolfgang Pauli's last piece of scientific writing was an essay about the history of the neutrino, based on the talk he gave that jarring evening in Zürich. He was greatly pleased not only that Cowan and Reines had proven his long-ago intuition true, but also at the excitement and new physics that his little particle was generating. Upon completing the essay, in September 1958, he sent a copy to Lise Meitner as a gift for her eightieth birthday.

He died unexpectedly on December 15, a few months short of his own fifty-ninth birthday, not many days after being diagnosed with pancreatic cancer. This brilliant and unusual man was haunted by strange coincidence up until the moment of his death: another of his obsessions was what he and many others saw as the fundamental significance of the so-called fine-structure constant, a ratio of fundamental physical constants that is expressed as a simple number with the approximate value 1/137. Pauli once wrote that the "theoretical interpretation of its numerical value is one of the most important unsolved problems of atomic physics." Charles Enz adds that "the number 137 also had an irrational, magic meaning for Pauli."

He was assigned to room 137 in the Red Cross Hospital in Zürich, and in that room he died.

Part II

The Dream of
Neutrino Astronomy

4. Wisconsin-Style Physics

I'm waiting for ignition, I'm looking for a spark.
Any chance collision and I light up in the dark.

—PETER GABRIEL

FRED REINES ONCE TOLD FRANCIS HALZEN THAT AFTER HE AND CLYDE Cowan showed "that the particle actually existed, literally everybody came up with the idea that one could do astronomy with neutrino beams."

Very few people put the idea in writing, however. It first appears in the literature in 1958, the year Wolfgang Pauli died, in the diploma thesis of one Igor Zheleznykh, a student of the respected particle theorist Moiseĭ Markov at Moscow State University. (Zheleznykh readily admits that it was his mentor's idea.) Markov first presented it in public at a high-energy physics conference in Rochester, New York, in 1960, and at about the same time, Kenneth Greisen at Cornell, a former Manhattan Project physicist, proposed a similar idea at a conference in Berkeley.

Although Markov and Greisen's concepts were based on the same basic principle of operation, they had a fundamental difference that has led to two separate lines of experimentation over the years.

In his talk in Rochester, Markov proposed "setting up apparatus in an underground lake or deep in the ocean in order to separate charged particle directions by Cherenkov radiation," and in a journal article published in

January of the following year, he and Zheleznykh expanded upon the idea: "All known particles with the exception of neutrinos are absorbed by scores of kilometres of [earth or rock] and thus entirely screened by the planet. . . . It is noteworthy that not only μ-mesons [muons] (from the reactions involving neutrinos) produced in the detector itself, but also the μ-mesons from the adjoining layers of the ground ('the cushion') are detected in the experiment." This is a description of a telescope designed to observe up-going neutrinos that have passed all the way through the planet, and it works quite well for AMANDA and IceCube.

In his remarkably prescient talk in Berkeley, meanwhile, Greisen prophesized the future development of "high-energy neutrino astronomy. . . . [T]he neutrinos will convey a type of astronomical information quite different from that carried by visible light and radio waves." And at the end of a now-classic review article on cosmic ray showers, published in December 1960, he specifically proposed "a large Cherenkov counter, about 15 meters in diameter, located in a mine far underground." Since Greisen's concept has very poor angular resolution, it's more detector than telescope, but it would prove easier to put into practice than Markov's and bear fruit much more quickly.

Both of these visionaries emphasized the detection of μ-mesons, or muons, and they were ahead of the game in that respect as well, since there was still no proof in 1960 that muon neutrinos actually existed. They would be detected two years later by a team that included the same Leon Lederman who had played a role in the parity discovery. The experimental method, which required an accelerator that was very powerful for the time, had been conceived by Bruno Pontecorvo in Russia before it occurred to Lederman's team, but the Soviets never built the accelerator, while the Americans did, and Lederman's team went on to win the Nobel Prize in Physics in 1988— the first Nobel based on his own thinking in which Pontecorvo did not share.

The principle of operation behind both Markov and Greisen's concepts— a third method for detecting the neutrino, actually—is that when a muon neutrino collides with a nucleon and produces a muon through inverse beta decay, the newly born muon will speed away from the scene of its parent's demise in nearly the same direction that the neutrino was going—much in the manner of a billiard ball hit head-on by a cue ball—and will emit a pale blue light known as Cherenkov radiation as it does so. By arranging a set of light detectors in some clever way in or around the specific volume of whatever clear medium they have chosen to monitor, neutrino enthusiasts can then determine the direction of the muon and its parent neutrino in turn. Both concepts are thus known as Cherenkov detectors.

This form of radiation is named for the Russian physicist Pavel Alekseyevich Cherenkov, who won a share of the 1958 physics Nobel for discovering it. It is produced whenever a charged particle, such as a muon, speeds through a refractive medium faster than light can. The most common example is the eerie blue light given off by pool-type fission reactors or spent nuclear fuel rods immersed in water. In this case, the light is produced by electrons emitted in the beta decay of the many radioactive by-products of the uranium reactor fuel.

If you've heard of the postulate in Einstein's special theory of relativity that nothing can travel faster than the speed of light, don't worry; we're not violating that rule here. For Einstein was speaking of the speed of light in a vacuum. In a refractive medium such as water, ice, glass, or even air, light will travel less quickly, so that it is possible for a different kind of particle to go faster than light can in that medium and not break the rule.

Cherenkov radiation is the optical equivalent of the sonic boom that occurs when a jet "breaks the sound barrier," or goes faster than the speed of sound: since the sound can't keep up with the jet, it is dragged along behind it in the same way that the waves in the wake of a speedboat ride off at an angle because they can't keep up with the boat. In a three-dimensional situation like that of the jet or a speeding muon, the sound or light waves take the shape of a cone, rather than the V of a boat wake. The muon, in other words, drags a cone of Cherenkov light along behind it. If it were to pass through a projection screen, a spot of light would appear on the screen at the moment it passed through, and the spot would change immediately into a tiny circle, which would expand and gradually fade as the muon zoomed away.

The difference between Greisen and Markov's concepts is basically one of geometry. Markov's idea was to place a three-dimensional grid of light detectors in a natural body of water and watch neutrino-born muons pass through the grid. We'll call this the Markov or "plum pudding" design, because the detectors are located *inside* the so-called detection volume. Greisen's idea was to surround a manmade tank of water with a shell of detectors, in which case the detectors would obviously be *outside* the detection volume. This is the "shell" or Greisen design. In both cases, the bigger the detection volume, the more sensitive the detector, since this increases the likelihood that a neutrino or neutrino-born muon will pass through it.

Stepping a few decades ahead for a moment, let's visualize the cone of Cherenkov light being dragged along by a muon as it passes through Ice-Cube. Think of the cone as a three-dimensional version of the waves produced by that boat as it skims along the surface of a calm lake. The three-dimensional grid of light detectors that the scientists have placed in the ice is analogous to a two-dimensional grid of pontoons floating on the surface of the lake. As the boat drags its wake through the grid, the pontoons will bob up and down. If you know the speed of the waves, then with the help of a little algebra and geometry you can reconstruct the speed and path of the boat using the timing of the wave fronts as they hit the different pontoons. The IceCube scientists reconstruct the direction and speed of an invisible muon zooming through their 3D detector in precisely the same way. Francis Halzen says, "It's like flying over the lake in an airplane. You might not be able to see the boat, but the waves tell you where it is and where it's going."

The reason Markov and Greisen focused on muons rather than electrons was that they knew they would be easier to detect. For the muon has enough mass and therefore momentum to travel in a straight line through whatever medium it finds itself in; whereas an electron, being two hundred times lighter, will be swayed by the electric fields of nearby nuclei and begin to zig and zag within a few meters of its birth. With each zig the electron produces bremsstrahlung radiation: photons, which, if they are energetic enough, will produce electron-positron pairs. These secondary pairs will in turn zig and zag and produce more bremsstrahlung, which will produce more pairs, and so on and so forth. The result, in IceCube parlance, is a cascade: a short, cigar-shaped flash of light, proportional in volume to the energy of the electron neutrino that created it and pointing in the direction that the neutrino was traveling in.

Not only is the long, straight track of a muon much easier to see than a cascade, its direction is much more precisely defined, so that it yields a more accurate measure of the direction of its parent neutrino. This makes a track much more useful for astronomy, since it points back more accurately at whatever cosmic object the neutrino came from. And highly energetic muons (which come from highly energetic neutrinos) have the added advantage of flying through ice or bedrock for several kilometers before they decay. This is what Markov and Zheleznykh meant by a cushion: the plum pudding design can detect a muon even if it's born a good distance outside the detector grid. Remember that we're looking for up-going muons created

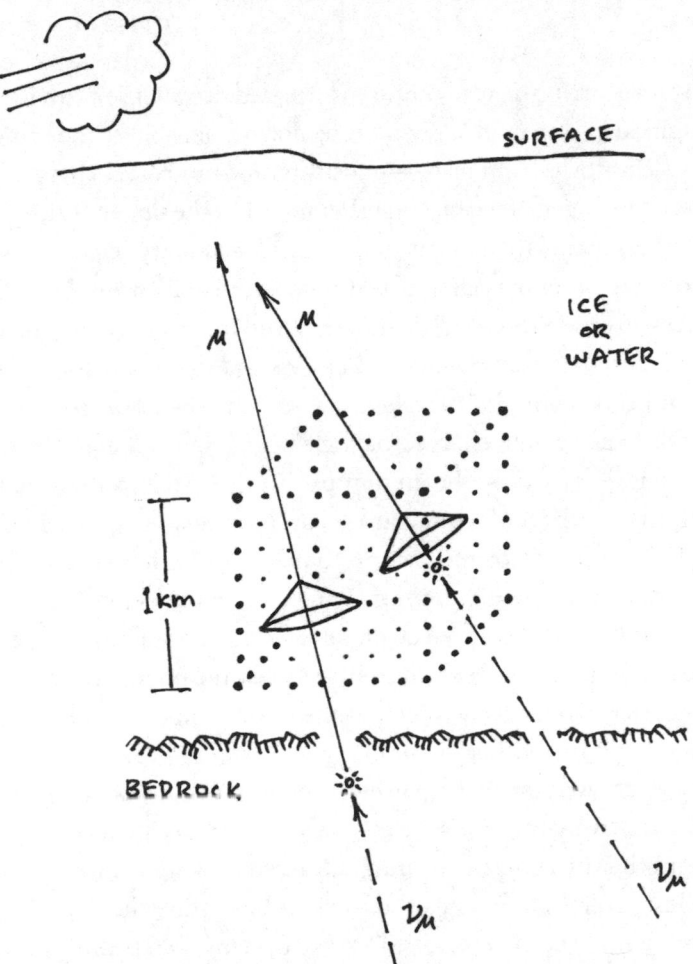

Two up-going muon neutrinos (v_μ) detected by an instrument of the Markov or "plum pudding" design. The neutrino on the left interacts with a nucleon in the bedrock or "cushion" below the instrument, giving rise to a muon (μ) that passes through it dragging along a cone of Cherenkov light. The neutrino on the right interacts inside the detector. The tracks of the muons indicate the directions of the corresponding neutrinos.

by neutrinos that have entered the planet somewhere north of the South Pole. IceCube will detect a muon born in the ice or bedrock to its side or beneath it as long as the muon's path eventually takes it into the grid. This enhances the effective volume of the detector and makes this type of instrument more sensitive to muon neutrinos than electron neutrinos, generally speaking.

✦ ✦ ✦

The reason a neutrino detector needs to be placed deep in the earth (or water or ice) is to shield it from cosmic rays raining down from above. The primary cosmic rays racing toward our planet—mostly protons and other charged nuclei—collide with nitrogen, oxygen, and other nuclei in the upper atmosphere and create showers of down-going pions and other "secondary" cosmic rays, which then decay into other particles, such as muons, or collide with the atmosphere themselves, to create so-called air showers. And this serves as a good example of the similarity between cosmic accelerators and the manmade variety.

The principle behind all accelerators is to employ powerful electromagnetic fields to accelerate charged particle beams to high energies and then slam them into targets or "beam dumps." In the case of a manmade accelerator this might be a block of carbon, and for cosmic rays it's our atmosphere. The "primary" cosmic rays, protons and nuclei, have been accelerated by some cataclysm in the cosmos, launched into interstellar space, and then steered by interstellar electromagnetic fields in a winding course to our planet. They hit the beam dump of our atmosphere and produce secondary particles in the same way that the beam in an earthbound accelerator does when it hits its beam dump.

The way accelerator physicists generate rare or new particles is to start off with a relatively easily produced charged particle, such as a proton; accelerate it either in a circle or a straight line, employing focused electromagnetic fields; divert it into a beam dump; and examine the resulting shards and debris with special detectors. In some designs, two beams are aimed at each other. (Richard Feynman compared this to smashing together two Swiss watches in order to find out what's inside.)

The standard unit of energy in particle physics is the electron-volt (eV), the amount of kinetic energy acquired by an electron when it is accelerated across a potential difference of one volt. This is a miniscule quantity by everyday standards—a one-hundred-watt light bulb puts out almost 10^{21} (one billion-trillion) electron-volts every second—but it's a convenient unit for expressing the masses of elementary particles. The electron, for example, has a so-called rest mass (its mass-energy in its "rest frame," exclusive of its energy of motion) of about 510 eV. And since Einstein has shown that mass and energy are equivalent, the way to generate more massive particles in the debris of a particle beam that has just hit its dump is to generate higher and higher beam energies, either by making the accelerator larger, so that it can accelerate the beam over longer distances, or by employing stronger electromagnetic fields.

The present record holder is CERN's $10 billion Large Hadron Collider, where the Higgs boson was discovered in 2012. The LHC is designed to cycle opposing beams of either protons or lead nuclei around in a circular tunnel twenty-seven kilometers, or seventeen miles, in circumference and smash them together at energies, in the case of the proton, of about fourteen trillion electron-volts.

Since cosmic accelerators are unconstrained by international science budgets and real estate considerations, however, such as the circumference of the Earth or even the Solar System, they generate energies far greater than anything humans can dream of replicating. The record thus far is held by the so-called Oh-My-God particle, which was observed by an instrument called the Fly's Eye in a Utah desert in 1991. (Kenneth Greisen, incidentally, invented this instrument.) This single subnuclear spec packed as much punch as a baseball going about sixty miles an hour—about three hundred thousand times the capability of the LHC. It seems to have been either a proton, a heavy nucleus, or even a neutrino; but it's impossible to know, since it died when it hit the atmosphere, giving birth to a shower of about *two hundred billion* secondary particles and decay products.

The mile of frozen overburden above IceCube serves as a shield from such down-going cosmic rays; however, many still punch through, penetrating deeply enough to reach the array. In fact, about a million atmospheric muons streak down through the detector for every more interesting muon, born of a neutrino, that streaks up from below. Thus one of the major challenges for this technology is to distinguish up-going from down-going muon tracks. Picking the up-going needles from the down-going haystack is a tricky business, as you might imagine.

Greisen's shell design is typically realized as a large tank of some ultra-pure, clear liquid, usually water, completely surrounded by walls of cheek-by-jowl light detectors, situated a mile or more deep in a mine of some sort. The idea here is to detect only those muons that are born inside the tank, and to aid in this purpose Greisen suggested that the primary detector "be . . . enclosed in a shell of scintillating material to distinguish neutrino events from those caused by [muons]." In other words, there were actually two shells: an inner one of primary light detectors and an outer one of scintillators, which would be used to exclude or "veto" atmospheric and

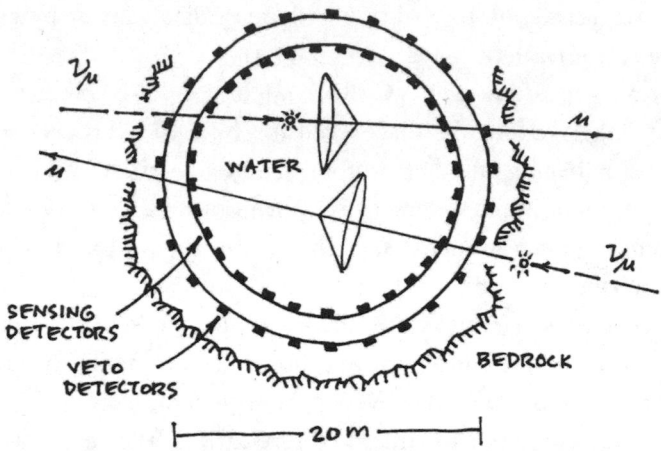

Muon neutrinos (v_μ) passing through an instrument of the Greisen or shell design. The upper neutrino, entering from the left, interacts inside the detector, so that the light cone from the resulting muon (μ) illuminates sensing and veto detectors only on its way out. The lower neutrino, entering from the right, interacts outside the detector, giving rise to a muon that illuminates detectors both as it enters and as it exits. Thus the shell of veto detectors helps reject muons born outside the detector.

other muons born outside the detector. Since these particles will pass all the way through the instrument, they will light up scintillators as they enter and as they leave, while muons that are born inside the volume will only cause a signal on their way out.

Any muon born in the tank (*or* passing through it) will light up a ring of primary detectors on the outer wall as it exits: one detector at first, followed by an expanding ring around it. And the muon's direction can be inferred from the shape of the ring: if it's passing through at an angle, it will produce an oval rather than a circle.

The resolution of the shell design is limited to twenty or thirty degrees—good enough to tell left from right and up from down, but not much better than that—while the plum pudding design (as embodied in IceCube) can resolve to less than half a degree, roughly the angular spread of a full moon.

John Learned, who would lead DUMAND twenty years later, the first attempt to make these dreams a reality, observes that "Greisen never did anything in this direction. Reines grabbed the ball and was running with it in the States, Reines and his gang from early on."

In a 1960 review article, Reines discussed the detection of both "neutrinos produced extra terrestrially (cosmic) and in the earth's atmosphere (cosmic ray)." But he comes across as being much more conservative than Greisen and Markov, possibly owing to the scars from his vexing attempt to detect manmade neutrinos in the first place: he doesn't even touch the first problem, and he pronounces the second "most formidable." Nevertheless, he began visiting mines to scout out locations for an instrument in the Greisen tradition as early as 1963 and was considering a Markov instrument in the deep ocean at least as early as 1966.

He must have concluded that Cherenkov detection was too much of a long shot, however, since he ended up attempting to detect atmospheric (or, in his words, cosmic ray) neutrinos by a different method. This was still a big step: the first attempt to detect neutrinos produced by nature, rather than a manmade reactor or bomb.

Atmospheric neutrinos are created in the same way as atmospheric muons: through the decay of secondary particles, including muons, born in the collision of primary cosmic rays with the atmosphere. They're useful for some kinds of particle physics, as we'll see, but have nothing to do with astronomy, because they're born so close by.

Sometime in 1963, Reines became aware of a doctoral thesis at the University of Bombay, suggesting that certain mines in India's Kolar Gold Fields might be deep enough to provide sufficient shielding against atmospheric muons to permit the detection of atmospheric neutrinos. He visited the Indian scientists who had come up with the idea and the mines themselves, but eventually chose the deepest mine in the world, the East Rand Proprietary Gold Mine near Johannesburg, South Africa. By this time, he had left Los Alamos to become head of the physics department at the Case Institute of Technology, now Case Western Reserve University, in Cleveland, Ohio.

In collaboration with a group from the University of Witwatersrand in Johannesburg, Reines installed the largest particle detector ever built at the time, comprising twenty tons of liquid scintillator, in a laboratory two miles underground. Evidently, the local miners referred to the scientists as "goggafangers," meaning "bug catchers," and to Reines himself as "makulu bass goggafanger," or "big boss bug catcher."

The idea was to detect muons traveling horizontally through two parallel walls of scintillators. Since there was significantly more than two miles between the detector and the surface of the Earth in the horizontal direction, any muon created in the atmosphere would decay before reaching the detector from that direction, so those that did reach it must have been created by a neutrino interacting somewhere in the earth between the detector and the surface.

But this time around, Reines had some direct competition. The Indian group whose idea he had borrowed, led by M. G. K. Menon, had gone ahead and installed their own detector in one of the Kolar gold mines. The Case-Witwatersrand group detected their very first naturally born neutrino on February 23, 1965 (a date that recurs in neutrino astronomy), and the Indians detected theirs about a month later. But Menon's group is accorded formal priority, because they published their results about two weeks before Reines's did. (Publication is important, as it usually indicates that a rigorous analysis has been done and uncertainties and potential misinterpretations addressed.) The two sides tended to argue about it, of course (discord tended to follow Reines), but pretty much everyone else chalked it up as a tie.

Such pettiness aside, the neutrino again demonstrated its shyness: all told, the Case-Witwatersrand group detected only 167 atmospheric neutrinos over the six years that they took data.

✦ ✦ ✦

Meanwhile, the seeds planted by Markov and Greisen gestated in fertile ground. Around the world, particularly in the United States and Russia, solitary individuals began visiting remote lakes in order to drop short strings of light detectors into the water and fish for muons. One was the young John Learned, a grad student at the University of Washington.

John was born in 1940 in Plattsburgh, New York, on the edge of the vast Adirondack State Park, which takes up a large portion of the northern part of the state. His grandparents lived in a sort of ancestral home there, which had been in the family for about a hundred and fifty years. When he was six, his journalist father moved the family to Staten Island in New York City, but John and his brother kept in touch with their rural roots by spending all their summers "pretty much unsupervised" in the north country: "We hunted, fished, hiked around, and camped out as we pleased. We built projects in Grandpa's shop (including cannons which we shot at each other), built models and tied flies . . . Indeed it was quite idyllic." Thus he developed a passion for exploring, tinkering, and messing around in the wilderness, which is basically what cosmic ray physicists do for a living.

John's years at Brooklyn Technical High also proved useful, as they gave him access to printing, pattern making, tin and machine shops, and even a foundry. He majored in physics at Columbia, and then worked in astronautics and aerospace engineering for a few years, first in the east at General Dynamics and then at Boeing in Seattle. After a few years, when Boeing began grooming him for management, he left, because he liked the technical stuff better. Too bad. A little management experience might have come in handy later on.

John first heard about the odd sport of neutrino fishing in his very first class at the University of Washington. He ended up pursuing the idea for his Ph.D. thesis, with the teacher of that class, Howard Davis, as his adviser.

"It was just wonderful fun, wonderful fun, because essentially I worked alone. I built a detector, which I took to Lake Chelan in the Cascade Mountains, and built a raft and had a boat and went out there and sank detectors down and counted cosmic rays as a function of depth and so on. Had a lot of interesting experiences. Didn't do any science of any particular merit. . . .

"You know everything has a history," he says. "Listen to this:

"One time I was there standing on the dock. I got my little barge pulled up, and there's this old guy with a German accent. Comes and says, 'Vat are you doing?' heh, heh. So, I tell him. He says, 'Oh! We did that in Lake Constance in nineteen thirty . . .' three, or something—before World War II. And I eventually looked it up. It was an old guy by the name of Regener. He actually made the first measurements in the lake all those years before. I don't remember what the hell he was using . . . I think maybe he lowered Geiger counters down or something [correct] . . . But it was just a marvelous moment, because . . . [mimicking the old man] 'I did this back in the thirties!' [and himself] 'C'mon!' "

This was indeed a marvelous visitation. Erich Regener had been both a "diver" and an "air rider" in what Pierre Auger called the heroic period of cosmic ray physics. He had begun dropping his "Bodensee Bombe" into the Bodensee, which is the German name for Lake Constance, in 1928. He wasn't fishing for muons; they hadn't been discovered yet. He was trying to answer the central question of the day, which was whether cosmic rays were particles or electromagnetic waves, that is, light, and he cast his vote in favor of the second hypothesis by christening his small research vessel the *Undula*. Regener also measured the intensity of cosmic rays as a function of height in the atmosphere, but he wasn't an air "rider" exactly. He came up with the innovation of attaching automated recording devices to his scientific instruments, so that he didn't need to ride into the air with them in the way that Victor Hess had when he had discovered cosmic rays sixteen years earlier.

Learned had no idea where his career was going in his school days. He thinks the best analogy would be not "a mountain campaign," which is usually focused on a specific objective, "but an exploration of new and ever-exciting territory"—an attitude that would help very much as he went on to pioneer neutrino astronomy. At Lake Chelan, he made his first baby step by being the first to count atmospheric muons in an open body of water, rather than an enclosed tank or detector.

As his time at the University of Washington was coming to an end, the head of his lab, an old cosmic-rayer named Bob Williams, asked him what he wanted to do next, and John responded with three experiments that intrigued him. In those good old days, you could get a job without a formal application. Williams called Fred Mills, a professor at the University of Wisconsin, and Mills simply created a position at a high-energy cosmic ray experiment on Mt. Evans in Colorado, the closest 14,000-footer to Denver.

Learned was connecting with the heroic age there again, since one of Auger's mountaineers, Nobel Laureate Arthur Holly Compton, had established a laboratory near the summit of Mt. Evans shortly after a road had been built to it in 1927. There, Compton had proven that cosmic rays are made up mainly of charged particles, thus proving Regener wrong. A few years later, the Italian physicist Bruno Rossi (who, as it turns out, was Kenneth Greisen's doctoral adviser) used data he obtained on Mt. Evans to make the first accurate measurements of the muon's lifetime.

The outward—and outdoor—aspects of John's new position made it a natural fit, of course; but he was also a natural for the research group in Wisconsin that he had unknowingly joined. It was known as CCFMR, after the names of the professors involved: Dave Cline, Ugo Camerini, William F. "Jack" Fry, Bob March, and Don Reeder.

In the late sixties and early seventies, the physics department at the University of Wisconsin–Madison was the crossroads for a surprising number of people who would play leading roles in the invention of neutrino astronomy later on—and in AMANDA and IceCube in particular. CCFMR was the energetic center of it all.

One of the many dreamers who passed through Madison during these years was Leo Resvanis, who would eventually become the flag-bearer for a still unrealized attempt to build a kilometer-scale neutrino telescope deep in the Mediterranean, off the coast of his native Greece. In my first conversation with Leo, he introduced himself in tones of surprising dignity with, "I am a Wisconsin-style physicist." He was speaking of the atmosphere that swirled around CCFMR.

"These guys were as good as they come!" Leo writes. "Very smart and very creative, but with a fantastic nose/gut feeling for physics. In a spirit of total anarchy (i.e. antidiametrically opposite [*sic*] to Committee Physics) they put their finger on fundamental questions. . . . There is a proverb you must have heard: 'a Camel is a Horse designed by a Committee.' They were driven by a pioneering spirit. This you do not find in today's experiments."

Bob Morse, who would become one of the main pioneers of AMANDA, is also a CCFMR alumnus. He did his graduate work in Madison with a different group, then left to pursue opportunities in Colorado for about six years, and returned to become a senior scientist with CCFMR in 1975. A happy and inspired expression lights up his face when he thinks of those days. "There was a lot of disciplined thinking, but as far as actually staying on message and on target, it was a little loose," he recalls. "And so what happened is, is you said, 'Oh Jesus Christ, we've got a test run coming up in three weeks! We better start to think about this!' you know. . . . We weren't as scared as we should have been is the best way to describe it."

"Style: fast, smart, intuitive, irreverent," writes John Learned in characteristically unmediated prose. "Willing to try many new techniques and tricks to get the physics answers . . . go for broke in physics and when not doing physics, party hard. Except for Fry and Reeder, lots of womanizing. . . . Emblem, the UW[†] red flying-cock-and-balls, as was painted on our research trailer at Argonne[‡]. . . . The style [stood in] particularly . . . strong contrast to the other boring uptight HEP[§] folks at UW, who all scoffed at our image but looked on with envy at the fun we were having. And in the end, despite a few miscues, we did better physics."

Jack Fry, the official leader of the group, had a knack for staying ahead of the curve. In 1952, he took the bold step of switching from cosmic ray detection to the accelerators, several years before the latter began to produce high enough energies to be of significant use in particle physics. An aficionado of Italian culture, he was best known outside the academy for

---

[†] University of Wisconsin.
[‡] Argonne National Laboratory, outside Chicago.
[§] High-energy physics

his work on the acoustics of the Stradivarius. He built and conducted research on violins.

Learned believes Ugo Camerini was "the real center of action" and sometimes refers to CCFMR as the "Camerini group." The son of a Milanese Jew who had moved his family to Brazil to escape Mussolini, Camerini had a long and splendid résumé in cosmic ray physics. After completing his undergraduate work in São Paulo, he re-crossed the Atlantic to join the renowned Powell group at the University of Bristol, where he assisted in the discovery of the pion, high in the Pyrenees. Cecil Powell later won a Nobel for this discovery. Camerini never actually obtained a doctorate, because he couldn't be bothered to submit a typed version of his thesis. He was in too much of a hurry to start on his next project: building the highest cosmic ray station in the world, at about seventeen thousand feet on Mt. Chacaltaya in Bolivia, where, in 1947, the discovery of the pion was confirmed.

"Ugo never finishes anything; his students finish things," says Morse. "Ugo would just throw these gems out and would hope that the students caught them, or caught most of them, and then he was off to something else." Learned suggests that he "gave off sparks."

Bob March had studied at the University of Chicago under Fermi. As the story goes, the FBI once asked Fermi to fire March for refusing to spy on his own parents, who were members of the Illinois Communist Party, and Fermi nobly refused. Along with most if not all of the CCFMR group, March opposed the Vietnam War—during which era, according to Learned, "there were huge political divisions in the Department, ranging from the wild eyes in HEP to real John Birchers in low temperature physics (more or less scaling with physics energy range!)." In other words, the higher the energy they were studying, the more leftist their politics.

Learned describes March as "a most interesting character, one of the most brilliant people I know." He was a renowned teacher and popularizer of physics—he hosted a radio show in Madison for many years—and he took a special interest in the culture of the field. One of his causes was to increase the number of women in graduate student and faculty positions. In 1970, he wrote a book named *Physics for Poets*, which led to a course on the subject at Madison. The idea caught on; similar courses are now offered at virtually every college in the United States, and other scientific fields have followed suit as well.

Although one doesn't hear stories about Don Reeder's extracurricular interests, one does hear that his colleagues thought of him as a solid physicist and a genuinely decent person. Dave Cline once admitted that Reeder probably helped Fry counterbalance the craziness of Camerini and March—and especially of Cline himself.

Cline is still remembered as the most remarkable member of the group, and from the late sixties to the mid-eighties he was also at the top of his game. The sheer number of cutting edge projects he engaged in, all over the world and all at the same time, simply boggle the mind. He was one of the main players in the breathtaking drama that has been particle physics for the past half century, and he stayed at the forefront until the moment of his sudden death, at the age of eighty-one, in 2015. John Learned describes him as "perhaps the most maniacally physics-focused person" he ever knew. Cline was a founding member of one of the collaborations that discovered the Higgs boson, and his main interest in the last few years of his life was the cold dark matter.

His tendency to take pretty much everything to the extreme may be explained by a tinge of fanaticism in his upbringing: Cline was born in Kansas City, Kansas, to accomplished and loving evangelical parents. "The psychology of religion is powerful, and it had me in its grips," he once told me. In his youth he always thought he would be a philosopher, but during a short stint in the army at a guided missile facility near El Paso, Texas, he developed an interest in physics. "Then I started realizing . . . that philosophy can't teach you anything either. So, actually the only way you can learn anything is physics. [It's] the only thing that controls the universe. God must have started physics, period, you know," although he added that he had taught too much astronomy to believe in God.

He majored in physics and minored in philosophy at Kansas State University, got a masters in physics at the same school, and then proceeded to Madison to pursue a doctorate under Jack Fry. He ended up doing his thesis research in Berkeley on a new method for detecting elementary particles called the bubble chamber. (Fry had again demonstrated his foresight by collaborating with Berkeley on the bubble chamber work.) After earning his degree in 1965, Cline returned to Madison and proceeded to rise through the ranks faster than anyone else ever has in the physics department, becoming a full, tenured professor in only three years.

Those were the days when the standard model of particle physics was coming into focus. At the center of this advance was so-called electroweak theory, a construct that unified the physics of the electromagnetic and weak nuclear fields, the latter of which is closely tied to the neutrino. The standard model could be used to make detailed predictions about particles that had yet to be discovered, including the carriers of the weak force, the W and the Z. So Cline, being an experimentalist, got right on the case.

In 1969, at the tender age of thirty-six, he was the lead author of the "E1A" proposal for the very first experiment to be run at the national accelerator laboratory, not yet named Fermilab, that was just being built in Batavia, Illinois. (Until the end of his life, Cline proudly pointed out that he wore visitor badge number one for Fermilab.) The most important goal of E1A was to discover the W using a beam of high-energy neutrinos. Thus "neutrino physics was one of the major justifications for building the Fermi National Accelerator Laboratory." But Bob Morse believes the proposal had even wider implications than that: "Other people have said that it outlined every important aspect for the next twenty years in neutrino physics. I mean the E1A proposal is a remarkable document—Dave Cline! It's just hidden away as a proposal, but it's gold. It's absolute gold."

The EIA experiment had been conceived by the comparatively sober-minded Alfred Mann of the University of Pennsylvania, but Mann had been so impressed with Cline's work that he recruited him early on in order to "add more weight to his proposal." They then brought in the brilliant and hotheaded Italian physicist Carlo Rubbia, who was a professor at Harvard at the time.

The story of the E1A experiment is told elsewhere. While it did not produce the W, it did result in the discovery of the so-called neutral current interaction in which a neutrino interacts with a proton or neutron without dying, as it does in inverse beta decay. This was an important advance, but the problem was that first they thought they'd discovered it, then that they hadn't, and finally that they actually had; so the joke was that they discovered "alternating currents" and Cline earned the nickname "A.C.D.C." His caution may also have cost him a Nobel Prize, since a group at CERN made the discovery while he was still dithering, and they did win one.

But Cline and Rubbia were not the sort to look in the rearview mirror. They moved on to a new experiment at CERN, where they detected both the W and the Z in 1983, and this led to Rubbia's winning a Nobel the following year. (It's impossible to know what might have gone on in the back rooms of the Nobel physics committee, but we do know that Cline was nominated by more than one Nobel laureate. It could be that his American citizenship was a strike against him. There was tremendous competition between Europe and the United States at the time, and this was the first Nobel that CERN ever won.)

Since the prize is awarded every year, it can become a bit ho-hum for the casual observer. But the W/Z discovery was a true watershed, as it provided dramatic confirmation of the standard model, which was still relatively young at the time. The only particle physics discovery that has rivaled it in excitement over the past thirty years is probably the Higgs, and Cline, of course, was involved in both.

The seventies were heady days. He and his collaborators used to sign off on letters with "the force is with us," alluding to both the weak force and the new movie *Star Wars*. Having passed through Berkeley at just the right time, Dave had also evolved into a true "hippy radical." In the book *Nobel Dreams*, he refers to himself and Rubbia as "radical, crazy people." He "once visited Rubbia in Geneva in 1972 dressed like Buffalo Bill, with a white suit and a white cowboy hat, and his shoulder-length hair." When he was reminded of this, Dave responded, "Not exactly, but, you know, it is possible. . . . In those days . . . we were euphoric."

Part of the reason it's possible is that he *was* into clothing. In the 1970s, while he was living in Madison, he and his second wife, whom he had met at Berkeley, were running a successful string of fashion boutiques in San Francisco and Hawaii, which was where she lived. He didn't mention this moonlighting job to his Wisconsin colleagues, "because I was already crazy enough, you know? This is one more level of craziness. This woman, by the way, was very beautiful." To give an idea of his dedication to physics, however, he made the conscious and difficult decision to dissolve this marriage in order to focus on finding the W, because he didn't think the added travel would be fair to his family. Until the end of his life, he questioned the wisdom of that decision.

The CCFMR group employed a Madison bar named the 602 Club, "painted a bilious green inside," according to John Learned, as a sort of ad hoc conference room. Bob March's holding of court there virtually every afternoon became a Madison institution.

"That was a weird club. That was one of my favorite places," Cline recalled. "Madison . . . was one of the most interesting and liberated cities in the world, and the 602 Club was our place to go. . . . You'd meet very crazy people."

"That's where future chancellor John Wiley used to drink a lot of beer," adds Bob Morse, "except on Sunday night when it was closed and we had to go to Glen and Anne's, heh, heh, heh. . . ." (Wiley, a friend and fellow grad student of Morse's, would play a huge role in both AMANDA and IceCube decades later.)

"You know, the 602 Club was a great place to drink, because they had sixteen-ounce schooners of beer for a quarter, and they didn't have a goddam juke box, which means that you were forced to actually talk to people instead of all this blaring music. We had people from the art department, the philosophy department. . . . Basically, it took all of nature's two-standard deviation misfits, and you threw them all into this one club. We all got along fine."

✦ ✦ ✦

Francis Halzen entered this milieu in its heyday, 1971, and his invitation to work in Madison was proffered, appropriately enough, in a bar. This was a very different bar from the 602 Club. It was located in a hotel in the resort town of Méribel in the French Alps.

He was attending—or, actually, about to leave—that year's Rencontres de Moriond, a prestigious, invitation-only gathering of particle physicists that took place annually at one French ski resort or another. They were purposely called "rencontres" or gatherings, rather than conferences, because the purpose was to create an intimate atmosphere "in beautiful and inspiring surroundings" in order to encourage the creative exchange of new ideas. The Moriond gatherings still take place, but they're more frequent and less exclusive than they used to be. In those days, the participants took a break to ski during the day and played music for each other at night, and it was considered bad form even to take a phone call. The talks were delivered in the hotel bar at Méribel. The speaker's blackboard was set up behind the bar itself.

Francis was a rising star in theoretical particle physics at the time. He turned twenty-seven just a few days after the gathering ended. He was finishing up a two-year fellowship in the theory group at CERN, and he had a permanent position waiting for him back home in Belgium, at the University of Leuven. He'd been invited to speak at Moriond on the basis of a paper he'd written with a theorist roughly his own age who had recently moved to CERN from a post-doc in Madison. Francis claims that this paper "would basically have gotten me a job anywhere," and it already had produced an offer from Caltech. He recalls looking out on the audience during his talk and seeing an Italian physicist sipping a cognac.

He had driven south from Geneva with his wife, Nelly, and they were planning to drive back the minute his talk was over. She was waiting outside in the "ugly hippie yellow/orange (excuse: we had no choice, we took the first model available)" MGB sports car that Francis had bought with a portion of the prize money he had won for his Ph.D. thesis at Leuven, two years earlier.

On his way out of the hotel, he was told that someone wanted to speak to him. "They better be important and be fast," he responded, "because my wife is waiting in the car."

The someone turned out to be Vernon Barger, the theorist who had hosted Francis's co-author in Madison. Barger explained that he didn't have a job to offer, but that he could probably find a way to support Francis for a visit of about six months.

"I was a bit of a dilettante at the time—probably still am," Francis observes. "I am not driven by plans or ambition. . . . I do what I want to do; you only live once. . . . This was better than going back to Belgium that particular day, so I go. . . . I mean what can go wrong in six months?" There wasn't much risk involved, since he had the job lined up in Belgium.

His visit began on October 6, 1971. ("Immigration made sure that I never forgot that day.") Almost half a century later, he's still there.

Another of Dave Cline's contributions to "Wisconsin-style physics" was to observe no clear boundary between theory and experiment. He often collaborated with Vernon Barger. The two had become "semi-famous," he recalled, by demonstrating that a new theory of so-called Regge Poles fit some new accelerator data—and they had also become infamous when Cline had once snapped a picture of a slide at a conference somewhere, brought it back to Barger to analyze, and the two published the correct interpretation of the data on the slide before the group that produced it did.

Dave had been toying with the idea of starting a "phenomenology" group to work at the interface of theory and experiment. When Francis arrived, Dave recognized his cleverness and broad knowledge right away and realized that Francis "fit perfectly into this picture." Dave was the principal investigator (PI) of the E1A experiment at that point, so he was bringing in roughly $1.5 million a year. When the money that Barger had found for Francis ran out, Cline hired him as a post-doc on the E1A project—a pure theorist working on an experiment, unheard of at the time—and made him an offer of a permanent position in the non-existent phenomenology group that he had in mind.

Cline was virtually unstoppable in those days. The phenomenology group was born in a meeting with the dean of the Madison graduate school and Cline's grant manager at the Department of Energy, when the dean promised $1 million from a remarkable fund at the University of Wisconsin called the Wisconsin Alumni Research Fund or WARF (about which more later), and the grant manager offered to match it. The group consisted of two experimentalists, Cline and Don Reeder, and three theorists: Barger, another professor named Martin Olsson, and Halzen, who observes that Cline was "by far the most imaginative of us. I mean, he led in every respect." Cline's insight about phenomenology turned out to be right on target. It was soon recognized as a legitimate occupation.

Cline was a synthesizer. Bob Morse says "he saw all things as being doable." He also disregarded the border between accelerator and cosmic ray physics. "Didn't matter," says Bob with a laugh. "Dave said, 'I'm interested in a particle of a certain energy.' If it meant he had to go to cosmic rays to find them, he'd go to cosmic rays. . . . He didn't see any boundaries."

The two most influential papers that Francis wrote during his first few years in Madison were co-written with Cline, and one of them dealt with cosmic rays—the first time Francis ventured into the field in which he would do his most important life's work. That paper, incidentally, is also considered one of the most influential of Cline's illustrious career. Francis recognizes Cline as one of the three or four important mentors in his life. In fact, he named his son, David, after Cline and another mentor, David Speiser, one of his advisers in graduate school.

And so, CCFMR trained its protean anarchy on experiments at accelerators and high-energy cosmic ray stations all over the world—usually a handful at a time. But the cosmic ray work took place mostly on mountaintops, so John Learned's thesis research was relegated to the background. After his stint in Colorado, he worked for a few years on traditional high-energy experiments at the Argonne and Stanford Linear accelerators, while his freewheeling mentors gave him "a good deal of freedom to padiddle around" in the nascent field of neutrino astronomy. Meanwhile, Francis Halzen remained pure, merrily writing papers as theorists will, and learning something about experimentalism by way of phenomenology, but not sullying himself on any true experiments quite yet. It would be almost twenty years before AMANDA would draw his first blood.

5. Peaceful Exploration by Interested Scientists Throughout the World

JOHN LEARNED FINALLY FOUND AN OPENING IN 1973, WHEN HE JOINED several scientists from the United States, the Soviet Union, and Japan in an informal discussion of Cherenkov detection at the biennial International Cosmic Ray Conference, which was held in Denver that year. Fred Reines was there. The Soviets were represented by Georgii Zatsepin, a brilliant experimentalist and theorist (not as unusual in Russia as it was in the west). Zatsepin had begun his scientific career in the high Pamir in the 1940s and conceived of Cherenkov-based neutrino astronomy at about the same time as Greisen and Markov, but had not gone to the trouble of writing it down. Saburo Miyake, a respected cosmic ray and neutrino specialist who had collaborated on the Kolar Gold Fields experiment, represented Japan.

The meeting consisted of "just a few people sitting around saying we should start to take this seriously and do something," Learned recalls. Reines coined the name DUMAND, for Deep Underwater Muon And Neutrino Detector, although the details of any such instrument were entirely nebulous at the time, and an ad hoc committee was formed. The first formal

DUMAND workshop was convened at Western Washington State College in Bellingham in July 1975.

A diverse array of experts was invited to the Bellingham workshop, including oceanographers, ocean engineers, submarine cable specialists, and biologists who might illuminate the physical scientists as to the potential effect of deep-sea life on their experiment. "For all we knew, a giant squid would try to have sex with the thing," says Learned, not entirely in jest.

While there is no, ahem, hard data on that particular possibility, it is well-known that many sea creatures give off light in a process known as bioluminescence, a complication well worth investigating, as it will create a disruptive background in a Cherenkov detector. At a subsequent DUMAND workshop, a specialist in marine invertebrates discussed the problem in a paper entitled "Serpents in an Astrophysical Eden."

Fred Reines and John Learned were voted chair and vice-chair of a formal DUMAND steering committee at the Bellingham workshop, and Reines then went about inviting others to join, including Dave Cline, who believed he was brought in for the knowledge of neutrinos he was gaining in the E1A experiment. Dave's CCFMR colleagues Ugo Camerini and Bob March could hardly keep themselves away from such a crazy idea, so Wisconsin became one of the principal institutions in the DUMAND collaboration.

Reines also brought in Arthur Roberts, who was John Learned's "grandfather" in a sense, since he had been thesis adviser to Learned's adviser, Howard Davis. This foray into cosmic ray physics would be the coda to Roberts's long and distinguished career. His quirky "personal history" of DUMAND would be his last academic paper. Like Reines, he had faced a career choice between physics and music in his youth—in his case piano— and all through his career, he wrote satirical songs in the vein of Tom Lehrer about physics and the academic life. One of the great benefits of having him around was that he knew about as many Gilbert and Sullivan songs as Reines did, so they would often break out in song together.

Another physicist who fit in handsomely was David Schramm, a renowned particle physicist-cum-cosmologist at the University of Chicago, who is said to have approached science in the same way that he approached his favorite sport, wrestling. Schramm was red-haired, six-foot-four, and had once nearly qualified for the Olympics. For many years, until his knees gave out, he kept fit by wrestling members of the Chicago Bears football team. Story has it that when he was advised as a graduate student that his

1. Participants in the DUMAND Summer Symposium in Honolulu in 1980. l to r: Fred Reines, Francis Halzen, John Learned, Andrew Szentgyorgyi, Bob March, Leo Resvanis, and Ugo Camerini. Szentgyorgyi, then a grad student of Camerini's, now works at the Harvard-Smithsonian Astrophysical Observatory. (Leo Resvanis)

2. An AMANDA hot tub party, sometime in the early 1990s, in one of the tanks holding pre-heated water for the drill. That's Bob Morse on the left with the beard and the shovel sticking out of his head. At the far left is Nigel Smith, a distinguished particle astrophysicist who later headed up SNOLAB, an underground neutrino physics laboratory in Canada. The other three naked individuals, as well as the fellow at the upper left who is about to become one, are students. The boxes in the background are cosmic ray instruments. The air temperature could not have been much above, say, minus twenty Fahrenheit. (Bob Morse, AMANDA/NSF)

3. *McMurdo Station, hub of U.S. activities in Antarctica, at the edge of the Ross Ice Shelf.* (Georges Kohnen, IceCube/NSF)

4. *The floating ice pier at McMurdo, with Hut Point and the hut built by Robert Falcon Scott's ill-fated 1911 expedition beyond.* (Mark Bowen, AMANDA/NSF)

5. *The Transantarctic Mountains through the window of a Lockheed LC-130 Hercules transport plane on the flight from McMurdo to the South Pole.* (Jim Haugen, IceCube/NSF)

6. *A river of ice flows through a gap in the Transantarctics.* (John Jacobsen, IceCube/NSF)

7. *The old dome at South Pole Station. The rectangular tower, left of center, was known as Skylab.* (Georges Kohnen, IceCube/NSF)

8. *The tunnel leading down into the dome.* (John Jacobsen, IceCube/NSF)

9. *The cramped buildings inside the dome. Steam escapes from the galley building at center, and "crapsicles" hang from the dome's ceiling.* (Mark Bowen, AMANDA/NSF)

10. *Driller Michael Jayred is warmed by the sun as he babysits the drill hose while it drops into the Ice. His job on that shift was to use cable ties to bind the hose to the electrical cable running down to the drill head every once in a while.*
(Mark Bowen, AMANDA/NSF)

11. *Master driller Bruce Koci taking the pulse of the AMANDA drill. Bob Morse says this may be his favorite picture of all time, since it gives a sense of Bruce's "genius of being able to see and feel logic, sanity, and process in an environment of seeming chaos and insanity."*
(Bob Morse, AMANDA/NSF)

12. Drillers Bruce Koci and Sven Lidström dressed in their finest for Thanksgiving dinner, 1996, in the galley in the dome. (Sven Lidström, AMANDA/NSF)

13. Head cook Sally Ayotte, outside the galley building in the dome, deep-frying one turkey and smoking another in preparation for Thanksgiving dinner, 1999.
(Mark Bowen, AMANDA/NSF)

14. *The Peacock Event, as first visualized in August 1998. Each dot or disk represents a photomultiplier tube in AMANDA-B10, and the size of a disk indicates the number of Cherenkov photons that hit the corresponding tube. This image was originally rendered in color, with the colors of the disks indicating the arrival time of the photons. In grayscale it still gives some idea of the arrival times: the muon entered the array at the lower left and exited at the upper right.* (AMANDA Collaboration)

15. *The view from the dark sector toward the dome in December 1999. The heating plant for the AMANDA drill, in the foreground, consisted of two Jamesways housing hot water boilers and the two open water tanks between them, which could be replenished with snow when necessary. The building on the left is "MAPO," the Martin A. Pomerantz Observatory.* (Mark Bowen, AMANDA/NSF)

16. *Members of the AMANDA drilling crew in the millennial season of 1999-2000. Bruce Koci stands at the far right.* (Mark Bowen, AMANDA/NSF)

17. The international crew of AMANDA scientists gathered at the hole, eight minutes into the new millennium. The black horizontal cable at the upper left, which runs to the pulley on the drill tower, would be string fifteen.

Sprawled in front: Albrecht Karle (Madison), Robert Schwarz (South Pole Station), Gary Hill (Madison). 2nd row: Dave Ross (Irvine), Michael Solarz (Berkeley), Katherine Rawlins (Madison), Kurt Woschnagg (Berkeley) Michael Boyce (U.S.A., winterover). 3rd Row: Darryn Schneider (Australia, winterover), Till Neunhofer (Mainz), Jean-Paul DeWulf (Brussels), Carlos de los Heros (Uppsala), Doug Cowan (University of Pennsylvania), Adam Bouchta (Zeuthen), Dave Besson (University of Kansas). Tall man in back: Bob Stokstad (Lawrence Berkeley Laboratory). (Bob Stokstad, AMANDA/NSF) thesis topic might be too chancy and contentious, he responded with an exposition of "jugular science."

Realizing that ocean engineering would make or break the project, Reines recruited Howard Blood, director of the Naval Ocean Systems Center in San Diego, to the cause. This was also a strategic move, since Blood's center had a branch in Hawaii, and when the ocean experts in Bellingham had been charged with finding a site three miles deep where the water was clear and advanced facilities were located on a nearby shore, they had recommended two sites in the Hawaiian Islands.

One of the stranger individuals to attach himself to DUMAND was a certain Peter Kotzer. He directed the Bellingham workshop and edited its proceedings. Learned is convinced that he worked for the CIA. Whether or not this is true, he *was* working quite openly on military applications. In fact, to give some idea of how all-seeing the military could be in those Cold War days, he had already been working for about three years on a project called UNCLE (Undersea Cosmic Lepton Experiment), in collaboration with Fermilab and the Naval Research Laboratory in Washington, D.C.

This was the era of the reassuring nuclear deterrence strategy known as mutual assured destruction or MAD. The Soviets and the Americans had reached nuclear parity, in which it was assumed that neither side could obliterate the other without being annihilated in return, and the final ace in the hole was the nuclear-powered submarine, armed with intercontinental ballistic missiles carrying nuclear warheads. The notion was that even if one side managed to destroy all the land or air-based nuclear weapons on the other, the submarines could still rise to the surface, wherever they might be, and retaliate—if they could be reached in time. Thus both sides were working on the difficult problem of communicating with distant submarines lurking at depth.

UNCLE was a feasibility study for communicating with the subs with neutrinos. The idea, believe it or not, was to outfit a submarine with a Cherenkov detector and use a particle accelerator to send pulses of neutrinos to it, in Morse code presumably. The great advantage of neutrinos, of course, is that they will penetrate anything, so the pulses could be aimed directly at the sub, no matter where in the oceans it might be. The usual sort of radio communication won't work, because the usual radio frequencies can't penetrate earth or water; only very low frequencies will, and this requires very long antennae, buried underground. (The military was working on that technology, too.)

This strange business was a bit of a growth industry in the sixties and seventies. Even Clyde Cowan, Fred Reines's early collaborator, seems to have been involved. He had moved from Los Alamos to The Catholic University of America in Washington, D.C., where he often collaborated with scientists from the Naval Research Laboratory, several of whom published a paper on "telecommunication with neutrino beams" in *Science* in 1977. According to Christian Spiering, who would later lead the German contingent in AMANDA and IceCube, the Soviets were interested, too. The notion never went anywhere, because the subs would have had to sprout enormous and ungainly wings of light detectors in order to receive the signal.

To their credit, the decision makers in the U.S. Navy dismissed this notion out of hand. Dave Cline remembers going to a meeting in Washington to discuss funding for DUMAND with "the biggest shot admiral in the Navy." He had taken his seat next to the admiral and Fred Reines was just walking in, when the admiral turned to his assistant and said, "I hope this isn't like the goddamn Kotzer thing," referring to Kotzer's UNCLE idea. (The Navy wasn't much interested in DUMAND, incidentally, because it was much too deep. Submarines only go down about eight hundred feet.)

It is hard to overstate the significance of DUMAND. At first, everything about it was as big and bold as Dave Schramm. For the first five years, it survived on the motherly love of a small group of pioneers. Arthur Roberts was the only full-time contributor, thanks to the support of Robert Wilson, the Wyoming cowboy who ran Fermilab, while the rest supported themselves with other grants, such as E1A, and nurtured the project along in their spare time. Although the name was later attached to a specific instrument, DUMAND was, and still is, basically a concept, encompassing underwater and underground neutrino astronomy in all its forms. It is fair to say that nearly every advance that has been made in neutrino astronomy over the past four decades was prefigured in some way by this collaboration. They even considered ice, but rejected it on the grounds that the logistics would be insurmountable in the polar regions, which are the only places on Earth with ice sheets thick enough to provide the requisite shielding from cosmic rays. These many years later, those few IceCube scientists who are even aware of DUMAND still appreciate the debt they owe to it. Bob Morse points out that you can tell how much influence DUMAND had by the fact that AMANDA even copied some of its mistakes!

Arthur Roberts presented their first thoughts at an international neutrino conference in Aachen, Germany, in June 1976. They had come up with three fundamental concepts and endowed them with mythological names: UNDINE (UNderwater Detection of Interstellar Neutrino Emission), ATHENE (ATmospheric High Energy Neutrino Experiment), and UNI-CORN (UNderwater Interstellar Cosmic-Ray Neutrinos).

One of their first insights was to realize that neutrinos with different energies called for different instruments and that higher energies called for larger instruments. This followed from the fact that low-energy muons travel a shorter distance in the detection medium than high-energy muons will, and give off less light altogether, since they have less energy to expend on that light in the first place. The size of a cascade induced by an electron neutrino will scale with energy as well.

UNDINE, the smallest of the three concepts, was aimed at detecting low-energy neutrinos, in the range from a few million to about a hundred million electron-volts, emitted by the Sun and nearby supernovae—the latter being much more interesting, but exceedingly rare. It is believed that a supernova takes place in our galaxy, the Milky Way, just once every fifty years on average, but that most go unseen by optical telescopes as they are obscured by interstellar dust. This is no problem for a neutrino detector, of course, since neutrinos pass through dust as if it isn't there, but fifty years is longer than the lifetime of most scientific instruments—not to mention the careers of most scientists—so the case for UNDINE was not compelling. The scientists estimated that they could observe a supernova about once a week if they extended their reach to the nearest cluster of galaxies, the Virgo Cluster, but that that would require a detector weighing 100 million tons, which seemed out of the question.

ATHENE and UNICORN were aimed at high-energy neutrinos, with energies above a billion or a trillion electron-volts. ATHENE would focus on atmospheric neutrinos, while UNICORN was aimed at the most exciting scientific frontier: cosmic particle accelerators in deep space. In other words, it was actually a telescope. The only real difference between these two concepts was their size. UNICORN was larger and therefore more sensitive, and this was the concept that would evolve into DUMAND the instrument, AMANDA, and IceCube.

The DUMAND pioneers realized the magnitude of the challenge from the outset: in order to see objects outside our galaxy, they calculated,

UNICORN would need to comprise at least a cubic kilometer of water. But the idea was exciting, nevertheless. "Adherents began to gather."

What is now seen as the seminal meeting in the field of neutrino astronomy took place in Honolulu in September 1976, when scientists from the United States, Japan, Switzerland, Germany, and Russia converged for a summer workshop hosted by the University of Hawaii.

The three mythological concepts were refined. UNDINE, the low-energy concept, solidified into a design in the Greisen style: a shell of tightly packed light detectors enclosing a tank of water. The standard detector for ultra-sensitive light detection is the photomultiplier tube or "phototube," which can detect even a single photon under the right conditions. But these are expensive items, roughly $3,000 apiece with their associated electronics, so a large shell-type instrument can be rather expensive to build. One of the current realizations of UNDINE, for example, is Super-K, the Super-Kamiokande detector, situated deep in a zinc mine in the Japanese Alps. Super-K contains "only" about 50,000 tons of purified water, so it's 20,000 times too small to detect a supernova in the Virgo Cluster, yet it still takes 13,000 phototubes to cover the walls of the tank. That's $40 million for the tubes alone. A simple scaling argument shows that if you wanted to make a shell-type instrument capable of seeing into the Virgo Cluster—weighing 100 million tons, that is—you would need about 2 million phototubes, which would cost about $6 billion.

ATHENE and UNICORN manifested as plum pudding designs in the Markov mode, in which the detection volume is filled with a grid of phototubes. Since this design requires fewer detectors per unit volume, it is less expensive than the Greisen variety. It also has better resolving power, so it's more suited to a telescope.

Since UNDINE would have had to have been prohibitively large in order to have a reasonable chance of detecting a supernova before the conferees died, she was rejected at the Hawaii workshop. Arthur Roberts writes that "UNDINE was returned to her sunless and solitary abode, there to be wooed and won by other suitors." The winning would not take long, and the union would prove remarkably fruitful.

The scientific legacy of the Hawaii workshop is carried on today by numerous experiments around the world, and two Nobel Prizes have so far resulted. However, its most important legacy is probably the vast network of connections it spawned among people and nations. The Soviet presence was especially meaningful. In a concluding resolution, the conferees proclaimed their collective view of "DUMAND as a vehicle most appropriate to collaboration on the peaceful exploration of this scientific frontier by interested scientists throughout the world."

This in spite of the numerous "spooks" from both sides of the Iron Curtain who lurked in the background at the workshop. Since the scientists refused to take them seriously, however, their presence only added to the sense of international camaraderie. John Learned remembers an amusing vignette involving the erudite Russian theorist Veniamin Berezinsky:

"There were some Russia-watchers, U.S. Navy people, as ever there were at those sorts of meetings, and Venya was chairman of one of the sessions. Having a Russian chairman was a big deal. He spoke quite good English and read English and read English literature and so on, was clearly much more cultured than the party apparatchik who was along. So Venya stands up to chair his first session, and he says, [fake Russian accent] 'Okay, now you have Russian chairman, the meeting will run on time. And remember, Big Brother is watching you!' And we all go 'What?!' Later we took him aside, saying, 'Ven, where did you get that line?' He said, 'Oh, I read Orwell, of course.' We said, 'Ah, this book you can get in Russia?' He said, 'No. No.' So that was great fun. And then there was clearly a KGB guy along who was taking pictures of everything and so on."

On the American side, they had Peter Kotzer and his friends.

Right around the time of this meeting, Fred Reines, who had now moved from Case to the newly established Irvine campus of the University of California, offered John Learned a position as a visiting scientist in his group.

Neutrino astronomy was off and running.

The friendships born in Hawaii were cemented in several more gatherings over the next few years. There were meetings at the Scripps Institution of Oceanography in La Jolla, California, and in Moscow, where the Russians expressed a strong interest in collaborating and offered "several thousand phototubes for DUMAND," an outlay that would have cost about $10 million in the west.

The Institute for Nuclear Research of the Soviet Academy of Sciences had been pursuing its own program of natural neutrino detection since the early sixties, and none other than Moiseĭ Markov, a senior member of the academy, had been directing it. The Russians already had two neutrino detectors running deep in a tungsten mine in the Baksan Valley in the northern Caucasus, one employing scintillation à la Cowan and Reines and the other the radiochemical method of Pontecorvo and Davis. In 1977, Markov chaired an international neutrino meeting at this laboratory, which occasioned an unauthorized and unsuccessful attempt to climb nearby Mt. Elbrus, the highest mountain in Europe, by John Learned and Dave Schramm. (We won't go there.)

But the most memorable of the early meetings took place in 1979 in the Russian far east, where the hosts had a location of their own in mind that had a different ambience altogether from Hawaii and combined both water *and* ice.

It had been identified by Aleksandr Chudakov, yet another Russian who had realized the potential of underwater Cherenkov detection in the fifties. Shortly after returning from Hawaii, he proposed a telescope of the Markov variety in the waters of Lake Baikal, the largest, deepest, and probably oldest freshwater lake in the world.

Baikal lies in a rift in the Eurasian tectonic plate that formed about twenty-five million years ago. It is crescent-shaped, four hundred miles long, an average of fifty miles wide, and in some places over a mile deep. Comprising about 20 percent of the Earth's liquid freshwater, it supports one of the most diverse assortments of aquatic life of any lake in the world, including the only known freshwater seal. In 1996, the United Nations listed it as a World Heritage Site, giving it the protective status of the Grand Canyon or Australia's Great Barrier Reef.

Relevant to a neutrino telescope, Baikal's waters are exceptionally clear and unpolluted, and its surface freezes in winter. It seems that Chudakov was first to realize that this seasonal covering would provide a convenient natural platform for the deployment of a neutrino telescope on the bottom of the lake. This sort of pragmatism and attention to the art has characterized the Russian branch of this business all along.

The 1979 Pacific Science Congress, a meeting encompassing many disciplines that had been held since the 1920s, took place in Khabarovsk, the second-largest city in eastern Russia. The DUMANDees joined everyone else there, and then shifted more than a thousand miles northwest to hold their own meeting at Baikal itself.

Outwardly, it seemed that there could have been a serious clash of cultures. The Russians were exceedingly formal and hierarchical, and the Americans, especially this bunch, utterly egalitarian. But it seems that everyone saw past these differences and got along famously. The Russian delegation was led by Moiseĭ Markov, who was in his early seventies by this time and one of the most prominent scientists in the USSR. Aside from being a distinguished member of the Soviet Academy, he was Secretary of its Department for Nuclear Physics, which oversaw the entire Soviet effort in both accelerators and cosmic rays. Meanwhile, the Americans didn't see any need for a leader at all until they found out that they had to have one in order to enter the country. They got together and asked Learned, who casually agreed.

"In those days, I was wearing Levis and a Levi jacket and had a ponytail and, as usual, a beard, and looking like the revolutionary. And here's this fine old silver-haired gentleman [Markov], who is dressed well. So I'm thinking, 'Uh-oh, we're probably not gonna hit it off.' . . . It turns out he was the nicest, nicest gentleman, and he paid no attention to my uppity-ness."

John expressed his disapproval of the class separation in the Soviet system by resorting to various antics, such as riding in buses "with the people," rather than joining Markov in his limousine, while his host responded with grace and tact and even, perhaps, quiet appreciation.

Markov had two lieutenants in neutrino astronomy, both theorists (who, I am told, have never gotten along). One was his former student, Igor Zheleznykh, who had written the seminal doctoral thesis in the late 1950s and was a bit of a dabbler. The other was Grigorii Domogatsky, who had a longer attention span and was taking the lead on the Baikal project.

Domogatsky seems to be the sole representative of the sciences in a well-known, highly cultured, upper-class Russian family, populated with artists of all kinds. He's a down-to-earth, no-nonsense individual, who used to insist on keeping his desk in a busy office with the students and post-docs in his Moscow institute, so as not to lose touch with "real" physics. He met the tired western delegation when they arrived in Khabarovsk, late at night after the usual travel delays.

"He was one of these guys who would smoke his cigarette held backwards, you know? Leather jacket and looking like a Soviet Mafioso for all the world," says Learned. "So we meet him in the dark of the night in this hotel and he's going around to each of us handing out stacks of rubles, because they gave you walking around money in those days; you couldn't easily convert dollars to rubles. It was just a very strange scene, this strange

Russian character handing out money to all the Americans in the dark of the night."

The physicists had a grand time in Khabarovsk—there was plenty of partying. And when they moved out to the lake, had a chance to have some fun at the expense of Peter Kotzer, the suspected spy, who had turned up uninvited.

"The Russians said 'What shall we do? We have to let him talk,' " Learned remembers. "I said, 'Okay, fine, but I'm not going to attend.' So they scheduled a trip out on a boat at the same time that Peter was going to talk. We all went out on the lake and drank vodka, and the crew got drunk and ran the ship up onto the shore and smashed up the dock on the way back, and . . . anyway, lots of stories."

Another of Kotzer's annoying habits was to add the other scientists' names to his research proposals without telling them in advance. Eventually, he was drummed out of the corps.

Fueled by such camaraderie and what seems to have been a good deal of alcohol, the science progressed famously. In 1979, a collaboration of institutions from the United States, Japan, Germany, and Switzerland submitted a successful proposal to the U.S. Department of Energy, and on the first day of 1980, the Hawaii DUMAND Center was established at the University of Hawaii. John Learned and Arthur Roberts moved to Honolulu to work full-time on the project, with John as technical director.

Synchronistically perhaps, Francis Halzen happened to visit Honolulu in the summer of 1980 to work on a problem involving quarks with fellow theorist Sandip Pakvasa. There were other Madisonians, past and present, in town, including Learned, of course, and Ugo Camerini, who was there to work on DUMAND. Joining them regularly for lunch, Francis began learning about the emerging field of neutrino astronomy. He still thought of particle physics as his "profession," and says that no respectable particle physicist would have touched cosmic rays in those days. But he wasn't overly concerned with that kind of respectability. He had dabbled in the field before, with Dave Cline, and now he began to move into it in a bigger way. Although he wasn't aware of it at the time, there was a reason for this change in orientation, and he would discover the reason as a direct result, more or less, of his visit to Hawaii.

The Hawaii physics department had figured out a creative way to fund

Francis's visit by asking him to teach what they represented as a "fake" weekly seminar. This seemed manageable, but when he showed up for the first class he found "the whole faculty . . . sitting there, heh, heh. So, ah, 'What do I do now?'"

Taking what he thought would be the path of least resistance, that is, work, he decided to review the current status of particle physics. But it turned out to be more work than he had imagined. As he proceeded to roll out the lectures, he realized that he was conducting this review at a critical moment. So many discoveries had taken place so rapidly over the previous decade that the particle physics community had not had a chance to assimilate them. "What I was about to do was to put a range of topics together which we now refer to as the standard model. It's not like I invented the standard model—it was clear; it was obvious—but [this] had never been done."

When he returned to Wisconsin, Francis taught the course again mostly to faculty. Then he discovered that one of his longtime collaborators, Alan Martin from Durham, England, had taught a similar course over the same summer that he had, and done a better job of it, to his way of thinking. The two got together and collaborated on a textbook, *Quarks and Leptons*, which remains, thirty-five years later, the most popular introduction to graduate-level quantum mechanics in physics programs around the world. It has been translated into many languages.

The book has never been updated, and there is a message in that. The sad fact is that it has not needed a second edition, for there have been no fundamental advances in particle physics since it was written. All the particles predicted by the standard model, except one, the Higgs boson, were discovered by the time the book was published.

Halzen and Martin completed the final draft in Durham just before Christmas 1982. Francis then drove to Belgium, with the draft lying beside him on the passenger seat of his Volkswagen Sirocco, to spend the holiday with his family. After the holidays, he flew to Japan to start a fellowship at the University of Tokyo, and when he got off the plane he was told that Carlo Rubbia, Dave Cline, and their colleagues at CERN had discovered the W and Z weak intermediate bosons. (Halzen, Martin, and Vernon Barger had done some important phenomenological work to facilitate the discovery, and Rubbia went out of his way to acknowledge this work in his Nobel lecture.)

The W and Z were the last undiscovered particles in the standard model, except the Higgs, and they—and the Higgs, too, for that matter—were already in the book.

At the beginning of 1983, then, more than three decades ago, accelerator physicists realized that they were entering what they came to call "the desert": the Higgs was the only likely discovery on the horizon, and it would require an accelerator orders of magnitude more powerful—and more expensive—than the one Rubbia had employed in winning CERN's first Nobel Prize. An unhealthy competition developed between the U.S. and European physics communities. The former began designing the Superconducting Super Collider, which ended up as a $2 billion hole in the Texas prairie, while the latter, led by the newly beatified Rubbia, started in on the Large Hadron Collider. Rubbia's original plan was to finish it in 1991. It went online about twenty years behind schedule.

In the thirty-plus years since the W/Z discovery, the particle physics community has been searching desperately for experimental evidence of any kind of physics beyond the standard model. The most visible theoretical forays have been supersymmetry, which posits a heavy sister or brother for each of the particles in the model, and string theory, which makes few if any experimental predictions. Rubbia, Cline, and their friends were on the lookout for supersymmetric particles even in the early eighties as they searched for the W and the Z—indeed, they were more interested in supersymmetry than in the particles they ended up finding—but they came up empty-handed. So far, the operators of the LHC are coming up empty-handed as well. Their main hope, as it was for their predecessors more than thirty years ago, remains supersymmetry.

While the discovery of the Higgs in 2012 was a tremendous triumph, as of this writing, nothing in its behavior or in any other data from the LHC has come as a surprise. It all conforms to the standard model. As Gary Taubes, the author of *Nobel Dreams*, wrote in 1986, "If there is in fact no life in the desert, no new particles, then there will be no more evidence forthcoming with which to build better theories. Progress will be at an end. The standard model will remain standard for the duration."

As I pointed out in the first chapter of this book, only one chink has so far been discovered in the armor of the standard model, and it has to do with the behavior of the neutrino. Moreover, it was discovered not by an accelerator, but by an underground neutrino detector in the mode of Greisen and UNDINE. More on that in a moment.

✦✦✦

One of Francis Halzen's nicknames is "the fastest pen in the west." He made his first move into neutrino astronomy in his spare time during his 1980 visit to Hawaii by contributing a theory paper to a major DUMAND symposium that John Learned organized in Honolulu that summer (see photograph 1). The paper gave estimates for the number and energy spectrum of the muons that the proposed instrument might find in high-energy cosmic rays.

Other theorists participated, including the leading Russians. One of Learned's aims was to build interest in the community, and theoretical work related to an experiment generally helps with that. But John believes Francis gave DUMAND an especially strong shot in the arm not only with his phenomenological work, but also by including DUMAND in the many talks he gave as he resumed his extensive travel around the world. This is a form of scientific pollination, since one picks things up on these travels as well.

"Francis would be absolutely scandalized if I made this analogy," says Bob Morse, "but . . . Francis is our new Dave Cline. Dave used to go around and visit all these flowers and take that little pollen on his legs and deposit it here and there and let us know what was going on. Francis does that now."

✦✦✦

Unfortunately, the delicate shoot of Cold War cooperation that was just taking root in 1980 was swiftly yanked from the ground. In December 1979, less than a month before DUMAND was funded, the Soviets invaded Afghanistan, and in the U.S. presidential election that took place in November 1980, Jimmy Carter was replaced by the more confrontational Ronald Reagan. Arthur Roberts observes that "the severing of the Russian link was done with elegance and taste. We were told, confidentially, that while we were perfectly free to choose our collaborators as we liked, if perchance they included Russians it would be found that no funding was available for us."

"We kept channels open and stayed friends, but we couldn't work together," adds Learned.

This may have been a blessing in disguise for the Russians.

6. Science at Its Best

THE DUMAND COLLABORATION STARTED OFF THINKING BIG. TOO BIG, most likely. The original vision called for an array comprising 1.6 cubic kilometers, three miles deep in the Pacific Ocean, nineteen miles from the nearest shore. There would have been more than 1,200 "beaded strings," holding 18 light detectors each, adding up to more than 23,000 detectors in all. Each bead would consist of a glass sphere capable of withstanding at least 500 times atmospheric pressure, encasing a photomultiplier tube and its associated electronics, and the phototubes alone would have cost $70 million. The strings were to be anchored to the ocean floor and held taut and upright by buoys at their tops. The collaboration planned to use fiber-optic cables to carry the electrical signals to shore, even though the technology had not been invented yet. Arthur Roberts observes that "the oceanographers were amazed—this project was larger than any other peacetime ocean project by a factor of the order of 100."

In the end, they never managed to place a single functioning string on the ocean floor. The first was lost in 1982 when the cable lowering it into the water broke, despite being rated at twenty times the actual load.

This was eventually ascribed to "snap loading": if the string was some distance down and the deployment vessel rose rapidly with an ocean swell, the string couldn't follow because the bulbous optical modules would resist being dragged through the water in the manner of a sea anchor. A second string was lost in 1985, "when the explosive bolts, designed to release it after a planned sojourn on the ocean floor, failed to release when fired."

As the realities of ocean engineering and cost reared their heads, the array diminished in size. (One section of Roberts's history is named "The Incredible Shrinking Array.") The collaboration came up with MICRO, MINI, and MIDI DUMANDs, the smallest consisting of only nineteen strings.

In its wisdom, the Department of Energy had funded only a feasibility study. In 1982, the department turned down a follow-on proposal for a thirty-six-string array, and offered to fund just one string, which would hang from the deck of a ship rather than sit on the ocean floor. Seven years into the project, in 1987, the collaboration finally deployed this so-called Short Prototype String, comprising seven optical modules, and ran what was basically a glorified version of the experiments Learned had run for his Ph.D. thesis twenty years earlier. They took a total of thirty-eight hours of data at five depths between two and four kilometers and measured the drop in the intensity of down-going, atmospheric muons the deeper they went. They also produced a rough profile of the muons' angular distribution. DOE deemed this enough of a success to justify the funding of a nine-string array named DUMAND-II in 1989.

The Russians, meanwhile, made steady progress. At about the time DUMAND was funded, the Soviet Academy of Sciences dedicated a laboratory at Moscow's Institute for Nuclear Research to the Baikal project and named as chairman Grigorii Domogatsky, the theorist who had handed out rubles in the night to the hippies from America.

He and his lead experimentalist, Leonid Bezrukov, took a conservative approach, beginning with their choice of site. They were dealing with a shallower body of water to begin with, only about seven-tenths of a mile, but they also chose a spot only two and a quarter miles from shore— about a tenth the distance to DUMAND. They would be plagued by more down-going muons at this depth, but the scale of their project would be manageable. Also, and crucially as it would turn out, for the two months of the year that the lake would be frozen over, they would be able not only to walk on their experiment, but drive heavy cranes over it.

Science is still seen as a noble calling in Russia, and Baikal, in its time, was probably the most soulful experiment in physics. Its field headquarters are still located in an abandoned station on a dead-end spur of what used to be the Trans-Siberian Railroad.

There are no roads to this spot. In summer it can be reached only by train or boat. In winter, when the lake is frozen, one can drive across the ice from the town of Listvyanka on the north side of the Angara River, about thirty miles northeast. Until 1997, when their German collaborators installed a satellite link at headquarters, there was no internet. A 1930s vintage railroad telephone provided the only connection to the outside world.

The researchers live in small wooden cabins heated with wood stoves, which burn mainly birch. There is no running water. There's a wood-heated sauna, furnished with a hand-operated water pump, and every Saturday night the residents gather for the weekly *banya*, or bath. The whole community strips down and bathes, in shifts, pouring water on the heated stones in the sauna to make steam. Many nights there is singing and guitar playing, and while the food does not receive high praise, generally speaking, delicious bread is baked on the premises daily.

Deployment takes place in March and early April, when the ice is at its thickest. Holes are cut in the frozen surface, the detector strings are pulled up from the bottom of the lake for repairs, tests are run, new hardware is added, and the telescope is re-committed to the depths. The last step is to gather the cables that have been used to lower the strings, attach them to a buoy, and release the buoy into the water, where it will hover safely below the surface through the warm months while the instrument takes data. (Data is transmitted to shore by an electrical cable, which snakes along the bottom of the lake.) At the start of each deployment season, divers in dry suits drop into a hole and swim down to the buoy with a line that is used to pull the strings up.

The scientists commemorate the release of the buoy each spring with a ceremony that borrows something from traditional Russian funerals. One person stands on the ice next to the deployment hole with a bottle of vodka and a small shot glass as the others file past. One at a time, they fill the glass, pour a few drops into the hole for Burkhan, the god of the lake, and drink the rest. This solemnity is followed by a final banya and a party.

(I once asked Ralf Wischnewski, one of the German collaborators, whether they take a dip in the lake before the sauna. "Some do, but only by mistake," he responded.)

Christian Spiering once showed me a scrapbook of photographs from one of these closing ceremonies. In one, the group poses naked and grinning in the snow by the sauna. In others they stand on the flat ice, with brown, wooded hills in the distance, surrounded by heavy machinery and large spools of cable, or lie on their bellies on the ice, peering through boxy, homemade, handheld "goggles" into the black water. They wear wooly winter clothes and the quintessential Russian fur cap with folded up brim and earflaps, as they wrestle with electronics or hack at computers inside boxlike structures that resemble ice-fishing huts. In one picture, Grigorii Domogatsky sends a piercing look toward the camera, under a fur cap with dangling earflaps, looking rugged and wiry even in his winter clothes. Lines of sharp intelligence radiate from his eyes, and a Papyrossi, the loosely rolled Russian cigarette, dangles from his mouth.

The Russians proceeded methodically. In 1981, they ran preliminary tests with individual phototubes. The next year, they lowered a detector nearly to the bottom and measured the natural luminescence of the lake water—a source of spurious light that limits the ultimate sensitivity of the detector. (This confounding background does not exist in ice.) In 1983—four years ahead of DUMAND—they lowered the equivalent of a Short Prototype String from their stable ice platform and took a few days' worth of data. The following year they placed a string on the bottom, connected it to shore, and detected their first atmospheric muons. They suffered a couple of mishaps that year, however. There was a leak in the buoy that held the string vertical, so that it sunk to the bottom over the course of about fifty days, and ten days after that a lightning strike fried the electronics.

In 1998, at a workshop on computer simulations for neutrino telescopes that was held in Zeuthen, Germany, I spoke with the leader of the Baikal group at the University of Irkutsk, a warm, soft-spoken man named Nikolaij Budnev. Drawing sketches in my notebook, he described the ingenious method they had used to recover the string that had sunk and been struck by lightning.

They designed a device that looked something like a window fan. It held a propeller inside a small frame that was fitted with fins, canted at an angle so that the unit would "swim" in spirals when it was dropped into the water. The autumn after they lost the string, they sailed a research vessel out to the spot where they had lowered the string the previous April. Attaching a flexible cable to their invention, they switched on the propeller and lowered it into the water by means of a winch. As their mechanical fish swam into the depths, the cable traced out the path of its spiraling descent, and as it approached the bottom it wound the cable around the collapsed string. The scientists then reversed the winch, the cable knotted itself around the string, and they pulled it to the surface.

The stark contrast between the simple, low-tech approach of the Russians and the complicated, high-tech approach of the Americans probably explains why one succeeded and the other failed.

In 1986, the Baikal collaboration produced its first physics. They set a limit on the existence of magnetic monopoles using a string equipped with improved buoys and protected against lightning. (The monopole, first postulated by Dirac in 1931, is an exotic particle whose existence has yet to be confirmed: a subatomic magnet with only one pole. In theory, if such a creature exists, it should produce a huge signal if it happens to pass near a Cherenkov detector. This was one reason the Baikal group thought it worthwhile to search for monopoles with just one string.)

These early studies resolved basic concerns about the feasibility of the lake for a kilometer-scale instrument, the most basic of which had to do with its shallow depth. Since more down-going muons punch through to this depth, it's harder to pick the up-going muons from the down-going haystack.

In 1987, the year the Americans deployed their Short Prototype String and got the go-ahead for DUMAND-II, the Soviet Academy of Sciences approved Baikal as a long-term research project and increased its funding. The following year, the Russians received a great shot in the arm when a group led by Christian Spiering from the East German Institute of High Energy Physics, in Zeuthen, just south of Berlin, agreed to join the collaboration.

A race was on.

Nineteen eighty-seven was a watershed year for neutrino astronomy, and, quite unexpectedly, the biggest splash was made by UNDINE, the shell-type Greisen design that wasn't supposed to have much promise astronomically at all. She had been wooed from "her sunless and solitary abode" about ten years earlier, for an entirely different purpose.

Back in 1974, two theorists from Harvard, Howard Georgi and Sheldon Glashow, had proposed one of the first credible theories of grand unification, the quest to unify the four fundamental forces of nature. This is a lofty goal. Albert Einstein spent the last thirty years of his life trying to unify two of them, electromagnetism and gravity, and failed more or less completely.

Georgi and Glashow's theory, known as SU(5) in physics parlance, was an attempt to unify the weak force, the strong force, and the electromagnetic force; it ignored gravity. The name is shorthand for "special unitary group of degree five," a term from group theory, which is a branch of mathematics that makes elegant use of all forms of symmetry and is one of the most powerful and aesthetically pleasing tools in modern physics. Wolfgang Pauli employed group theory, for example, in proving his theorem of CPT or "strong reflection" symmetry in the 1950s.

It took a few years for the implications of SU(5) to sink in. The most surprising was its prediction that the proton, which had hitherto been thought to be absolutely stable, might actually decay after some unimaginably long time—or, as Glashow once put it, that "diamonds might not be forever." This would also have the larger implication that all the matter in the universe might eventually dissolve into a vast sea of elementary particles. By 1979 (the same year in which Glashow, Steven Weinberg of Harvard, and Abdus Salam of Imperial College, London, shared the Nobel Prize in Physics for their previous success in unifying the weak and electromagnetic forces), the new theory had been used to estimate the proton's lifetime at somewhere between 10^{27} and 10^{32} years. This poses little threat to the quality of your diamond ring or your health and well-being, in case you were wondering, since even the lower bound works out to about 100 million billion times the age of the universe. Amazingly, however, the entire range was within reach experimentally.

"This was accessible," says John Learned. "If it was right, we would have found it with a low-energy detector. So we pulled the [UNDINE] designs out, and that's why there's this interesting tie."

Over the next few years, several people who had attended the seminal 1976 DUMAND workshop in Hawaii submitted proposals to build UNDINE or Greisen-style detectors to search for proton decay.

"Now, Saburo Miyake claims that he told the design for that detector to [Masatoshi] Koshiba, who then proposed the Kamioka detector in Japan," Learned continues. "So, the amusing thing is that DUMAND, or the idea for generic DUMAND, was in some way the mother or father of this dual chain of experiments that goes on 'til today. . . . So, that's kinda neat. I like to point that out to people."

Ahead of the game, as usual, Dave Cline held the first-ever meeting on proton decay in Wisconsin in 1978. Fred Reines, who attended the meeting, took his eggs out of DUMAND at that point and put them in this basket.

In May 1979, Reines's group in Irvine joined forces with groups from the University of Michigan and Brookhaven National Laboratory to propose a 10,000-ton detector in the shell-type or UNDINE style, two thousand feet deep in a Morton-Thiokol salt mine on the shores of Lake Erie. They named their collaboration IMB, after the names of the institutions involved.

As with neutrino detection, the size of the detector was key: if the proton had a lifetime of 10^{32} years, then 10^{32} protons, which is roughly the number in a hundred tons of water, should produce an average of one decay event per year, and a 10,000-ton tank should produce a hundred events per year. Theory held that the proton would decay most often into a positron and a neutral pion, which would fly off in opposite directions. The positron would send a cone of Cherenkov light toward one wall of the detector, while the pion would swiftly decay into two gamma rays, each of which would produce electron-positron pairs, and these would produce multiple Cherenkov rings on the opposite wall. This characteristic light pattern would serve as a signature of proton decay and help distinguish this kind of event from atmospheric muons passing through the detector from above (which were minimized by placing the detector underground) and from the occasional neutrino interaction inside the tank. Neutrinos, in other words, were a hindrance to this experiment; they added to the background. At one point Reines compared the teasing of the signal from the background in IMB to "listening for a gnat's whisper in a hurricane."

The Japanese project, led by Masatoshi Koshiba, was named Kamiokande, the Kamioka Nucleon Decay Experiment. It was located one kilometer deep in a zinc mine owned by the Kamioka Mining and Smelting Company, in the center of a mountain in the Japanese Alps. In the end, it was smaller and therefore less sensitive than IMB: 3,000 tons versus 8,000. IMB also got off to a faster start.

But Dave Cline, Carlo Rubbia, and Jim Gaidos of Purdue (another Madison alum) beat them both to the punch with a smaller, easier-to-build experiment, which they placed two thousand feet deep in a silver mine in

Park City, Utah. Like IMB, this effort was named for the institutions involved: Harvard, Purdue, and Wisconsin. All of Cline's colleagues in CCFMR participated in HPW at one point or another . . . including Bob Morse, who would later become the principal investigator of AMANDA. In fact, this was Bob's introduction to cosmic ray physics. He would carry the DNA and even some of the equipment from HPW into AMANDA. (He also carried the lineage of Kenneth Greisen. For Bob's doctoral adviser, William D. "Bill" Walker, had obtained his Ph.D. at Cornell under the great man. "That's important, you know, your lineage in this business," says Bob. "So Morse is traceable to Walker, traceable to Ken Greisen at Cornell, and Greisen, of course, walks on water.")

"In typical Carlo/Dave fashion," Bob continues, "they were going to build a small, cheap and dirty detector, about eight hundred tons of water, and if the proton decay lifetime was hovering around ten to the thirty-two [10^{32}] years, we would've nabbed it. It was a high risk/high payoff experiment. We took a chance on it, and we missed."

Part of the reason they missed was that their detector had a serious design flaw. The phototubes weren't placed only in a shell around the detection volume, they were placed inside too, and there were mirrors in there as well. This was supposed to help catch every bit of light from a decay event, in order to measure its total energy, but since it also made it nearly impossible to reconstruct Cherenkov light cones, all directional information was lost.

I once asked Bob if HPW had done anything useful. He thought for a long time . . . "I don't think so," he finally said, "other than figuring out that there's a lot of goddam thorium in that mine. Thorium is radioactive, hah! hah! . . . Between the wire chambers and the mirrors on the inside of the tank, the reconstruction was a mess!"

None other than Sheldon Glashow pronounced the experiment "an unmitigated failure."

Neither HPW, IMB, nor Kamiokande ever detected a single credible proton decay. (Maurice Goldhaber from Brookhaven once remarked that IMB "had some candidates, but they weren't elected.") They took a step forward, however, nevertheless.

On a philosophical level, it is impossible for science to prove that something does not exist—"absence of evidence is not evidence of absence," as Carl Sagan used to say. But if you run an experiment and obtain a null result you can use the sensitivity of the experiment to estimate just how unlikely it is that whatever you were looking for does exist or will occur. So, while these experiments did not find proton decay, they still made progress by setting new limits for the minimum lifetime of the proton.

Limit-setting is a common game in particle and astrophysics. We will see a lot of it when AMANDA and IceCube begin generating data. The code for the setting of a new limit is to use the word *search* in the title of your journal article, meaning you have searched but not found. You would use *evidence* or *observation* if you had.

Limit-setting can also be used to rule out theories, and in fact, IMB and Kamiokande set high enough limits for the proton's lifetime to rule out SU(5)—a significant contribution. Then, sure enough, SU(5) was extended by supersymmetry, and the theoretical limit for the proton's lifetime was bumped up to 10^{36} years—beyond the reach of those experiments. Kamiokande was then replaced by the 50,000-ton Super-Kamiokande, or Super-K, which holds the present record of 10^{34} years. ("The Kamiokande people have just beat that one to death," notes Morse.) But an UNDINE-style detector would need to weigh in at about a million tons to cover the next two orders of magnitude and put supersymmetry to the test. So there's a Hyper-K in the offing . . . and so it goes.

A proton decay experiment also becomes more sensitive the longer it runs. IMB and Kamiokande started searching for proton decay in 1982 and 1983, respectively, and ran all through the eighties.

This was a marvelous coincidence, because in 1987 a near-miracle occurred: the brightest supernova in almost four hundred years.

✦ ✦ ✦

On February 23 (twenty-two years to the day after Fred Reines and his collaborators detected the first naturally occurring neutrino in South Africa) an insignificant star in a nearby galaxy burst into the first supernova visible to the naked eye since 1604. The remnant of the earlier event is known as Kepler's Star, because Johannes Kepler himself followed its course in detail. He had no choice but to observe it by eye, since it preceded the invention of the telescope by about four years.

No one has ever been lucky enough to be observing a star at the moment it exploded, and such was the case with Supernova 1987a. When the excitement had died down and events were sorted out, it was determined that its first light must have reached Earth in a window of about seventy-eight minutes, the time between the last "non-observation" and the first observation.

On the evening of February 23, his time, an amateur astronomer named Albert Jones made a routine scan of the sky with a homemade telescope from his driveway in Nelson, New Zealand. Jones knew the sky pretty well. He has been called the greatest amateur astronomer of all time. At his death at the age of ninety-three in 2013, he had made more than half a million variable star observations and received many degrees and awards for his work, including the Order of the British Empire from the queen.

That night he was checking on some stars in the Large Magellanic Cloud, an irregularly shaped dwarf galaxy, visible only in the southern sky. It orbits the Milky Way at a distance of about 160,000 light years, which makes it one of our nearest neighbors. In galactic terms it's basically right next door.

Jones noticed nothing unusual. It was about eleven a.m. universal or Greenwich mean time.

When he rose at his usual predawn hour to make observations the next morning, he scanned a different portion of the sky since he'd covered the Large Magellanic Cloud the night before. This was a shame. Otherwise he would have been the first person on Earth to observe one of the most spectacular events in our region of the cosmos in 383 years.

The following night, he was monitoring stars in another portion of the sky when some clouds moved in, so he "poked the telescope" back to his targets of the previous night in the Large Magellanic Cloud. There he "was quite surprised to see a bright stranger." He quickly called his friend Frank Bateson, director of the variable star section of the Royal Astronomical Society of New Zealand. "Frank," he said, "there's a star in the Large Magellanic Cloud where there was no star before."

Meanwhile, almost five hours before Jones observed the event and a little more than seventeen hours after he made his crucial non-observation (we are now at about 4:20 a.m. universal time on the twenty-fourth), a professional astronomer named Ian Shelton, who was also searching for variable stars, completed a three-hour exposure of the Large Magellanic Cloud with a camera on one of the telescopes at the Las Campanas observatory in Chile. At first he thought the bright spot on his photograph was a flaw on his photographic plate, since it had not appeared in a photograph of the same galaxy he had taken the previous night. After convincing himself that it was real, he walked from his telescope to the control room of one of the observatory's larger telescopes to discuss the new development with his colleagues. They concluded that an object that bright and that far away had to be a supernova.

At one point during the conversation, the night assistant on the larger telescope, Oscar Duhalde, chimed in to say that he had seen the object by eye about an hour and a half earlier when he had stepped outside during his coffee break. This would make him the first person who actually saw the event. Duhalde had taken many plates of the Large Magellanic Cloud, so he knew it well. Evidently, he had been intending to tell the astronomers about his sighting when he returned from his break, but one of them was telling an off-color joke when he stepped inside, and by the time it was told and the punch line translated from vernacular English into Duhalde's Chilean Spanish, he had lost his train of thought. Anyhow, they all walked outside the dome to take a look, and, sure enough, there it was.

The scientific community, which may be overly precise in such matters, tends to recognize Shelton as the official discoverer of Supernova 1987a. The more generous members of the community give Jones and Duhalde credit as co-discoverers.

The news traveled quickly. In New Zealand, Bateson phoned Jones's message to an observatory in Siding Spring, Australia, where everyone dropped what they were doing to focus on the bright stranger. One of the Siding Spring observers, Rob McNaught, then realized that he had included the Large Magellanic Cloud in a wide-field photograph he had taken at 12:20 p.m. universal time on the twenty-third, seventy-eight minutes after Jones had seen nothing. When McNaught developed the negative, the supernova was there. This observation was quite valuable scientifically, as not only did it serve to bracket the time of the event, it also evidenced "the swift rise to brilliance" of the supernova, which strongly constrained the current theoretical models.

Within a day, virtually every professional and amateur astronomer in the southern hemisphere was marveling at the sight, either through massive instruments on remote mountaintops and plateaus or small store-bought or homemade telescopes in backyards and on other driveways, in every wavelength range from the infrared to the ultraviolet. Over the following months the range was extended to the x– and gamma ray bands, and Supernova 1987a was observed with telescopes and numerous other sorts of detectors, mounted on balloons, rockets, satellites, and at least one airplane.

Searching their charts, astronomers identified the star that had exploded as a so-called blue supergiant named Sanduleak-69° 202a (after Nicholas Sanduleak who had catalogued it two decades earlier). This was the first supernova to arise from an identified star—and identification brought surprise, because Supernova 1987a turned out to be the most violent type of supernova, a type II, and theorists had previously believed that only *red* supergiants, which are about ten times larger than the blue variety, could produce that type.

For many, it was the experience of a lifetime. In a review that appeared in *Science* about a year later, two leading supernova theorists, Stan Woosley of the University of California at Santa Cruz and Mark Phillips of the Cerro Tololo Observatory in Chile, offered a "personal observation."

"Especially during the first few weeks following the supernova, observers and observatories of all nations and from all continents shared data, speculations, and the sheer exhilaration of the moment. Little was held back. In the process some mistakes were made, but were quickly subjected to test and the errors freely admitted and corrected. It was science at its best. The data we have now will occupy theoreticians for at least a decade, but the memory of the shared experience will last even longer."

The messenger for all this excitement was the photon, which has delivered pretty much all the news that our ancestors have ever received from the cosmos, since they first raised their eyes to the twinkling stars.

A new cosmic messenger delivered the news of Sanduleak-69° 202a's death about four hours before the photon did, with exquisitely precise timing, and it told a story from the heart of the star that was inaccessible to the photon. A six-second flash of neutrinos began passing through the IMB detector in Ohio at 7:35:41.37 a.m. universal time on February 23, plus or minus five one-hundredths of a second. The indeterminacy arose from the uncertainty of the clock on the instrument. Kamiokande saw the flash at the same time within its clock uncertainty, which was about one minute. The flash was also observed by the large liquid scintillation detector, similar to the one Cowan and Reines had used to detect neutrinos in the first place, at the Baksan Neutrino Observatory in the Caucasus. All three of these detectors happened to be located in the northern hemisphere. With neutrinos it doesn't matter.

The IMB and Kamiokande teams may have been looking for proton decay, but as physicists will, they had kept the door open for neutrinos all along.

Kamiokande was first out of the gate. February 23 fell on a Monday that year. Their data tapes were sent by bus from Kamioka to Tokyo according to their standard weekly schedule, they were analyzed through the night on Friday, the discovery was revealed to the collaboration at large on Saturday morning—and then to the rest of the world.

The Americans weren't as well organized.

The IMB tapes were in the possession of the M in the collaboration, Michigan, and their leader, Jack Vander Velde, didn't think the instrument was sensitive enough to see the neutrinos. Most of the Michigan folks also happened to be out of the country that week, attending that year's Moriond conference at a ski resort in France.

But John Learned, an active member of IMB even though he was up to his eyeballs in DUMAND, knew they must have seen it. He had written the supernova section of the IMB grant proposal, so he had the numbers at his fingertips. A few days after the event, John cajoled the Michigan group into sending a copy of the tapes to his former student, Bob Svoboda, who had recently moved to a post-doc in Irvine, and he and Svoboda made a plan for scanning the tapes. Not long after he put down the phone from that conversation, John got a call from one of the Japanese scientists, who told him that Kamiokande "had it." Particle physicists were way ahead of the curve on e-mail, since it was invented at CERN, so John was using it regularly. He shot back a request for the time that they had seen it.

"As near as I can recall," he says, "just about simultaneously, I got the time in Hawaii, I called up Svoboda in Irvine, he's just run the tape, and he says, 'We got it!' And I say, 'Okay Bob, what time?' And we check the times with each other, and they're within two seconds. At that moment we knew we had it. . . .

"The little amusing inside story is that we then called Michigan to say, 'Hey, we have it!' and the first response from Jack Vander Velde was, 'Oh, bullshit, John.'

"I said, 'Fine. You get the tape, and you scan down to event number so and so. Then call me back.' And so then things went wild."

IMB announced their results to the world about ten days after Kamiokande did, and the two collaborations published back-to-back papers in *Physical Review Letters* on April sixth.

✦ ✦ ✦

"When all is said and done," wrote Woosley and Phillips, "the most exciting and unique aspect of SN 1987a will remain the detection of the neutrino burst that signaled the collapse of its iron core into a neutron star. The numbers are awesome."

They estimate that the neutrino luminosity during the first second of the star's collapse was more than four times the photon luminosity of the entire observable universe ("all that matter from which we could have received light since the big bang"). The observable universe is about twenty billion light years across, while this neutrino explosion came from a region thirty miles across. Expressed another way, in that first second, the neutrinos given off by Supernova 1987a carried off about one hundred times the total energy that will be given off by our Sun in its entire ten-billion-year life. "All the nuclear weapons in the world, on the other hand, could only power the sun for a few millionths of a second. Supernovae are by far the most violent events in the universe."

Think of the neutrino flash as a thin spherical shell that expanded at about the speed of light. By the time it passed through our planet it had a radius of 160,000 light years (which means that the star actually died that many years ago). Since the surface area of a shell grows by the square of its radius, the intensity of the neutrino flash drops at the same rate. Nevertheless, even this far from the center of the blast, an average of fifty billion supernova neutrinos passed through each square centimeter of our planet each second for between ten and twenty seconds.

The neutrino being a shy creature, only twenty-four were detected in all. Kamiokande picked up eleven in a span of about thirteen seconds, IMB eight in six seconds, and Baksan five in ten seconds. And luck played a big role: Kamiokande almost missed the flash, because the detector had switched itself into calibration mode, which it did for roughly two minutes out of every hour, just one or two minutes before the neutrinos passed through. And IMB was partially crippled: one of its power supplies had failed several hours before the flash, and a quarter of its light detectors were down.

The most important thing about these twenty-four neutrinos was that they provided the first direct link between a supernova and the birth of a neutron star. As theorist Adam Burrows put it at a neutrino conference in 1988, "Core collapse has been studied for thirty years and neutron stars for fifty years in blissful theoretical isolation. . . . These detections provide us with the first definitive tests of the basic theory connecting stellar death, supernovae, and neutron star birth. . . . Within [about] 10 seconds . . . that theory was transformed into an astronomy."

Not all stars end in supernovae. Lighter stars, such as our Sun, pass quietly into white dwarves. Stars weighing eight or more times the Sun will explode into neutron stars, and above about twenty-five solar masses, they probably leave black holes or exotic and so far unverified creatures named quark stars behind.

Black holes and quark stars are the only cosmic objects that are believed to be denser than neutron stars. Theory dictates that neutron stars should be about 40 percent more massive than the Sun, and every candidate whose mass has been measured comes in close to that value. They should be only about twenty kilometers across, which would make the force of gravity at the surface more than a hundred billion times stronger than the force we feel on Earth.

The towering Russian theorist Lev Landau first postulated the existence of neutron stars in 1932, just a few months after James Chadwick discovered the neutron. Two years after that, none other than Walter Baade, who had placed a bet involving a case of champagne with Wolfgang Pauli, joined fellow astronomer Fritz Zwicky in proposing the link to supernovae. "For many decades after these original contributions, and without any observational evidence, the idea of the neutron star was kept alive only by stalwart theorists." Their existence was finally confirmed in the late 1960s with the discovery of radio pulsars. One of the first pulsars to be discovered was the one at the center of the Crab Nebula, a diffuse, blue, vaguely crab-shaped object in the constellation Taurus, which seemed likely to be a supernova remnant, since the Crab supernova had been observed historically. But this was circumstantial evidence. The neutrinos from Supernova 1987a sealed the deal.

Radiant stars are basically a balancing act between gravity and nuclear fusion. In the core of a young star, the fusion of hydrogen nuclei, that is, protons, into the second-lightest element, helium, generates enough outward pressure to keep the core from collapsing under the pressure of gravity. Thankfully for life on Earth, fusion also produces light and energy.

Sanduleak-69° 202a was probably about eleven million years old when it died, and the reason for its demise was that it had run out of fuel.

It had burned hydrogen for about ten of those eleven million years. This is known as the main sequence in the life of a star. When its main sequence ended, Sanduleak-69° 202a consisted of a core of helium, the so-called ash from the hydrogen burning, surrounded by a large spherical envelope of unburned hydrogen, which was now removed from the life-giving production of energy and pressure at the core.

Stars are the furnaces that produce all the atoms in the universe heavier than hydrogen. Without them, even planets could not form, much less life. We are stardust, as the saying goes.

The helium burned to produce carbon and oxygen over a span of something less than a million years. At the end of that stage of its life, the star consisted of a core of carbon and oxygen "ash" enveloped by a shell of helium, enveloped in a large cloud of hydrogen.

And so on and so forth: lighter nuclei fused into heavier nuclei (carbon and oxygen produce neon, sodium, and magnesium), the remaining ash burned into still heavier nuclei, the leftovers retreated to non-participating shells outside the core, and the star took on the structure of an onion: a central, ever shrinking core, surrounded by concentric shells of progressively lighter elements. The death spiral also spun faster: each successive stage took less time than the previous.

The final fuels were silicon and sulfur. It took them only about a week to burn into a core of solid iron. The process stopped there, because iron is a stable nucleus. There is no energetic advantage to burning it; energy is required either to fuse it into heavier nuclei or split it into lighter ones.

The core of Sanduleak-69° 202a was extremely dense at the moment it was poised to explode. Its diameter was about half the Earth's, but it weighed just what it had to in order to turn into a neutron star: 40 percent more than the Sun. The entire star, with its concentric shells of lighter elements and its outer envelope of hydrogen (which was blue on account of its temperature), had a diameter of about thirty million kilometers, about one-fifth the distance from the Earth to the Sun.

With no fusion left to produce outward pressure, the iron core collapsed in about a tenth of a second. The outer surface fell in at about a quarter the speed of light. Neutrinos began to emerge during the collapse, as increased pressure and the breaking up of some iron nuclei promoted inverse beta decay: protons captured electrons, turned into neutrons, and gave off electron neutrinos, which sped away.

As the core shrank to a diameter of about sixty miles, pressures and densities rose unimaginably and the cataclysm began: nuclei were ripped apart, free electrons combined avidly with protons to produce more electron neutrinos; electron-positron pairs emerged from the high-energy void to provide the leptonic material for more neutrinos and antineutrinos; other, similar reactions produced muon and tau neutrinos; ultimately all flavors of neutrino and antineutrino were produced in roughly equal measure. But the collapsing core was so dense and so opaque that even the neutrinos couldn't escape. They and the light were trapped inside.

At the very center of the collapsing star, densities increased by a factor of a million, to a few times that of an atomic nucleus: about 10^{14} grams per cubic centimeter. This is like packing a pool of water one kilometer square and a hundred meters deep into a cube one centimeter on a side. The nuclear material was packed so tightly that the strong force, which normally binds protons and neutrons together, reversed sign and began pushing them apart. This outward pressure, originating at the very center of the core, presented a "brick wall" to the collapsing matter that was still rushing in, and this material bounced off the wall in a so-called shock wave, which triggered the stellar explosion.

As the shock slowly "broke out" of the star, hitting progressively less dense layers of in-falling material and knocking them back out, more nuclei were ripped apart, generating more material for neutrino production, and the neutrinos that were trapped in the core were instantly released. This produced an ultraluminous neutrino burst lasting one or two hundredths of a second—the brightest phenomenon known to astrophysics.

The theoretical models encounter a problem at this moment, because the release of the neutrinos should take enough energy out of the shock—wind from its sails, so to speak—to cause it to stall and the star to begin to collapse again. But the star does explode, obviously, and one idea is that a stream of neutrinos from the embryonic neutron star at the center of everything keeps the shock going. The ten-second duration of the flash seen by IMB, Kamiokande, and Baksan lends credibility to this notion.

(You can thank neutrinos, incidentally, for the quality of your teeth. A recent study indicates that some of the antineutrinos given off by the core during these ten seconds undergo inverse beta decay with neon nuclei in that shell of the star and change them to fluorine, and that this seems to be the primary mechanism in the universe for making this element.)

However the shock was kept alive, it continued to work its way out through the rest of the core and the progressive layers of the onion—which were oblivious to the goings on inside until the shock reached them—and gradually ejected everything but a core of pure neutrons, a neutron star, into the surrounding interstellar medium. Only when the shock reached the surface of the star did it release the light that was seen by eye and telescope 160,000 years later.

As grand as the optical pyrotechnics may have seemed to the many astronomers watching them from the southern side of our planet, they were as nothing to the neutrino flash and the shock wave. The neutrino flash was about thirty thousand times as energetic as the optical display and two to three hundred times as energetic as the material explosion. Put another way, about 99.5 percent of the energy released in the gravitational collapse of the iron core escaped in the form of neutrinos, and more or less all of the remaining half a percent was carried away by flying material. Only a few thousandths of a percent was given off as light.

The reason the first light from Supernova 1987a reached us after the neutrinos did was that it took time for the shock to work its way to the outer surface of the star. It traveled at only about a fiftieth the speed of light, while the neutrinos sailed out at close to the speed of light. And the fact that the delay was only a few hours added support to the identification of this supernova with a blue supergiant: it would take the better part of a day for a shock to escape from a red supergiant, because they're that much larger. One of the great triumphs of Supernova 1987a was that the energy of the neutrino flash, as measured by Kamiokande and IMB, lined up almost perfectly with theoretical estimates for the gravitational energy that should be released in the collapse of an iron core weighing 1.4 solar masses.

In 2002, Masatoshi Koshiba, the leader of the Kamiokande collaboration, shared half the Nobel Prize in Physics "for pioneering contributions to astrophysics, in particular for the detection of cosmic neutrinos." No one from IMB was cited, probably because the only real choice would have been Fred Reines, who had already won one Nobel and had died in 1998. Baksan was ignored altogether.

Part III

Touching the Mystery

Neither the application of the discoveries of science nor even their achievement is to be compared with the struggle in winning them.

—FREDERICK SODDY

7. Solid-State DUMAND

The idea was kind of a non-idea.

—FRANCIS HALZEN

FRANCIS WAS BORN IN THE TOWN OF TIENEN, IN THE FLEMISH REGION OF Belgium, in March 1944. The story in his family is that he was almost born by the side of the road. His father had to drive his mother about seven miles to get to the hospital, and the Germans were bombing the road at the time. It was about a year before the end of World War II.

Tienen was his father's hometown. It had a population of about twenty thousand, relatively large by Belgian standards. Francis's parents and older sister had lived there before the war, but had moved to his mother's small village when the war began, because it was easier to get food in the countryside. His father owned a road and bridge building company that had been in the family for several generations. As such, according to Francis, his choice was to "work for the Germans and be killed later, or not work for the Germans and be killed immediately." Louis Halzen hid in the Pyrenees in the south of France for most of the war, and figured out a way to return and live anonymously in his wife's village for the last two years or so.

Francis was too young to remember the war, and as a child he wasn't particularly aware that he was growing up in its aftermath, although he does remember many stories of suffering, mostly for lack of food. Even for a few years after the war, the family ate chicken only once or twice a year. They moved back to Tienen when he was about four.

He remembers his childhood as a placid one. He ranked at the top of his class all through his education in every subject except religion. ("I don't know what went wrong there. It's not like it's a big deal. It was not intentional. I tried.") But he claims that his excellence in school had no effect on him. "I never really realized it. Neither did my parents. In fact, I had to be pushed to go to university." Looking back, he realizes it was probably a lack of confidence.

He remembers the Belgian school system as being "absolutely superb, I have no other word for it." He was good at science and math, and could speak Dutch, French, Greek, and Latin in addition to his native Flemish by the time he graduated from high school. He was also reasonably proficient in Elizabethan and Middle English from reading Shakespeare and Chaucer. He remembers being laughed at by British tourists when he tried to make small talk with them in these antiquated dialects during his family's yearly vacations to the shore near Bruges.

It was a classical education, not a practical one, and there were no elective courses—a good thing in his view. He believes his high school education was "pretty much the equivalent of what you would get at a very good liberal arts college" in the United States and that he could have gone straight into a Ph.D. program from there.

No one in his family had ever gone to university. He was destined for his father's construction business. Ironically, however, it was his father who encouraged him to go on in school. Louis had an active mind and a great interest in literature. When Francis remembers him, "it's with a book."

Being a diffident young man, interested in just about everything, it was unclear what he should study, so his father had him speak to a counselor, who suggested pharmacy of all things. He was steered away from that by one of his high school math teachers, who stood him in front of a pharmacy one day, pointed through the window at a man inside mixing pills, and asked him if that was what he wanted to do for the rest of his life.

Since Belgian education followed the French, Cartesian tradition, in which mathematics is the most revered of the disciplines, he made the non-choice and majored in math. He wasn't really free to choose his university either. The Belgium university system was sliced into four more or less rigid segments, dictated by the country's bedrock provincialism: the official state religion, Roman Catholic, versus secular, and Flemish versus Walloon- or French-speaking. Being Flemish and Catholic (though non-observing), he went to the Catholic University of Leuven, eighteen kilometers from Tienen, and commuted by train.

He remembers the formal instruction at Leuven as being vastly inferior to his instruction in high school. For one thing, the school was weak in science. Had his family been more aware of the world of higher education, he probably would have chosen the University of Brussels, a non-sectarian school that was (and still is) much stronger in the sciences. In fact, Brussels now participates in IceCube. Oddly, the classes Francis found most stimulating at Leuven were the ones in religion that he was required to take once a year. Theology was an active field at the time owing to the debates engendered by Vatican II, and there were a few left-wing theologians on the faculty with remarkable intellects. "Even if I wasn't very interested in what they were telling me, being exposed to people that are the top of their field is very important."

Francis believes that one always lives with the blind spots in one's upbringing, but that one can erase them, at least partially, through education and the broadening of one's worldview. He believes this is how he discovered his true calling professionally, over a period of about twenty-five years.

He majored in math at first and slowly shifted to physics as he realized that that was what he "really wanted to do." "You have to get rid of your upbringing slowly, right? . . . I think deep down I always wanted to be a physicist and an experimentalist; it just took me a long time to move from mathematics to real physics. [I] wandered for twenty years doing theoretical physics, and it's clear from my period as a theoretical physicist that I'm a very poor mathematician." He obtained intermediate degrees in both subjects and his final degree in physics alone.

Most people would probably agree, almost half a century later, that Francis has had a stellar career in science and academia. But he is more proud of being the first member of his family to graduate from university and that in being so he might also have set an example. Almost every family member who has come after him has earned a Ph.D.

He never figured he'd continue in physics. He was planning to teach high school for a year or two and then join the family business. His father figured the same: his son's education had gone on long enough as far as he was concerned; it was time to start earning a living.

Unbeknownst even to himself, however, Francis caught the research bug during his third year in university, the first in his entire education in which he was allowed to take an elective course, and the first in that formal milieu in which he ever spoke directly to a professor. The physics department required a thesis, and by pure luck he did his thesis work in the one corner of the department that had a sense of what true research actually was. Since it's an international game, the professors he began to work with also had contact with the wider world.

The luminary of Leuven's physics department in those years was Georges Lemaître, who is now recognized as the father of both big bang theory and the theory of the expanding universe (the latter is often mistakenly attributed to Edwin Hubble). Lemaître was a Catholic priest and, for a time, head of the Pontifical Academy of Sciences. A superb mathematician, he had developed his two cosmological theories in 1925 as solutions to Einstein's field equations for general relativity, which had appeared only nine years earlier. At first, even Einstein didn't believe in the physical reality of Lemaître's solutions, telling him famously, "Your calculations are correct, but your physics is abominable." But Einstein and the rest of the scientific community came around when Hubble provided the first observational evidence to support Lemaître's theories, about two years after he proposed them.

Lemaître was an old man by the time Francis crossed paths with him; he died the year Francis graduated from university. And although Francis knew him by reputation, there is vivid evidence that he did not know him by sight: The great man could no longer walk, and he was driven to work in a limousine. Francis and another student sometimes had the job of meeting the limousine, transferring this old gentleman in his ecclesiastical robes to a chair, and carrying him up the steps to the institute where they worked. Francis only learned who he had been carrying twenty years later, from the other student, who had become a professor at Leuven.

Lemaître was particularly devoted to numerical calculation and as a scientist of some renown, had access to the most powerful computers of his time. One of Francis's early accomplishments was to use a machine that had been built under Lemaître's direction to carry out the first computerized calculation of a Feynman diagram. ("It doesn't matter what this means, but I'm very proud of this," he says.) He undertook this project under the direction of Martinus "Tiny" Veltman, at the University of Utrecht, who would win the Nobel Prize in 1999 for a fundamental theoretical triumph that also required computers. So Francis found himself in the finest international company the moment his research career began.

Luck has played an extraordinary role in his career—mostly for the good, it seems, but sometimes for the bad—and his undergraduate thesis is an example of both. One day his supervisor handed him a copy of the paper, fresh off the press, in which George Zweig of Caltech first proposed the existence of quarks. (It was Zweig who later offered him a position at Caltech.) It was an odd paper, because Zweig didn't have the math to describe his new model; he employed strange geometric constructions to do the calculations. But Francis, with his rigorous training in mathematics, saw right away that they could be done with group theory and recast the model in elegant group theoretic terms. ("Imagine the luck involved!" he says.) It was a good piece of work, but the subject seemed so highly theoretical that he dropped it when he finished his thesis. Looking back, this was an obvious mistake, but how could he have known that quarks would become the greatest discovery of the next few decades in particle physics?

And so, having done estimable work even as an undergrad, Francis got a job as a high school teacher in a town near Tienen—but never took it up. His work had caught the eye of a Belgian theorist named Edward Verboven, who invited him to study with him at the University of Nijmegen in the Netherlands (to which Francis refers as "the land of great physicists north of the border"). Again it took a wise teacher to make him realize what he really wanted to do. This also happened to be a promotion, since graduate students were paid more than high school teachers. When the physics department at Leuven realized that Nijmegen was trying to steal him away, they offered him a position, so he followed the path of least resistance and stayed where he was—a theme in his career.

Francis looks back on his doctoral thesis as a total waste of time. It had to do with S-matrix theory, an idea that was sexy at the time but has now disappeared "because," he points out, "it had nothing to do with real world." He calls it the string theory of the late sixties, since he suspects that string theorists "are going through a similar period now." (This is one of his milder pronouncements about string theory. "It's the ultimate form of bad physics to think that your imagination is more important than data and evidence," he once told me. "It's not just arrogance, it's not just bad taste, it's bad physics.")

He earned his doctorate over the course of only three years, defended his thesis a few days before Christmas 1968, and was married that same week. His wife, Nelly, was training to be a teacher at a school near Tienen where his sister taught, and she and Francis had met at a school dance. Most women of an intellectual bent studied to become teachers in Belgium, as it was rare for a Belgian woman to attend university back then.

Francis has made surprisingly few career moves. They have all either been passive, as we have seen in his schooling, or completely spontaneous. And he has *never* applied for a job. The right offer has always appeared just in time to allow him to "transition from place to place following the science I wanted to do."

Even though he doesn't think much of his Ph.D. thesis, it won him, as I have mentioned, a monetary prize from the Belgian government for being the best doctoral thesis submitted in the country that year. It must have been a hefty amount, since it allowed him not only to buy the "ugly hippie yellow/orange MGB" that fate would have him and Nelly drive to Méribel, France, it also allowed them to move to Geneva and support themselves while Francis volunteered in the theory group at CERN.

It was a large theory group, a hundred or so, and it was an unsupportive, sink-or-swim environment. The group was led by a famously tough Belgian theorist named Léon Van Hove. One might guess that this would have given a fellow countryman a leg up, but this guess would be wrong. Within a few weeks of Francis's arrival, Van Hove strode into his office and said, "Halzen, the people we've had here from Belgium have been so bad over the last few years, you may be the last one."

"Well, at least you get me cheap," Francis responded, pointing out that he wasn't being paid.

He had learned to work independently at Leuven, and he wasn't intimidated by CERN's milieu. He found some collaborators, began producing papers, and had a paying job by Christmas 1969. "So it was lucky that Van Hove had come into my office to insult Belgian theoreticians. Otherwise, I may not have had a salary for much longer."

Two years later, he had the chance encounter with Vernon Barger in Méribel that led to his spontaneous relocation to the American Midwest.

This was a big jump, culturally. Madison was a sleepy country town compared to Geneva. Francis sums up the new state of affairs with his usual pithiness: you couldn't get a decent cup of coffee in Madison, and he and Nelly used to drive all the way to Chicago to shop. (Francis is a dedicated gourmet. He tends to sample the best restaurants in the innumerable cities he visits in his travels, and his stories often revolve around memorable meals in the world's most famous establishments.)

But Nelly liked Madison immediately, thanks mostly to the educational opportunities. She began taking classes at the University of Wisconsin, earned a bachelor's degree in the history of art, went on to a doctorate in French literature, and now teaches in the department of French and Italian.

As for Francis, as important as gustatory considerations may have been, the quality of the research and teaching environment came first. As Dave Cline has observed, he fit perfectly into the Madison picture. He got tenure almost as fast as Cline did, within two or three years.

✦ ✦ ✦

Francis was still pure in the watershed year 1987, sixteen years into his American "visit," when the three underground detectors sensed the neutrinos from Supernova 1987a. He'd been dabbling in the dark side since his visit to Hawaii seven years earlier—he and Tom Gaisser of the Bartol Research Institute at the University of Delaware wrote an influential paper on cosmic rays in 1984—but he still worked mainly in particles. And he continued to travel widely, visiting collaborators, giving talks, picking up pollen on his knees, and spreading it around.

In the fall of that year, he was invited to speak to the physics department at the University of Kansas. He guesses that the topic was "cosmic accelerators and the (now defunct) evidence that Cygnus X-3 emits mysterious particles." He might well have talked about Supernova 1987a, and considering what happened next, he undoubtedly mentioned DUMAND.

The Cygnus story demonstrates, incidentally, how science can be advanced even by false leads. In 1983, groups from Kiel, Germany, and the University of Leeds presented evidence that Cygnus X-3, an X-ray source in the constellation Cygnus, was emitting very high-energy gamma rays. Subsequent studies proved these studies wrong, but the "damage" had been done. Some of the more adventurous members of the particle physics community, driven in part by the fact that their field was just entering the desert (the W and Z were discovered that same year) began migrating into cosmic rays. This led naturally to progress, and in 1989 a specialized telescope in Arizona detected the first true high-energy gammas from the Crab Nebula. Gamma ray astronomy has since flourished, and dozens of gamma ray sources have been discovered in the meantime. And since gamma ray and neutrino astronomy borrow many of their techniques from particle physics and share a lot of science with it as well, their ascendance has given rise to a new field that combines astronomy and cosmic ray physics. It's known as either particle astrophysics or astroparticle physics, take your pick.

Anyhow, it so happened that an unusually versatile scientist by the name of Ed Zeller was sitting in the audience in Kansas. Zeller had been trained as a glaciologist, also had an appointment in physics, and was an old Antarctic hand. During the discussion after Francis's talk, he mentioned a feasibility study that was taking place at the Soviet Union's Vostok Station in Antarctica, in which a small array of radio antennae was being used to "listen" for neutrinos interacting with the ice. It was known as RAMAND, for Radio Antarctic Muon And Neutrino Detector, and it was being led by Igor Zheleznykh, the second of Moiseĭ Markov's neutrino lieutenants. RAMAND had not detected any neutrinos; in the end, it never would.

The radio method had been conceived by the Russian physicist Gorgen Askaryan in 1961. It is based on the fact that at very high energies both muon neutrinos and electron neutrinos will generate cascades, and these cascades are essentially sparks consisting of millions of positrons and electrons, which produce jagged radio pulses.

"My interest peaked with [Zeller's] comment that the Russian physicists had been unable to compute the power in the signal," Francis later recalled. "With an enthusiasm reminiscent of graduate electromagnetism, [two collaborators] and myself solved the 'hardest Jackson problem ever,' using superior computer power not available to our Russian colleagues." (J. D. Jackson's textbook, *Electrodynamics*, provides a bittersweet rite of passage for many a physics student.)

The problem was interesting, but the answer was disappointing. Francis, John Learned, and Bartol theorist Todor Stanev later estimated that with the technology of the day, even if RAMAND had been monitoring a cubic kilometer of ice it would only have detected about one neutrino every hundred thousand years.

Francis believes theorists write two kinds of papers: "the ones they really believe in, that they feel passionately about," and others simply to show how smart they are. This one fell in the showing-you're-smart category. He wrote it, put it in "the stack that you submit every year to the university for your salary," and moved on.

But Zeller's remarks prompted him to put two and two together on another front. He was quite aware of DUMAND, of course, and now he had heard about neutrino detection in ice. "You don't have to be a genius," he says. "Eventually you're going to hit the idea: if the Hawaii stuff doesn't work in water, maybe it works in ice. And forget the radio detectors; do just what [DUMAND does]. . . . That was all there was to it."

He sent an e-mail to his theorist friend Sandip Pakvasa in Hawaii, asking him to ask John Learned if he'd "ever thought about whether you can do this in ice."

He and Learned exchanged frantic e-mails for several weeks. They queried Zeller, who assured them (naïvely in retrospect) that the deep ice in Antarctica ought to be clear. And the more they considered the idea, the more compelling it became: Ice would be free of the dissolved radioactive potassium in natural bodies of water, which gives rise to a confounding background luminescence. The detectors would remain stationary; they wouldn't slosh around with changes in water currents. Neither would they accumulate sediments, which was also a worry so close to the ocean floor. And finally, as at Lake Baikal, the ice would provide a stable platform for deploying the instrument.

But Francis wasn't ready to take the plunge into experimentalism quite yet. He still thought of himself as a theorist. He was happy to write a few papers about "DUMAND on ice" and present the idea at some conferences, but that's as far as he was willing to go. Learned, on the other hand, was at the height of what might be called his neutrino megalomania at that point in time. He was still leading DUMAND, he was a member of IMB, he was about to join the successor to Kamiokande, Super-K, and he had another line in the water in Arkansas. Nevertheless, he was more than willing to take a shot at building a second DUMAND on the bottom of the planet.

There are many political angles to doing science in Antarctica. One is that the continent is governed by a surprisingly enlightened international treaty, which encourages science and environmental protection but prohibits military activity and territorial claims. For this reason, Learned's primary funding source, the Department of Energy, isn't particularly welcome, since it is deeply involved in the U.S. nuclear weapons program. The purely scientific National Science Foundation (NSF) oversees the entire U.S. presence in Antarctica.

The Madison connection crops up again here. For as it happened, John Lynch, the administrator in charge of Polar Aeronomy and Astrophysics at NSF, in both polar regions, north and south, knew John Learned and

Francis Halzen well. He had been a graduate student in the Madison physics department back in Learned's post-doc days and Francis's early days as a professor. He was also a contemporary of Bob Morse—the two had met on their very first day of graduate school. Perhaps most importantly, Lynch had been a regular at the 602 Club, the bar where Dave Cline, Bob March, and the rest of the CCFMR gang held court.

As Lynch remembers it, Learned sent him an e-mail "with no capital letters in it, demanding that I, you know, let him drill six hundred holes five thousand feet deep at South Pole, next year." Although he dismissed this half-baked request out of hand, he appreciated the concept well enough to show the e-mail to his boss, Peter Wilkness, director of the foundation's Office of Polar Programs, whom he describes as a "genuinely larger than life guy," who "just loved big things."

In May 1988, Learned got more formal, but not formal enough. He sent Lynch an eight-page letter of intent, based on a paper he and Francis had written for a cosmic ray conference that would take place, in Lodz, Poland, in September.

Francis observes that "research is when you don't know what you're doing." The most naïve remarks in this paper relate to the ice: "A crucial question is how deep we must go to obtain clear, bubble free ice because optical scattering from bubbles can ruin . . . muon detection. . . . It appears, based upon conversations with Prof. Edward Zeller of the University of Kansas that we will obtain good optical clarity below about 150 meters near the pole, but that we may have to go below 500 meters to find bubble free ice."

An insight that has held up, on the other hand, was that it would be much less expensive to build a neutrino telescope in Antarctica, surprisingly enough, than virtually anywhere else on the planet, assuming a well-equipped base nearby. Demonstrating a charming lack of political savvy, Learned even suggested that the polar project would be "an order of magnitude more cost effective" than his own project, which was already beginning to founder in the warm waters of Hawaii.

Learned figured the Antarctic project would attract collaborators from other underground neutrino projects, and "given an acceptable political climate . . . that our Soviet colleagues would want to participate as well." Indeed, a few months after he wrote these words, he received a telex from his friend Moiseï Markov indicating that "Prof. G. I. Marchuk, President of the Academy of sciences of the USSR . . . considered this suggestion as interesting."

It's too bad the Russians didn't get involved, since they would have brought an extensive knowledge of ice to the table. (Wherever there is ice, one will tend to find Russians.) Markov's telex contained remarks that could have saved the soon-to-be-born AMANDA collaboration years of effort and anxiety, had they been aware of them. Based on the analysis of a 2.2-kilometer-deep ice core that the Russians had drilled at their Vostok station, Markov pointed out that there were "only sporadic bubbles in perfectly clear ice at 1300–1400 meters, so the holes in ice for [phototubes] have to be rather deep"—an insight that would prove dead-on.

At the end of his letter of intent, Learned requested $100,000 for the upcoming Antarctic season, which was about six months away. In a friendly response, Lynch and Wilkness turned him down.

Learned's involvement came to an end when the Department of Energy told him he had to make a choice (although that may be a polite way of putting it). "Basically, I was told by the DOE that I had better stick to—I had to choose between DUMAND and playing these other games," he says.

One of the first signs that a field is gaining momentum is when conferences develop around it. And the first regular conference dedicated to neutrino astronomy was one of the most civilized one can imagine. It was conceived by Milla Baldo Ceolin, an elegant, brilliant, and beautiful woman, who was in fact the first woman ever to become a professor of physics in Italy. She was Galileo Professor at the University of Padua. There was a Madison connection there, too: Milla, as she was widely known, had begun collaborating with Jack Fry when he had made his early shift to the accelerators in the 1950s.

Until it became more popular and started focusing on things besides neutrino astronomy, this was Francis's favorite conference, and he's been to a few. It was held in the Istituto Veneto di Scienze, Lettere ed Arti in Venice, in noble sixteenth-century halls once graced by Galileo himself. At first, it was very small and by invitation only.

"In the early Venice meetings we just gathered in a room at the Palazzo Loredan, the smaller building of the Academy," Francis writes. "Everybody was working on neutrino telescopes; we were there to discuss and help each other—we knew this was going to be difficult. . . . The meetings were small and I remember the room very well. There was a small transparency projector with the screen a foot from Tintoretto's *Madonna col Bambino*."

Milla may also have had a civilizing effect upon the competitive men in the field. Francis says that her concept for the meetings was that "with mutual help we could all succeed." At a meeting in the early nineties, for example, by which time he was engaged in AMANDA, he remembers asking a member of the DUMAND collaboration about the computer simulations they had developed for their instrument, and the response was simply to give him the code. This was a shocking act of generosity to someone accustomed to the sharp elbows in particle physics.

By the time of the first Venice meeting in 1988, DUMAND on ice had attracted the interest of Bob March, who was already working on the watery version. He delivered a paper, co-written with Learned and Halzen, entitled "Neutrino Detection in Clear Polar Ice." His literary flourish is evident in the prose.

> One might call the detector to be described in this talk "Solid State DUMAND." ...
>
> It must be emphasized that this is a "Dio Volente" detector, for to make it feasible three conditions must be met that are beyond our control:
>
> A natural body of clear ice, of area at least tens of thousands of square meters and thickness a few hundred meters, must exist somewhere within a kilometer or so of the surface;
>
> The light transmission of the ice in the Cherenkov band [blue] must be comparable to that of pure water, with attenuation (absorption and scattering) lengths of tens of meters;
>
> The ice body must be situated near an existing permanent research station, for the cost of constructing, maintaining, and supplying a base would be prohibitive. . . . The third condition effectively limits the choice of sites to the USA's South Pole Station and the USSR's Vostok Station. . . .
>
> In summary, this is a detector that requires a number of happy accidents to make it feasible. But if these should come to pass, it may provide the least expensive route to a truly large neutrino telescope. Exploratory studies may begin at the South Pole within the next few years.

Francis was now becoming passionate about this project. With Learned out of the picture, however, he needed to find a collaborator on the experimental side in order to move it along. The obvious choice would have been

March or one of his CCFMR colleagues, so Francis would occasionally ask them if they'd like to do something more than just write another paper. "After a while," he says, "whenever I showed up on the experimental floor of the building, people ran for their offices and closed the doors. You know, they just thought I was crazy. . . . The one exception was Bob Morse."

Bob was the perfect man for the job, for a couple of reasons. For one, he specialized in getting new projects off the ground. For another, he had been one of the earliest defectors from the accelerators to particle astrophysics. As he remembers it, at about the same time he joined Dave Cline and Carlo Rubbia on their HPW fiasco in Utah, "Jack Fry and Ugo Camerini convinced Dave that we ought to go into the gamma ray astronomy business. And I thought, 'Great!' So off to Hawaii, off to Mount Haleakalā!

"One of the things I'm accused of is that I'm fast out of the gun; I love to start new things, but I'm never around to help finish them up when all the messy details come around. And I think that's probably a quasi-legitimate indictment. I really do love planning the new things and getting stuff laid out and getting into the ground work and stuff. And so . . . Jack Fry came up, and like all good salesman he convinced us that there were probably stars out there . . . that were putting out lots of high-energy, pulsed—that was the magic word here, *pulsed*—gamma rays."

The gamma ray telescope that the CCFMR group erected on Mt. Haleakalā was the first particle astrophysics experiment of any kind to be placed among the more traditional telescopes on the summit of this 10,000-foot volcano. Unfortunately, however, like HPW, it didn't work out. Camerini and Cline went in with the arrogant attitude that they could scoop the astronomers by employing some tricks from the accelerator repertoire to detect what they assumed would be periodic pulses of gamma rays from sources that emitted other periodic signals, such as X-ray pulsars, and that the data would roll into their laps like coins from a slot machine. About ten years later, when the first true high-energy gammas were detected from the Crab Nebula, it turned out that they weren't periodic, and it wasn't until 2008 that a more sensitive, second-generation instrument finally detected pulsed gammas from the pulsar at the center of the Crab. So the particle physicists got a lesson in astronomy, rather than vice versa.

But the side benefit of Haleakalā was that it led Bob Morse to the South Pole. He and his friends soon realized that the equator wasn't a great place to look for periodic signals, because other periodic effects complicated the measurement: the daily cycle of the Sun and the rising and setting of the very stars they were trying to observe. Basically, it's more difficult to measure one periodic signal when it's riding on top of another, so these diurnal cycles increased the noise.

The South Pole experiences only one very long day a year. The Sun rises and sets at the equinoxes, on about September 21 and March 21, so for about six months it's below the horizon and won't complicate things. The stars in the southern sky don't rise and set either. They describe circles in the sky, centered on the zenith; they're always in view, although they're hard to see in summer when the Sun is up. If you set up your telescope to track a star, therefore, you can keep it in sight all winter long, barring clouds, blowing snow, and the spectacular auroras at the pole, of which there are many. The two main complications at Haleakalā don't come into play.

So, sometime in the late 1980s, Bob began talking to his friend John Lynch at NSF about trying the Haleakalā idea at the pole.

Martin Pomerantz, the recently retired director of the Bartol Institute, was by then already known as the father of South Pole astronomy. He'd been the first person to do any kind of astronomy there, and he was a great proponent of the pole's unique advantages for certain kinds of observation.

The South Pole resembles an ocean, except that it's white and you can walk on it. The surface of the ice is about as high as Haleakalā, 9,300 feet, owing to the fact that the East Antarctic Ice Sheet is about that thick there. It's dark all winter, the same stars are always in view, and owing to the incredibly cold temperatures, the air is extraordinarily dry. This makes it nearly transparent to infrared light, so the pole is particularly well-suited to studies of the cosmic microwave background. It's also an excellent place to observe cosmic rays, because charged particles like to travel along magnetic field lines, so they're steered by the Earth's magnetic field to the north and south magnetic poles. The auroras australis in the south and borealis in the north are simply the light given off by cosmic rays as they hit the polar atmospheres.

In the epochal year 1987, Pomerantz and a droll Scot by the name of Alan Watson (a member of the Leeds team that had made the false but fortuitous Cygnus X-3 "discovery") installed a cosmic ray detector named SPASE (South Pole Air Shower Experiment) at the pole. Pomerantz's Bartol colleague Tom Gaisser, who had previously collaborated with Francis Halzen, participated in this experiment as well.

Bob Morse points out how science begins to move in a new direction: "You get this sort of subcritical mass of stuff boiling around. . . . Everybody's sort of dabbling in everything." He was talking to Lynch about trying out the Haleakalā idea at the pole; Pomerantz and Watson started SPASE; John Learned was pretty much everywhere; Trevor Weekes, the leader of the team that would soon detect gamma rays from the Crab Nebula, was interested in setting up a telescope at the pole similar to the one he already had running in Arizona; and Francis Halzen was traveling the world trying to rustle up interest in solid-state DUMAND.

Economic politics also tilted in a helpful direction. In 1989, partly in response to a perceived economic threat from the advancing technological knowhow of Japan, the National Science Foundation began funding so-called Science and Technology Centers: centers of excellence in specific subject areas, usually run by universities, where the foundation would invest in infrastructure and advanced instrumentation that could be used by research teams from around the country. Part of the idea was to spin off any new technologies that might result into the private sector.

In June 1989, Pomerantz chaired a conference at Bartol for the express purpose of setting up a Science and Technology Center named the South Pole Astrophysics Research Center (SPARC) at his institute. Although SPARC never came to be, this was an epochal gathering. Many pioneers in particle astrophysics attended, including Learned, Watson, Weekes, locals Tom Gaisser and Todor Stanev, and, of course, a healthy contingent from Madison. Morse presented his idea for Haleakalā at the pole, and Francis gave what was now his standard talk on neutrino astronomy. Near the end, he suggested "that field investigations begin as soon as possible to examine the relevant optical properties of deep Antarctic ice. It is amusing to envisage the Antarctic ice sheet as a giant neutrino telescope, with the whole earth as its rotating neutrino bandpass filter."

This time the pollen found receptive flowers in two grad students from Berkeley named Doug Lowder and Andy Westphal. They had come to Delaware with their mentor, Buford Price, chairman of the Berkeley physics department, to present a balloon experiment they were planning to launch over the upcoming Antarctic summer, just a few months away.

Buford is a courtly southern gentleman with an unusually broad range of interests, scientific and otherwise. He speaks Russian, for example. He was trained in solid-state physics and early in his career invented new methods for recording the tracks of nuclear decay products in solid materials.

This led to investigations of the ancient tracks left by decay products in natural materials, most prominently mica—and thence into geophysics. (The tracks in ancient rocks act as signatures for radioactive constituents that have long since decayed away.) He was elected to the National Academy of Sciences at a young age and was one of the few investigators to be awarded samples of the "Moon dust" brought back by the Apollo astronauts. He, Lowder, and Westphal were planning to send a new glass-based track-recording device into the air above Antarctica in order to study the isotopic composition of heavy, iron-related elements in cosmic rays. This might help determine whether supernovae, which are essentially exploding balls of iron, are a source of the cosmic rays produced in our own galaxy.

It is unclear whether Lowder and Westphal actually spoke to Francis after his talk; if they did, he doesn't remember. He was still a theorist, after all. He jetted out of Delaware and continued on his merry way.

The seasonal cycle in Antarctica begins to pace the story.

The U.S. Amundsen-Scott Research Station at the South Pole ("Pole" in Antarctic-speak) is about 850 miles from the nearest station, McMurdo, the hub of U.S. operations, which is on the coast of the continent. And the safest and most convenient way to get to Pole from McMurdo is by air.

Since 1988, air support between McMurdo and Pole has been provided by the 109th wing of the New York Air National Guard (which sports one of the best acronyms of all time: NYANG). This is the only wing of the U.S. Air Force that flies the Lockheed LC-130 Hercules transport plane, aka "Skibird," a turboprop equipped with a unique wheel-ski combination that allows it to land on both tarmac and snow.

It goes without saying that an LC-130 needs good weather to make the trip to Pole—they frequently "boomerang": turn back in mid-flight. One specific requirement is that the temperature must be above −50° Celsius or −58° Fahrenheit,* because at that temperature jet fuel, which is basically kerosene, turns to a noxious sort of jello and hydraulic fluid becomes so viscous that the moving parts of the plane stop moving. This limits the working summer at Pole to about three and a half months, from sometime in late October or early November to mid- to late February. When the last LC-130 lifts off and makes its farewell flyby, it leaves behind a skeleton crew of "winterovers" to spend the next nine cold, dark, and to most, sublimely

---

* Quick rules of thumb: minus forty is minus forty, and minus eleven F is plus eleven C.

beautiful months in isolation. The winterovers maintain the science experiments and the station itself, and keep themselves safe, alive, and entertained, but they don't do much in the way of improvement to either the station or the experiments. Most of that sort of thing—and much else—takes place during the short, hectic summer.

Summer in Antarctica would be winter in Wisconsin. As Francis remembers it, he was sitting in his office one winter's day, about six months after the Bartol conference, when he "received a call from an irate NSF program officer telling me that 'two guys from Berkeley had been caught trying to sneak a string of photomultipliers into Antarctica to detect muons in the ice,' and asking whether I might have put any crazy ideas in their heads. I had in fact never heard of them. . . .

"You have to realize, I'm a theoretician; I'm funded by DOE. I'd never talked to anyone at NSF in my life. I get this phone call, a guy shouting at me, and he says, 'You know what it is to do science in Antarctica? You just don't run around there.' I thought, 'This guy is crazy. He has the wrong number.'"

It was Lowder and Westphal. Their balloon experiment had been piggybacked onto a field study led by an eminent Caltech geologist named Hermann Engelhardt, and they were shuttling around to various locations on a quickly moving ice stream on the West Antarctic Ice Sheet, drilling short, quick holes in order to study the movement of the stream. They had volunteered to help with the drilling in order to get a chance to launch their balloon experiment and had also taken along a simple apparatus built by another of Buford Price's students, Steve Barwick, in the hope of running one of the "field investigations" that Francis had outlined at Bartol. They planned to lower a light source into one of Engelhardt's drill holes, drop a couple of photomultiplier tubes into a second hole nearby, turn on the light, and measure the transparency of the ice. They weren't aware of having broken any rules, incidentally, since Engelhardt had told them it was quite alright.

The NSF officer soon decided that Francis really was clueless about this little foray, and Francis convinced him to let the renegades have their fun. Ironically, however, they never did run their test, because Engelhardt's drill broke down before they got the chance. And they had a frustrating season all around, because their balloon experiment failed, too.

As it happened, Bob Morse made his first visit to Pole that summer, to install a Haleakalā look-alike named GASP (Gamma-ray Astronomy at the

South Pole). And it may be emblematic of the soon-to-be-born AMANDA collaboration that, as far as I can tell, its first "pre-meeting" took place in the most popular bar on the continent, Willy Field.

"The first time I ever met Doug Lowder, I ran into him walking out of the men's room at McMurdo, at Willy Field," says Bob. "And so I introduced myself, . . . and he was not in a very good mood. He said he didn't want to be in this goddamn place, and the experiment wasn't going anywhere, and all he was doing was hanging around."

It is not unusual to find people in foul moods at McMurdo, which is by far the largest base in Antarctica. For the vagaries of air travel on the continent can leave you stranded there for a week or more on your way to or from your work site. Some say "Mactown" isn't part of Antarctica at all, since you don't have to relate to the out of doors much at all when you're there.

The coincidences mount. As it also happened, the biannual International Cosmic Ray Conference took place in Adelaide, Australia, that year, and when you come off "the Ice" you fly from McMurdo to Christchurch, New Zealand, and often connect through Australia. Bob Morse stopped in at the conference on his way home. Francis was there, Buford was there, Steve Barwick was there. And one sunny January afternoon, as these four sat together on the lawn by the main building at the University of Adelaide, they resolved to form a collaboration to build a neutrino telescope at the South Pole—Morse being the in-house "expert" on logistics now that he had been there all of once. As a badge of this exalted status he was sporting a large spot of frostbite on the tip of his nose. It seems that Alan Watson from Leeds participated in the discussion, but declined membership in the collaboration. Everyone remembers Doug Lowder as being there, but in fact he was not: "Neither Andrew or I went to that conference," he writes. "Those guys were sitting back, drinking Foster's and having a meeting while Andrew and I were down in Antarctica getting the work done." Or not.

The upshot was a request by the two professors involved, Buford and Francis, to meet with John Lynch at NSF.

John had been expecting this. "Here it comes," he said to himself. He asked his boss, Peter Wilkness, if he'd like to participate, and the answer was, "You bet!"

In the spring of 1990, Francis, Buford, Bob, and Steve Barwick met with their potential funders in Wilkness's office at NSF headquarters near Washington. The scientists presented an overview of the project and explained that their first, limited goal was to study the optical qualities of South Pole ice. When the discussion turned to how to do this, Lynch paused and said, "Wait a second. Let's get Zimmy down here": Herman Zimmerman, the director of polar glaciology in the Office of Polar Programs, who happened to be the man who had chewed Francis out several months earlier. Zimmerman oversaw all the ice drilling in NSF's portfolio, not only in Antarctica, but also in Greenland, where the summer season was about to begin.

The first step would be to drill a hole in the ice, drop some phototubes into it, and see if they could detect down-going muons—just as John Learned had done years earlier for his graduate work in the Cascades. Zimmerman said, "Well look, this is going to be hard at South Pole, but maybe we can get started in Greenland."

So they decided to run their first test on the two-mile-high summit of the Greenland Ice Sheet. The summit is that high for the same reason the pole is, the ice is that thick, and an American team was just commencing a multi-year effort to drill an ice core all the way to bedrock there.

But how to fund this sneak attack? According to Francis, Zimmerman thought the idea was "so crazy, he could never fund it if it needed to be reviewed." But Lynch, as a physicist, didn't consider it quite so crazy and also had a creative solution to the funding problem. A few weeks earlier, he had read about a new mechanism called a Small Grant for Exploratory Research or SGER ("sugar" in NSF parlance), which had a limit of $50,000, could only be used once on any single project, and did not require formal review. Wilkness happily signed off on a full $50,000, which John divided into two grants, one to Madison with Bob Morse as principal investigator, and the other to Berkeley and Buford Price. He is quite proud of having awarded what may have been the first SGER in all of NSF and definitely the first in Polar Programs.

So the physicists finally got a chance to encounter some honest-to-god ice—and they met Bruce Koci.

8. Enter Bruce

The miracle of AMANDA/IceCube is the clarity of the South Pole ice and the ice-genius of Bruce Koci.

—FRANCIS HALZEN

As previously mentioned, I first met Bruce in Bolivia, where he was engaged in his other life, high-altitude ice core drilling, with the climatologist Lonnie Thompson. He and Thompson then struck off on an expedition to the Himalaya, and the next one after that, about two years later, was to Mt. Kilimanjaro in Africa, where I joined them again.

One afternoon in February 2000, after a long day's drilling, Bruce and I sat together on the sand in the volcanic crater on the 19,000-foot summit of Kilimanjaro. As we leaned against our packs and watched the sun set, he reminisced about his career.

"I never was in the drilling business to be a driller. I hate machines. Maybe one of the few engineers in the world you'll ever find that feels that way about them. I hate them. . . . That's one of the few times I will ever fly into a rage is over a machine that does something that it shouldn't do.

"I'm here for the experience. I came into this thing as a canoeist. I walked out of good aerospace job and decided to go into ecology and then got back into engineering through glaciology, starting at Minnesota. I've always come for the place; I haven't come to do the drilling. I'll do my damnedest to make sure the drilling goes well, because that means I can go to another good place.

"I have only a love relationship with rivers or mountains and always ask their forgiveness for our trespass and delving into their inner secrets."

Growing up in Minnesota, Bruce dreamed of visiting the Canadian Arctic from the age of five. At sixteen, he took the first of about five long canoe trips in Canada, a 650-mile excursion on the Athabasca, Slave, and Hay rivers in Alberta and the Northwest Territories. He became especially enamored of Baffin Island, one of the more awe-inspiring landscapes on the planet, with enormous granite towers sprouting from its many glaciers, and took two long backpacking trips on the island in his late twenties, one of three hundred miles, the other of four.

After earning an undergraduate degree in aerospace engineering at the University of Minnesota, Bruce worked in the industry for a few years, and in 1972, returned to his alma mater to enter a graduate program in wildlife ecology. After obtaining a master's degree in that field, he switched to glaciology, and this took him back to Baffin, where he worked on the Barnes Ice Cap for four years in a row, occasionally punctuating his field work with kayak trips through the island's spectacular fiords.

One of his friends from that time recalled an experience he shared with Bruce on the Barnes in 1976:

> It was our first day on the glacier. We had ridden some miles north to the previous year's camp to dig out [one of] the snowmobiles buried last season. All was going rather smoothly until, just as we had the machine out and were preparing to start it, a storm blew in. Having never been quite this far north in a raging whiteout, I became mildly concerned with my future. Bruce continued working, not seeming to bat an eye at our impending demise (he was without a doubt one of the coolest customers I've ever known). That is, until the snow began to pile up and the machine we had labored so hard to retrieve wouldn't start. We tried and tried and nothing worked. Finally Bruce gave it one last mighty pull on the starter; the rope snapped in his hand; and Bruce lost it. What then began was the most eloquent rant I have ever heard to this day. He started with the combustion engine. Technology in a general sense took a broadside; man's hubris; ice; the weather; the meaning of science; and finally to sum it all up, "the sonofabitch who invented the wheel."

We laughed all the way back through the whiteout to camp, Bruce leading the way because I was utterly lost. Never has rage found a better poet, or wild places a more gentle champion.

I suspect that this was the sort of rage Bruce was confessing to on Kilimanjaro. I'm sure they were infrequent, I suspect they were usually laced with humor, and I'll bet they were rarely directed at a living thing. Lonnie Thompson once said that Bruce's distinguishing attributes were "his loyalty, endurance, constancy, creative craftsmanship, disheveled wisdom, and soft-spoken nature."

He had an extraordinary memory for numbers and dates. He could summon an obscure specification for a drill he had built decades ago, the number of days it took to drill a certain ice core, how much the ice weighed, or the flow rate of an old AMANDA drill without skipping a beat. His intuition was legendary, and his work sites, his desks, and his person were famously disheveled, but this hid the fact that he was well organized beneath it all and employed rigorous engineering methods in all of his designs and procedures. In the days before calculators and smartphones, he used to carry a slide-rule and tables of logarithms around with him. His intuition was based on a solid understanding of the physics and a firm grasp of the numbers.

He had not quite earned a degree in glaciology, and was, he recounts, "rapidly running out of time, when all of a sudden I got this call that the University of Nebraska was looking for someone with a degree in engineering and some understanding of glaciology. So I called them up, got hired over the phone in, like, mid-October, and was on a plane two weeks later for the Ross Ice Shelf."

The Ross Ice Shelf floats on the surface of the sea by McMurdo Station in Antarctica and provides a stable platform for the station's airstrip, Williams Field. It is the largest body of floating ice on the planet, about the size of France. Bruce first visited the place over the 1977–78 Antarctic summer.

The University of Nebraska had contracted with the National Science Foundation to drill a hole through the two-hundred-foot ice shelf so that a group of scientists could study the ocean beneath. The drill was a so-called flame-jet, which is conventionally used by the mining industry to cut crystalline rock. It consisted of two 10,000-pound compressors, feeding air at a thousand psi to a modified jet engine—a huge Bunsen burner, basically—which was lowered into a roiling, water-filled hole, spitting out flame and partially combusted diesel fuel.

"Well," said Bruce, "it makes an awful racket . . . Lotta noise; lotta smoke; it's real dirty . . . But it did drill through the ice shelf and relatively quickly. Provided a hole about eighteen inches in diameter, so the scientists were then able to lower their things down and do their experiments."

As with any engine that burns fossil fuel, however, the flame-jet produced prodigious quantities of carbon dioxide, which dissolved easily in the freshwater in the hole and the seawater beneath it, turning both into something like seltzer water. At one point during the drilling, a kink formed in the hose that delivered air to the drill, and when it un-kinked a large air bubble entered the freshwater column. This caused the pressure in the column to drop and the dissolved carbon dioxide to bubble out of the water like the fizz from a champagne bottle.

"We had four thousand pounds of hose, a fifteen-hundred-pound drill on the end of it, and we were lowering it down the hole with a bulldozer. All of a sudden it quit going down the hole . . . and then it started coming back out of the hole . . . and everybody took off! The drill, the hose and everything came out the top, and then we had a geyser about forty feet high and four feet in diameter. And we got salt water! The fact that it was salt water really blew us away, because it meant that we'd bailed the whole hole."

So that was his first Antarctic experience. Over the next decade or so, he helped a glaciologist named Charley Bentley from the University of Wisconsin at Madison drill numerous short holes with a hot water drill—basically a glorified garden hose—in various parts of Antarctica, so that Bentley could drop charges of dynamite into the holes for seismic testing. He became adept at the sophisticated art of ice core drilling, in which a hollow, tubular drill with a smooth interior, threads on the outside, and sharp cutting teeth on the bottom is repeatedly lowered into the ice to carve out core segments and pull them up, one meter at a time. He drilled ice cores in Greenland and many locations in Antarctica, including the pole—and also found time to invent the field of high-altitude ice core drilling with Lonnie Thompson. This took him on at least four expeditions to high mountain glaciers in the Peruvian Andes and the Qilian Shan, the range that defines the border between the Tibetan Plateau and the Gobi Desert.

At the beginning of the 1980s, the National Science Foundation made the dubious decision to turn the group that was carrying out the Ross Ice Shelf project into a small bureaucracy, which they named the Polar Ice Coring Office, or PICO ("pike-oh"). The organization remained at the University of Nebraska until 1989, when the contract was taken over by the

University of Alaska, Fairbanks. Bruce and his understanding wife, Ann, moved with it.

By the spring of 1990, when Francis Halzen and his colleagues met with the NSF officers in Washington, Bruce was pretty much at the top of his game. He had participated in about thirty remote drilling expeditions altogether and was arguably the most accomplished practitioner of ice drilling in all its forms on the face of the planet. He had spent two months the previous austral summer helping Lonnie Thompson drill an ice core on the high spine of the Antarctic Peninsula in generally appalling conditions, and he was about to make his tenth or eleventh trip to Greenland to help get the second Greenland Ice Sheet Project off the ground—or into it. GISP2 would succeed in recovering a two-mile-long ice core at the summit three years later.

When PICO's new director, John Kelley, walked into Bruce's office and asked him if he'd be interested in helping a group of physicists drill some holes in Greenland in order to explore the possibility of constructing a neutrino telescope at the South Pole, he jumped at the chance. "Absolutely!" he remembered saying. "This is the neatest project I've ever heard of in my life! I'll work at night if I have to and just not sleep."

In August 1990, Bob Morse and a fourth Buford Price student named Tim Miller embarked for Greenland to conduct the first known ice fishing for muons—their fishing line consisting of three photomultiplier tubes that Morse had scavenged from Carlo Rubbia and Dave Cline's failed HPW experiment.

Tim and Bob flew to the summit in an LC-130 operated by the same wing of the New York Air National Guard that supports the science work in Antarctica. They were met in that high, cold place by Bruce Koci and Bill Barber, a tall, good-natured, unflappable, and incredibly strong Brit. This is where the physicists had their first encounter with glaciology—a field they would know better than they had ever wished by the time their telescope was built.

The topmost, so-called firn layer of a glacier consists of opaque snow, which becomes increasingly dense the farther you go down. Below the firn, compression from the overburden and the passage of time have transformed the dense-packed snow into bubbly ice. Very deep in a glacier the individual air molecules in the bubbles migrate into the ice to form a crystal, known as a clathrate, in coordination with the ice molecules, and even the bubbles disappear. Ultimately, the array for a neutrino telescope must be placed in this lowest and clearest, bubble-free region, but since the goal in Greenland was simply to look for down-going muons, the bubbly layer just below the firn would do. Basically, since down-going atmospheric muons give off light, if all three of the phototubes were to light up within a very short time span—thirty billionths of a second or so—the odds would be very good that they had detected a muon rather than three simultaneous pulses of random noise.

Bruce had prepared an ice core hole he had drilled the previous summer for Ellen Mosley-Thompson, Lonnie Thompson's wife and research partner. Since ice core holes will collapse over time from shear forces in a glacier, he had reamed this one out to a depth of 217 meters, about a hundred meters below the firn/ice transition.

The physicists dropped their fishing line into the dry hole and took their first series of measurements. Then they decided they wanted to increase the optical coupling between the phototubes and the ice, so they asked the drillers if they had any liquid on hand that they could pour down the hole. Unfrozen liquid is scarce on the summit of Greenland, but there happened to be a tremendous amount of butyl acetate on hand, for use in keeping the GISP hole from collapsing. They poured enough of it into the hole to cover the string and took another round of data.

"I don't know why we felt like we had to haul the string back up," says Bob, "but we did haul the string up, and all of a sudden we saw this blue sludge all over everything. . . . The butyl acetate had completely dissolved the outer jacket of the wires, and it turned all the snow and all the liquid in the vicinity this beautiful purple-blue color. . . . We wondered if we had any light transmission at all, because everything seemed to be, heh. . . . And we took pictures of the tubes going down, and then we took pictures of the tubes coming up, and the picture of the tube coming up made it look like a nice big grape snow cone. [Bruce called it a "blue slushy."] There was blue in my gloves, blue in my clothes, blue in my face. . . . It was the dye or whatever it was that's normally in the cable. . . . Electrically, it worked fine. It was just, optically we weren't sure what the hell we had done."

Bob had begun suspecting that there was something strange about Bruce when he realized he was sleeping outside in an unheated tent, even though he was surrounded by numerous heated tents all over the summit encampment. On the day he, Bruce, and Tim flew out, Bob's suspicions were confirmed. Two LC-130s left the summit that day. The one Bob and Tim rode in was heated, while the other was not, because it was loaded with ice core segments that had to be kept frozen. Bruce elected to ride in the unheated plane.

"I thought Bruce was really insane then," Bob recalled, some years later. "Now I know he's insane."

Back in Madison, Bob showed Francis the picture of the blue slushy, and Francis told him to hide it and never, ever show it to anyone again! He couldn't have been all that serious, though, because not only did they not hide it, they showed it to their funding officer, John Lynch.

Just enough data had survived the slushy incident. The blue dye *had* blocked out the light, but it had seeped into the surrounding liquid gradually, so that enough light had reached the detectors at the beginning of the run to demonstrate that the ice in Greenland was transparent to the blue light given off by traveling muons.

Analyzing the data on the blackboard in Bob's office, he and Francis came up with estimates for the transparency of the ice and the effective area of their phototubes for sensing muons. This area is larger than the geometric size of the tubes, because a muon doesn't have to hit one in order to be detected. Since Cherenkov light can travel some distance in the ice before being absorbed or scattered, a muon could pass about a meter away and still be sensed. In later years, when they learned more about the ice, they would realize that these back-of-the-envelope calculations were off by a bit, but this was a detail. There was little doubt that they had detected down-going muons in polar ice.

By this time, Steve Barwick had become an assistant professor at UC Irvine (where there was a vigorous program in neutrino physics, thanks to Fred Reines) and had come up with the name AMANDA (Antarctic Muon And Neutrino Detector Array) for the new experiment. (Francis doesn't like giving instruments female names; he thinks it's sexist.) The fledgling AMANDA collaboration, consisting of Doug Lowder, Tim Miller, Buford Price, Andrew Westphal, Steve Barwick, Francis Halzen, and Bob Morse, submitted a letter to *Nature*, which was published the following September. Francis believes this "letter launched the experiment" by showing that the idea of using polar ice for neutrino detection "was still crazy but not that crazy."

One reads that "the hole was filled with butyl acetate, an organic liquid chosen for its low freezing point and optical clarity." There is no mention of blue slush. "We find these results very encouraging, and are planning more extensive experiments at the South Pole during the coming austral summer."

This, warts and all, is how experimental physics is done. As Bob Morse writes,

> Greenland is a really beautiful example of an experiment quickly thrown together to meet a window of opportunity—mistakes were made—and where flawed or less than perfect data is also very useful, as failure can be when in the right hands. . . . This little pre-AMANDA experiment has all of the features of the later AMANDA and IceCube experiments. The later successes . . . were simply a matter of getting the bugs out of the deployments and data retrieval systems—not a trivial task. . . . This is a rare example where the funding agencies had more faith in the data (flawed as it might be) than many of the experimenters pushing on the DUMAND-in-ice idea.

Francis adds that "it's pretty clear we had no idea what we were doing, and so this was real research, right?"

He suspected "that many people had had this idea, knew more about glaciology than I did, and obviously concluded it could never work." "If we really had [known] what we were doing we would probably not have done it. And, in fact, it turns out that a lot of the things we should have known turned out not to be true."

In lectures to young scientists today, he sometimes uses the early days of AMANDA as an example of the dictum, "Don't read books. Do things. There's nothing better than to be ignorant and lucky." (It goes over well.) It is usually the young, unaware of accepted knowledge, who make original discoveries. He believes that the only reason he managed to do something original in his late forties was that he was "like a young person again" in the sense of being naïve. "It's only when you're ignorant and you haven't read all the books yet that you can do something original and new."

He was now taking a clear step into experimentalism, which is not ruled by the pristine logic of theory. Not only do numerous practical and strategic considerations come into play, oftentimes the wisest course is to stop thinking for a while and just do it.

Had he read what was then the definitive textbook on the optics of water and ice, for example, he would have "learned" that the absorption length of blue light in pure ice—the distance over which about two-thirds of the light will be absorbed—was about eight meters. That would have been a show-stopper. They would have given the phototubes back to Cline and Rubbia and gone home. If Cherenkov light really was absorbed in that short a distance, it would take something like two million phototubes to fill a cubic kilometer of ice, and the tubes alone would cost about $6 billion. Luckily, the book turned out to be very, very wrong. They obtained an estimate of eighteen meters from the Greenland data, and even though it, too, turned out to be wrong, it was a step in the right direction.

A few years later, when they were still struggling to understand the ice but had seen signs that the absorption length was actually much longer than even eighteen meters, the library in Madison mistakenly delivered the textbook to Francis's office; it was meant for someone else. He started browsing through it, naturally, and when he came to the line about the eight-meter absorption length a chill ran up his spine.

So they were headed to Antarctica. John Lynch helped them again by finagling a way to get them started on the Ice with no need for a grant proposal. AMANDA became an official NSF project, with little need for direct money, because PICO paid for the drilling, travel to Pole was subsumed under the foundation's huge Antarctic program (standard operating procedure), and Morse was going south for his gamma ray project, GASP, anyway. Madison, Berkeley, and Irvine kicked in small sums for the fabrication of two short strings of phototubes, and the scientists earned their salaries through other channels. Whatever cash was required came from GASP, which was about to be shut down anyway. Morse was still arguing for it, but Lynch kept telling him that "GASP was on its last gasp" and he ought to put his energy into AMANDA. (At around that time, the GASP group discovered a fatal design flaw in their instrument and realized it would never work.)

"I'm not in favor of diverting funds to work on a NASA project or to send your daughter to college," says John, "but I am in favor of diverting funds from an iffy idea to one that [has] a lot of promise."

At this point, although they didn't quite know it, AMANDA became a project in applied physics and glaciology. Their lofty goals in astro- and particle physics receded to the background while they taught themselves how to turn the ice at the South Pole into an enormous particle detector. It would take about ten years.

The critical unknown was how deep to drill in order to get below the air bubbles, which would scatter the light from muons and erase all memory of their direction. And there were two obvious places to look for the answer: an ice core the Americans had retrieved in 1968 at the Byrd station in West Antarctica and the Russian core from Vostok, both of which had reached about 2.2 kilometers into the ice.

Antarctica may appear to be one big chunk of ice when you look at it on a map, but it is actually divided in two by the Transantarctic Mountain Range. Vostok and Pole are located on the larger of the two pieces, the East Antarctic Ice Sheet, on one side of the range, while Byrd is located on the much smaller West Antarctic Ice Sheet, on the other. One could argue, therefore, that Vostok is more relevant to Pole. But the AMANDA scientists did the convenient thing and talked only to their fellow Americans.

The expert on the bubbles in the Byrd core was Anthony Gow from the Cold Regions Research and Engineering Laboratory (CRREL), a division of the Army Corps of Engineers, located in Hanover, New Hampshire. His work suggested that the bubbles should begin to disappear about 800 meters down and vanish more-or-less completely at about 1,100 meters. Recall that in 1988 Moiseï Markov had informed John Learned by telex that the bubbles at Vostok disappeared at 1,300 to 1,400 meters. This was another case in which ignorance may have been a virtue, for if the AMANDA scientists had realized at that early stage that they had to drill the better part of two kilometers into the ice, they might have lost heart.

When he looks back, Francis sometimes wonders just what they were thinking. They resolved to drill the first holes to only a thousand meters, above which even Gow predicted the ice would have bubbles. Perhaps they compromised because Bruce Koci told them they'd have a hard time drilling even to that depth. Oddly, too, on their first few strings, they placed the detectors only a few meters apart—as if they *had* read the magical textbook. But they were basing almost everything they did on the pioneering work of the DUMAND collaboration, which had realized years earlier that the spacings need to be large in a Markov-type, plum pudding design. So, yes, what were they thinking?

The University of Wisconsin demonstrated its formidable capabilities from the outset. The Madison campus runs a facility in the farm country near town called the Physical Sciences Laboratory, or PSL, which builds large instruments for high-energy physics experiments and the like. Since PSL had some experience building hot water drills for glaciologist Charley Bentley, PICO contracted with PSL to build a drill for AMANDA.

Although Bruce could do nothing to prevent this (nor would it have been his way) he realized that this arrangement allowed PICO to duck responsibility. They had very little skin in the game, so they low-balled it. They charged PSL with modifying what he called "an existing *small* drill that we had left over from the Crary Ice Rise," a site on the Ross Ice Shelf where he had drilled several years earlier. Morse recalls that the specifications consisted of "a bunch of pencil scratches on napkins"—typical Bruce. They named the drill Bucky-1.

When you travel to the South Pole with the U.S. Antarctic Program you always pass through Christchurch, New Zealand, and McMurdo on your way south. And the only time you'd be doing such a thing would be in the Antarctic summer, which makes for an interesting contrast, since New Zealand will be experiencing spring or summer as you prepare to fly to the Ice.

Christchurch has been called the most English city outside England. Before it was hit by a major earthquake in 2011, the lovely old town in the center of the city featured several impressive neo-gothic landmarks of red brick with white trimmings and many fine old buildings made of rectangular gray stones. In spite of the devastation, which is still quite apparent, it is a wonderful town to walk in. One comes across clusters of Christ's College schoolboys in their black-and-white-striped jackets and ties. There are beautiful English gardens, many parks, and a world-renowned botanic garden, which seasoned Antarctic travelers tend to make a point of visiting on their way south, in order to take in the colors and smells of life one last time before heading to the white and blue, mostly lifeless landscape of the frozen continent.

One is more aware of the natural world in Christchurch than in most cities of its size, owing not only to the gardens and the rivers that flow through the town, but also its location. It sits on a plain between the Southern Alps and the sea, so the mountains afford a majestic backdrop and there is usually a salty smell in the air.

The town figured largely in the golden age of Antarctic exploration, because the South Island of New Zealand is the closest major land mass to

Ross Island, the site of present day McMurdo Station, which happens to be the farthest south—in other words, the closest to the pole—that it is possible to travel by ship. In 1911, both Robert Scott and Roald Amundsen made final landfall in Lyttleton, the port city five miles from Christchurch, before sailing to the Ice and engaging in their legendary race to the pole. When Scott finally reached what he mistakenly believed to be the exact geometric pole, slightly more than a month after Amundsen had come within a few hundred yards of the real thing, he wrote in his diary, "Great God! this is an awful place and terrible enough for us to have laboured to it without the reward of priority." He and his men then proceeded to lose their lives on the gloomy trudge back to Ross Island. New Zealanders, with their British heritage and all, tend to favor Scott over Amundsen. There's a statue of him in the town center, and there's a wonderful collection of historic Antarctic artifacts in the local Canterbury Museum.

On my way to Pole in 1999, I stayed at the Windsor Hotel, a comfortable bed and breakfast that was a favorite of Antarctic travelers. (Sadly, it was destroyed in the 2011 earthquake.) Having had some experience as a mountaineer and climber, I had heard a lot of adventure stories by then, so I approached Antarctica with a sort of ho-hum attitude, expecting more of the same. At breakfast on my first morning at the Windsor, however, when I noticed the gleam in the eyes of a man who had just come off the Ice, the place began to exert its magnetic pull. A kind of madness takes over—like an addiction. I have never met anyone who went to the Ice who didn't scheme to go back for a year or more after their return.

The gateway to the Ice is the International Antarctic Centre, a gleaming set of buildings that is part Disneyland for the tourists and has an adjacent set of runways, where you will catch your flight to McMurdo. Your first order of business is to pick up your "Extreme Cold Weather" or ECW gear: a big red down parka with a fur-edged hood, insulated Carhartt overalls, very warm boots, socks, long underwear, gloves, hats, face mask, goggles— enough to fill a large duffle bag. Then you slip into a pattern of waking at five a.m., donning your ECW paraphernalia, shuttling from your hotel to the Centre, doing the first of many "bagdrags" (dragging your luggage to your awaiting flight), finding your flight has been canceled, and returning to your hotel to change into spring or summer attire and spend the rest of the day sightseeing in Christchurch. On a lucky day you'll get the call that your flight has been canceled before you leave your hotel; on an unlucky one, you might board your plane and take off, fly for a few hours, and boomerang. The flight to McMurdo takes from five to eight hours, depending on the aircraft, so it's not unusual for conditions to deteriorate mid-flight.

All flights from Christchurch forward are in military transport planes, either LC-130s or other much larger aircraft that don't have skis and can only land on one of the sea-ice runways at McMurdo—if it has been made flat and hard enough. This is a hit-or-miss proposition nowadays, as the climate on the Ross Ice Shelf has warmed enough in recent years to create a surface layer of slush for much of the summer.

These planes are basic affairs for the passengers, who sit in canvas-slung seats along the walls of the huge, open cargo compartment in order to keep it clear for whatever equipment and supplies the plane might also be carrying. There are only a few small windows in the compartment, but the mood is quite relaxed, in contrast to the usual commercial flight. Folks sprawl on the floor to sleep, or chat together leaning on boxes of cargo. You're even allowed to enter the cockpit every once in a while to join the pilots for a panoramic view of the land or seascape.

You land and disembark in the middle of the vast, flat plain of the Ross Ice Shelf, five or so miles from shore. As you walk away from your plane, even it begins to look tiny in the huge and unusual, truly breathtaking landscape. White mountains rise sharply in the distance, at the edge of the icy plain. Mt. Erebus, a bulbous white dome dominating the skyline of Ross Island, behind McMurdo, is probably emitting white steam. It is the southernmost active volcano on the planet.

Everything is pristine white and blue, except for your destination. You ride a shuttle bus across the ice to a collection of metal and wood buildings clustered in random array among the low hills at the edge of Ross Island. McMurdo Station is mainly black and brown. It looks like a mining town (see photograph 3). The roads are dirt, and the dirt is volcanic. Little besides lichen grows in it. It doesn't have the rich smell of topsoil. In fact, your sense of smell goes more-or-less dormant in Antarctica. A farm of disc-shaped fuel tanks lends the aspect of an oil refinery. McMurdo is the fuel depot for the lion's share of U.S. operations on the continent, and fuel is what makes it possible to live in this alien environment. It keeps you warm, it cooks your food, it melts ice for your water, it generates your electricity, it enables you to go where you want to go. It is delivered by tanker once a summer on a route cleared by icebreakers, after the sea-ice has softened enough to let the breakers through. A large golf ball housing a satellite dish stands on a hilltop, and tiny Hut Point juts into the sea on the far side of the large rectangular pier, ingeniously crafted of floating ice covered with dirt, where the oil tankers and cargo ships dock and unload (see photograph 4). The hut that Scott built for his fatal 1911 expedition still stands on the point, the cold, dry conditions having kept the building and its contents remarkably well-preserved. You sit through some briefings and find your Spartan accommodations.

McMurdo has most of the amenities of a small town: bars, clubs, gyms, movie theaters, a chapel. No children or pets allowed. As at most bases on the continent, there is a rowdy party scene. It's a busy place in summer, with innumerable operations to run and scientists to support. It has a strong ethos of its own, and many people go back year after year, but you won't find too many McMurdo fans among the scientists who work at Pole. To them it's a place to get stuck on the way to or from your workplace. "Rotting in McMurdo," it is sometimes called.

You soldier through your period of waiting and abortive bagdrags, which may last as long as a week, and finally board the LC-130 that will take you to Pole, hoping not to boomerang. You're a little more careful about your ECW gear on this flight, since you'll be stepping from the plane directly into minus thirty or forty degree weather.

The flight is spectacular. Nothing but blue above and white below, except during the brief crossing of the Transantarctic Range, a band of gray sandstone nunataks jutting up through more than a mile of ice, which flows in great rivers around them (see photographs 5 and 6). This crossing occurs early in the trip, so for most of the flight all you see is the unblemished ocean of the East Antarctic Ice Sheet. A few minutes before the plane lands you discern a collection of black dots, tiny Amundsen-Scott South Pole Station, on the snowy plain in the distance.

The flight protocol calls for an LC-130 to remain at Pole for as short a time as possible and to keep its engines running. The passenger exit is near the bow of the plane on the left. Once you step onto the Ice you notice a couple of people standing there, directing you toward the station. (I had made the acquaintance of a South Pole veteran named Martin Lewis on my way south. He stepped off the plane just before I did and was hit by a fusillade of snowballs.) Meanwhile, a cargo crew, the "cargoids," swiftly empties and reloads the plane through the large access door at the rear, and a group of "fuelies" siphons off some of the plane's fuel, which will be used to power the station. Almost half a million gallons need to be stored in the huge tanks on-site before the last flight leaves in February and the winterovers are left to enjoy the nine-month winter alone. After the incoming passengers have debarked, the outgoing passengers board. If all goes according to plan, the plane is on the ground for less than half an hour.

In the early days of AMANDA, the nerve center of the station was a large, orange metal Buckminster Fuller dome that functioned as a sort of tent, enclosing a few small buildings that resembled refrigerators. Their doors were just like the locking doors on walk-in freezers in the real world (as they say in Antarctica), except that they were meant to lock the cold out rather than in. The dome was unheated—a few grimy icicles, known by the locals as crapsicles, hung from the triangular panels about fifty feet above your head—and it was entered by way of a tunnel. It has now been replaced by a gleaming, new, multi-hundred-million-dollar station and is widely missed. An AMANDA veteran once told me that moving into the new station was something like moving from a comfortable old bed and breakfast into a Best Western hotel. The dome was "the real Pole" (see photographs 7–9).

Aside from adjusting mentally to a new, surreal environment, you are immediately beset by acute altitude sickness, since you have just flown from sea level to about ten thousand feet: headaches, nausea, listlessness, and sleeplessness in varying degrees. You are also prone to dehydration: the air is extremely dry, owing both to the altitude and to the cold. Technically speaking, Antarctica is a desert; it snows less than three inches a year at Pole. But buildings are eventually buried nevertheless by blowing snow. (The dome proved to have an unanticipated advantage in that respect, since its spherical shape somehow minimized the drifting. It lasted about twenty years longer than it was expected to.)

At the briefing you attend first thing, you are advised to drink plenty of water, but not do much else with it, since it takes labor and fuel to make this valuable substance at forty below zero. You are allowed two two-minute showers per week and one load of washing. Your room will most likely be an eight-foot by twelve-foot canvas-walled cubicle in one of the Jamesways, a sort of insulated, canvas Quonset hut, out in Summer Camp, the berthing area for the "hordes" who descend in summer. There's a bed, a small side table or bureau, a large tin can for peeing in—and not much privacy. Conjugal relations in these quarters don't involve much pillow talk.

A vibrant culture has developed among the people who have maintained South Pole Station continuously now for about sixty years. The first station, which was known as "Old Pole," was built during the Antarctic summer of 1956–57. Until it was blown to smithereens during the summer of 2014–15, it lay buried beneath the snow about three-quarters of a mile from the new station, out beyond the "dark sector," where the telescopes are located. It was blown up for safety reasons. Some mini-crevasses had opened up around the buildings and the tunnels that entered them, and it also used to be a favorite forbidden pastime to explore the place—especially in winter. Imagine crawling around in an underground maze, partially filled with snow, by headlamp, in the dark, at minus fifty degrees. If you were caught, you'd be escorted off the continent forthwith.

It's a tribe really. They tend to be frontier types, cultists, and free spirits: Deadheads, rock climbers, Red Sox or Cubs fans. . . . In 1999, I met a guy who was writing a book about riding the rails through the American West like an old-fashioned hobo. Another fellow had just sold a bar he had owned in Thailand (which seemed to be a popular R&R spot for "Polies" after leaving the Ice). The tattooed bulldozer mechanic you strike up a conversation with in the galley, who looks like he could easily break you in half over his knee, happens to knit Icelandic sweaters for a hobby. There's a women's knitting club called "Stitch and Bitch." There's a fly-tying club.

The real hard core are those who have worked on the continent for years, many of whom have wintered over in the past and may be at it again this year. They are privy to the innumerable secrets of construction work and getting infernal gadgets to work and stay working in this brutal environment. This is an oral tradition. One trivial example is not to leave metal tools outside. If a cargoid thinks she might need a big box-end wrench to open the cargo bay or disassemble a large piece of equipment that needs to be removed post-haste from the back of a "Herc" (Hercules transport plane, aka LC-130), she leaves it in the cab of the forklift that she races up to the back of the plane with. If she leaves the wrench outside at minus fifty, her hands will freeze when she uses it, no matter how thick her gloves are.

The saying among the inner circle is that the first year you come for the adventure, the second for the friends, the third for the work . . . and the fourth because you don't have any friends left in the real world anymore. In other words, it's mainly for the work—or "mission" as they call it. The scientists find themselves in the hands of a seriously competent group of people, who take pride in their work and are inspired in return by the quality of the science they are supporting. The science talk given every Sunday in summer is usually well attended. There is a sense of trust in one another that was reminiscent to me of the people I used to share a rope with on rock or ice climbs. Competence and trust translate into survival in dangerous situations. The atmosphere at the station is infectiously positive and can-do, with a rough, unsentimental edge that makes it all the more effective.

Scientists are known as "beakers," and while it may not sound like it, this is a positive term. The derogative is "jafa," for "just another fucking academic," which you don't hear as often as "fingy," for "fuckin' new guy," a carryover from the Vietnam War that applies to the clueless scientist and non-scientist alike, there for his first "summer jolly" on the Ice. (A large fraction of the support personnel are first-timers, seeking only adventure, who won't be back next year.) The PICO drillers, who are doing skilled manual work in the service of science, liked to be called redneck beakers.

It seemed to me, incidentally, that the most competent heavy-equipment operators, that is, bulldozer, bucket loader, and forklift drivers, all happened to be women.

There is no place on Earth further south than the South Pole, obviously, so when you're traveling to the place you are inevitably headed south, and when you leave you're headed north. There is no longitude at the pole either; its coordinates are 90° South. While, technically speaking, every direction is north, it is necessary to be able to talk about directions, so at Pole they are defined by a grid. Grid north follows the prime meridian or zero line of longitude, which passes through Greenwich, England. Grid south (180°) passes through New Zealand. You can also step easily into any time zone, so for convenience the clocks are set to New Zealand time.

The endless white plain surrounding the station has been divided into pie-shaped sectors. The clean air sector lies up the prevailing wind, grid northeast between 340° and 110°; the wind blows from this direction 90 percent of the time. Although the South Pole has the cleanest air on the planet (upwind, at least), it is not immune to pollution. The U.S. National Oceanic & Atmospheric Administration runs an observatory in the clean air sector not only to observe the weather, but also to monitor air quality, and it was at the pole during the International Geophysical Year, 1957–58, that the anthropogenic or manmade increase in carbon dioxide, the greatest contributor to global warming, was first confirmed. (Old Pole was built under the auspices of the IGY.) It is ironic therefore that the U.S. Antarctic

Program runs almost entirely on fossil fuels. The snow in the downwind sector has registered the pollution from the station's power plant ever since it was erected. The snow is noticeably darkened, and in the immediate vicinity of the station you can even pick out annual layers of black soot.

Grid east lies the quiet sector, free from sound and radio noise, and grid west, across the airstrip and a little more than half a mile from the station, lies the dark sector, where the telescopes are located.

Also, incidentally, since everything we've talked about is sitting on top of an ice sheet, it's all sliding grid northeast, in the rough direction of the Transantarctic Mountains, at the rate of about thirty feet per year. Every New Year's Day there's a grand and motley procession, led by John Wright if he's in town, a miner and demolitions expert who also plays the bagpipes. The procession ends with a ceremony for placing a marker at the precise location of the geographic South Pole on January 1. There's a line of them stretching across the Ice in the direction of the dark sector (see photograph 18).

✦ ✦ ✦

In the austral spring of 1991, Bob Morse, Bruce Koci, Steve Barwick, and Tim Miller traveled south to do the first drilling for AMANDA. PICO also tasked a drilling crew that included Bill Barber to the job.

Since AMANDA had no infrastructure in place, Bob and Bruce decided to ride on the backs of SPASE and GASP. They drilled out by the "SPASE shack," so they could retreat to its warmth every once in a while during the drilling of the holes and the deployment of the strings, which took place in the wide-open in temperatures that were pretty much always below twenty below.

Drilling was always the key. Francis, the theorist, estimates that over the two decades that it took to build AMANDA and IceCube, he spent more than half his time thinking about drilling, directly or indirectly.

"Everybody knows that hot water drilling got started by a bunch of glaciologists standing around drinking beer, going to the Ice and putting two and two together," Bruce once told me. "Something like that happens after about a six pack of beer."

The idea is to use a hose with a nozzle on the end to spray a parallel stream of hot water into the ice, and let the nozzle free-fall as it melts its way down. In practice, this is a lot more complicated than it sounds. High technology has been applied to every detail, and the nozzle or drill head is especially sophisticated. Calipers are attached to the head to measure the diameter of the hole as it goes down; inclinometers are attached, so that the path of the head can be reconstructed as it works its way down and the straightness of the hole can be determined; etc., etc.

The technique was first employed in temperate glaciers in Alaska, where it wasn't difficult to reach depths of a thousand meters. Since the temperature of the ice wasn't that far below the freezing point, it wasn't that hard to melt, and the water used for the drilling didn't lose a lot of heat through the walls of the hose as it streamed down to the depths. The method was first attempted in the polar regions on the Ross Ice Shelf, where the ice is also relatively warm, to depths of about four hundred meters, and then at Antarctica's truly frigid, 10,700-foot Dome C, but only to about sixty meters. The ice at Pole sits at minus fifty Celsius, so the Amandroids were taking a step into the unknown by trying to drill to a thousand meters.

What you want in a hot water drill is a lot of very hot water at high pressure and a hose with a wide diameter and heavily insulated sidewalls, so that the water will stay hot as it goes down and carry as much heat as possible out through the nozzle and into the ice. (There's a bit of a balancing act here, because you do want some heat to escape through the sidewalls in order to keep the water in the upper portions of the hole from re-freezing as the drilling proceeds.) Bruce once pointed out that in the same sense that a neutrino telescope can never be too big, "you can't have too big a hot water drill either, 'cuz the best hot water drill is the one that drills the hole instantly; it's the most efficient." This is a mental construct, obviously—a drill that will deliver an infinite amount of heat in zero time—but it gets the point across: you want a lot of heat and a huge hose.

Bruce knew that Bucky-1 represented a serious compromise in both respects. The hose was only an inch in diameter, and the heating plant, which consisted of a gang of hot water boilers standing in the open out on the Ice, produced only half a megawatt of power. The first hot water drill he had used on the Ross had produced two megawatts. "We knew [Bucky-1] was limited, but we didn't know how limited, exactly, because it was the first time anybody really tried to go deep in ice this cold." He calculated that by the time they got to a thousand meters the heat that the water would lose on its way down would exceed the output of the heating plant. In other words, the bottom of the hose stood a good chance of freezing. He could play various games to try to keep it from freezing, such as raising and lowering the drill in order to reheat the water in the newly opened hole, but this would cost fuel, and it was a tricky business in any case.

"Hot water drilling is not for the faint of heart either," he says. "We're trying to keep water in some place that's minus fifty degrees, and that's not a good thing to do."

They had taken along a combination of detector strings, some of which had been constructed in Berkeley and the others jointly by Madison and Irvine. The phototubes on the Berkeley strings were only three inches in diameter, while the ones on Madison/Irvine strings were not only larger to begin with, they were encased in the same pressure vessels that DUMAND was using at the time: ten-inch glass spheres, designed to withstand the high pressures at depth. They were made by a company named Benthos and called Benthospheres.

The team took the conservative approach and deployed the smaller detectors first. It is unclear exactly how deep they drilled, but the process went well enough. Unfortunately, however, when they lowered the string into the hole, it got stuck at about eight hundred meters. According to Bruce, "they weren't jigging the thing, and the holes were pretty irregular, so it froze to the side." He had suggested to the physicists that they raise and re-lower the cable at least once every five minutes. They had let it sit for half an hour at one point, and that's when it froze in. (Bob Morse observes that Bruce "was a taciturn man, who would only offer his advice up to the listener once, and you ignored it at your peril.")

The string didn't work very well, either. Of the four detectors, only one ended up functioning, the suspicion being that water had leaked into the other three. This first string ever deployed for AMANDA was the last without Benthospheres.

After a few days at Pole in summer, once you've adjusted to the altitude, the brightness, and the surroundings, you begin to realize that everything you learned in grade school about the spinning of the Earth and its motion about the Sun is probably true. It's obvious that there's only one day a year down there, as the Sun circles counterclockwise overhead, at a constant angle above the horizon that depends on how recently it has risen or how soon it will set, in other words, how far you are from an equinox. At the solstice, around December 21, it will be twenty-three-and-a-half degrees above the horizon, which is the angle between Earth's spin axis and the ecliptic, the plane in which we rotate about the Sun. This is the latitude of the Tropics of Capricorn and Cancer.

Since summer is so short, three eight-hour shifts work round the clock all through the season, everyone works six days a week, and if your task allows, Sunday is your day off. This means that the place is on pretty much all the time. On top of that, it's hard to turn yourself off anyway, since you begin to buzz on the dazzling amount of light pouring constantly into your eyes. It's easy to get frazzled, and it can be difficult to get to sleep. There are those who fall into a pattern of staying awake all through the work week and sleeping all day Sunday—until they completely burn out. Furthermore, since it's always "nighttime" for one of the three shifts, there's usually a party going on somewhere, and people find creative ways to celebrate in what is truthfully, for most, a disconcertingly blank landscape.

The AMANDA team was keeper of a stellar attraction in the tubs that they used to heat water for their drill. Until NSF caught on and made them illegal, there were hot tub parties out at the drill site (see photograph 2). One time, the pilot of a Twin Otter, a relatively small plane that is ubiquitous on the continent, sighted a party in progress as he flew into the base. He landed on the snow nearby, taxied up, stepped out of the plane, stripped off his clothes, and jumped in. There were also naked rolling-in-the-snow contests. An Italian scientist set the record with twelve complete rolls.

It's good they were having some fun, because the drilling was not going well.

On the second hole they got greedy and tried to go to a thousand meters. This led to about the worst thing that can happen in hot water drilling.

"We stuck Bucky-1," Bob later recalled. "He's still there."

"Yup, radar marker," said Bruce.

Bob was asleep in his cubical in one of the Jamesways in Summer Camp when a PICO driller named Dave Kestor poked his head through the curtain and whispered the bad news. (Since one of the three shifts is always trying to sleep, silence is observed in Summer Camp twenty-four hours a day.)

"What should we do?" Dave asked. "Are we going to put the instrument down or not?"

"God, I don't know," Bob replied, "but I think we probably ought to. We've invested this much in the hole. We ought to stuff whatever we can down there."

He got out of bed and went to find Steve Barwick. "I said, 'Steve, ah, they've stuck the drill.' And Steve went ballistic at this point and went screaming—he didn't know where Tim Miller was, so he went walking into every Jamesway at about three in the morning, screaming at the top of his lungs, 'Tim Miller? Tim Miller? Where the hell are you? Where the hell are you?' . . . And I thought that there was gonna be some construction worker, some six-foot-four guy, was gonna get up and just *kill* Steve. I thought Steve was going to *die*."

PICO had detailed only two shifts to the drilling, each working twelve hours a day, and it was the night shift that had stuck the drill. Bruce was on the day shift, and although he didn't sleep much during drilling operations—he'd hang around to keep his finger on the pulse even when he wasn't on duty—he happened to be asleep when the disaster occurred. Not only had the drill gotten stuck, the flow in the hose had stopped as well. This is about the only situation you can't get out of with a hot water drill.

"Everybody who's ever done hot water drilling has made the mistake of going with not enough heat once in their life," Bruce later observed. He thought they were lucky to have gotten as far as they had.

They tried to pull it out with Caterpillar D7 bulldozer.

According to Bob, "That goddamn hose was like a violin string. . . . It necked down to about half its original diameter. . . . Bill Barber was the only one that had the courage to go over there and stand while the hose was down the hole and take a hacksaw and cut through it. And we saw this hose, or heard this thing, disappearing at the speed of sound down this hole, like *phewwwwww.*"

And so, with the drill head and a significant portion of the hose down the hole, they acted on Bob's first sleepy thought and deployed one of the Madison/Irvine strings anyway. For whatever reason—the hose in the way, the hole too narrow—it only went down 150 meters. Then they began worrying that light from the surface might wick down to the detectors and flood out any muon signals they might possibly detect. The photomultipliers were exquisitely sensitive; they were operating at what is known as the single photon level, which meant that they could detect individual particles of light.

"I just looked around and said, 'We've got to plug that hole. What can we do to plug it?'" Bob recounts. "There was asbestos and some green garbage bags around—the eco people would shit a brick if they heard this: I started stuffing asbestos insulation into the bags to give them bulk and flicking them down this hole, trying to make a light seal. . . . I think I threw down three or four. I threw down as many as I had."

Experimental physics, blemishes and all.

Bob remembers the date of the disaster quite accurately, since it was his wedding anniversary: January 17, 1992. He sent an e-mail to Francis back in Madison.

"It said, 'After a particularly difficult deployment, we've managed to put the phototubes exactly where we wanted them'—that was a touch of irony. . . . 'Oh, by the way, buy my wife some flowers.' Hah! Which he did! Francis can always be counted upon to do the right thing in that sort of situation. He sent her a beautiful bouquet of roses."

9. The Crossover

THERE IS NO SUCH THING AS A DISASTER IN BASIC RESEARCH. WHATEVER happens, you learn from it. They had stuck their collective nose into it and learned what to do next time. They had even made a significant, though not unexpected discovery: that Antarctic ice is optically inert; it doesn't give off light on its own. The few detectors they had succeeded in deploying proved to be extremely quiet when they weren't picking up light from the occasional muon zooming by. This demonstrated that the ice did not contain potassium-40, the radioactive isotope of potassium that is present in all natural bodies of water. The background light given off by dissolved potassium-40 limits the ultimate sensitivity of all water-based neutrino telescopes and always will. Ice, they had now confirmed, does not have that limitation.

From a practical standpoint, the drillers had learned the necessity of more heat and a larger hose, and the physicists had learned that phototubes housed in Benthospheres could survive the refreezing of the water in a relatively deep, hot-water-drilled hole.

The collaboration had also come face-to-face with a management issue that would plague them for the rest of the decade: that PICO, the Polar Ice Coring Office, was fundamentally incapable of doing the job it had contracted to do. The problem was structural in nature; it had nothing to do, necessarily, with the competence or good will of its employees. Since PICO was a separate organization from AMANDA and answered to NSF rather than the physicists, it had no institutional stake in the outcome; there was no pressing need to go the extra mile. The AMANDA contract was one of several in its portfolio, and a small one at that. As long as the organization stuck to the letter of the contract, it would win it again the following year, whether the project succeeded or not.

Bruce Koci and Bill Barber were exceptions to this bureaucratic mentality. Bruce was totally committed to the project and intrigued by the science, and he understood the importance of working with the scientists to understand their needs—even if they couldn't articulate those needs explicitly on their own. Research is when you don't know what you're doing. It also helped that Bruce and Bob Morse had become friends over the course of the season, and that Bruce was the most inventive engineer at PICO.

John Lynch and Peter Wilkness at NSF hardly batted an eye. They were used to failure the first time around, and the project had flown under the radar anyhow. Near the beginning of 1992, the AMANDA collaboration submitted a three-year grant proposal for $2.6 million that was greatly buoyed by the letter they had published in *Nature*. (This is a tried-and-true tactic in the grantsmanship game: publish in *Nature* or *Science*, get your grant. They would use it again at the next crucial juncture, ten years later.)

Funding commenced in July, and a few months later the collaboration picked up its first new members: a team from Stockholm and Uppsala Universities in Sweden, led by the distinguished high-energy physicist Per Olof Hulth.

Peo, as he was known to his friends (pronounced "Pay-oh"), was a modest and friendly man with a warm sense of humor, who didn't see a lot of difference between work and play. He would come up with several inventive and even entertaining ideas for AMANDA and IceCube over the years, and students or professors who took sabbaticals in Stockholm have fond memories of excursions with him in his beloved sailboat in the Stockholm archipelago. He supplied a steady hand experimentally and a wise, behind-the-scenes approach to resolving conflicts in the collaboration, of which there would be more than enough over the years. Like most converts to the new field of astroparticle physics, he was seeking refuge from the accelerator world at CERN, with its huge collaborations, rigid management structures, and brutal politics. Astroparticle physics seemed "new and exciting," he once told me. Its very existence demonstrated that physics was "dynamical"; it could change.

He and several of his countrymen had considered building a neutrino telescope in Torneträsk, a lake in northern Sweden that is reputed to be the clearest in Europe, but it turned out not to be as clear as they'd hoped, nor deep enough to provide the necessary shielding. They had then looked into DUMAND, but decided against joining, partially because it seemed too high tech. Peo told me that when he first heard of AMANDA, it seemed "like a joke . . . absurd." Evidently, the more he looked into it the less absurd it became. The Swedes were a perfect match and began contributing right away.

After a bit of back and forth, the collaboration decided against another deployment attempt the following season. Instead, they took a deep breath, redesigned the drill, and began designing a sizeable array, which they hoped to begin deploying over the 1993–94 Antarctic summer.

In retrospect, Bruce concluded that it would have been impossible to improve much upon Bucky-1, no matter how hard PICO had tried, because the drill was limited by the hose, which was the best available at the time. This would be a recurring theme, all the way through IceCube. As prosaic as it may sound, one could say that the success of this project, that is to say, the invention of neutrino astronomy, hinged upon the development of an adequate hose for hot water drilling.

Irrespective of its insulation within the bureaucracy, the first year's failure had been an embarrassment for PICO, and this gave Bruce the leeway to develop a larger hose, which he outsourced to a company that could make it to his specifications. It was an inch and a quarter in diameter, only a quarter of an inch wider than Bucky-1's, but since flow rate grows rapidly with diameter, he was confident it would allow them to reach a thousand meters the second time around. A team at Wisconsin's Physical Sciences Laboratory, led by a superb engineer named Bill Mason, built the first dedicated drill for AMANDA, again to Bruce's specifications (and again probably written on napkins). The Swedes paid for it.

In a vivid demonstration of the debt that AMANDA owed DUMAND, the first true array envisioned for the South Pole was almost a direct copy of the array that DUMAND was hoping to deploy at the same time. DUMAND-II, which had been funded in 1989, was to consist of eight strings arranged in a regular octagon, with a ninth in the center, while AMANDA-A, as it came to be called, was to have nine strings on the perimeter and a tenth in the center.

Bob Morse points out that the tenth string was added to break the high symmetry of the octagon, because computer simulations had shown that too much symmetry can confuse the reconstruction of muon tracks. It doesn't make it difficult to reconstruct the direction of the track; however, surprisingly, it *can* make it difficult to figure out which way the particle was traveling—up or down, for example, which seems kind of crucial. AMANDA-A would have twenty modules per string, spaced ten meters apart. And they would be deployed in thousand-meter holes, so they'd end up precisely in glaciologist Tony Gow's magical depth range of eight hundred to a thousand meters. The plan was to deploy six strings the first season and the remaining four the next.

As they prepared for this campaign, they fell into the annual build cycle that would drive AMANDA and IceCube for the next twenty years. You don't carry a hot water drill or two hundred optical modules to Antarctica in your luggage. Anything that goes to Pole, for whatever purpose, even as part of a building, must be "Hercable": it has to fit in the cargo bay of a Hercules LC-130.

The Herc is a standard coin for working at Pole. The drill, for example, with its numerous hot water boilers and thousand-plus meters of hose, filled almost two Hercs. When you submit your proposal for your working season at Pole, you negotiate with NSF not only about how many people you need to house at the station and for how long, but also how many Herc flights it will take to get your equipment and instrumentation down there. These items are transported to McMurdo on cargo ships that leave the continental United States on non-negotiable schedules from Port Hueneme, a naval base on the California coast, not far north of Los Angeles. If you miss the boat you may very well miss your season.

The plan was for each of the universities in the AMANDA collaboration to become expert in most of the hardware tasks, so that whatever collection of students, faculty, and other personnel happened to find themselves on the Ice at any given time would have a large enough reservoir of knowledge to deploy the hardware and get it running. The optical modules were assembled in Madison, Irvine, and Sweden, while Berkeley made other critical parts. Since Steve Barwick's institution, Irvine, was closest to Port Hueneme and he had access to large workspaces on campus, Irvine served as the staging area, where the hardware was integrated, packed in wooden crates—or, in the case of the hose and the cables for the strings, wound onto large reels—and shipped by truck to the port. Once it reached McMurdo, it would be transferred to Hercs and flown to Pole, on an unpredictable schedule that was highly dependent on weather. The trip from Port Hueneme to Pole usually takes between seven and eight weeks, so equipment for the start of the season needs to reach the port by the end of September. Ships continue to depart on a regular schedule until about the first week of January.

Barwick was a powerhouse. He became the hardware guru. Among other things, he designed a data acquisition system that would run on an Apple Macintosh, to collect and computerize the raw electrical signals as they streamed to the surface from the phototubes below. (This so-called DAQ was small enough to be carried south in his hand luggage.) He was in a strategic position, of course, and this became both a blessing and a curse—especially when money got tight, which it did right away.

The management of an academic collaboration is very different from that of a corporation. For one thing, it's more democratic. For another, its finances are subject to the whims of capricious elected officials—in this case, the United States Congress. Thus the annual funding cycle for U.S. science agencies resembles a rollercoaster ride. NSF, DOE, NASA, and the other science agencies prepare budgets in advance of the fiscal year; they are modified over some months, first by the White House, then by Congress (the "appropriators"); agreement is usually reached; budgets are usually approved; and money becomes available at some indeterminate date, usually in the late summer or early fall. Depending upon political winds, there can be excessive delays. The infamous budget stand-off in the mid-nineties between Bill Clinton and Newt Gingrich stretched into the following year and wreaked havoc on countless scientific programs. In the end, an agency's budget may actually exceed its original request, but it usually falls short—as it did for NSF in the first year of the first AMANDA grant.

Wisconsin was the lead institution, Berkeley and Irvine were listed as subcontractors, but the budget called for each institution to get about the same amount of money. Now, some of that money was earmarked for salaries: 50 percent of Bob Morse's salary, for example (since he was a re-search scientist, rather than a professor, he lived on such "soft money"), and two months of Steve Barwick's, to cover the summer. Students were par-tially covered as well. Without going into the arcane details of NSF money management, let's just say that Lynch and Wilkness changed the ground rules only a month or two after the grant was funded. They were happy to pay for any and all hardware and for salaries at Madison, but decreed that Berkeley and Irvine should pay their salaries from other grants. (Buford Price had a longstanding NSF grant for nuclear physics research that was more-or-less rubberstamped every three years.) This upset Buford and Steve tremendously, and they began making frantic requests for money, both to NSF and to Bob and Francis in Madison. Steve's were particularly emotional. In an e-mail written in July 1993, scant months before their sec-ond deployment season would begin, Francis wrote to Buford that "after talking to Steve (for hours) I have the impression that there might be a misunderstanding concerning the AMANDA funding. Steve reacts with his heart although his brains know the facts and, given the circumstances, this is hardly a criticism."

As the hard deadline for getting their equipment to Port Hueneme ap-proached, Steve upped the ante with what would become another annual rite: he held the project hostage. He threatened not to deliver the final two strings until he got more money.

Considering the University of California's reputation as the premier public university system in the world and its long history of running huge pro-grams like the Los Alamos bomb project, it is surprising how much more adaptable the University of Wisconsin proved to be in dealing with these vicissitudes.

The reasons go back to the beginning of the twentieth century, when the president of the state's university system, a man named Charles Van Hise, first enunciated what became known as the Wisconsin Idea: that the purpose of the university should be to serve the people of the state by pro-moting public service and the search for truth. "I will never be content until the beneficent influence of the university reaches every family in the state," he declared. The Idea is now enshrined on a bronze plaque on a gran-ite boulder on a hill on the Madison campus, and the people of Wisconsin support it across the political spectrum. It may sound lofty, but it has had immense down-to-earth ramifications.

In 1923, a Madison biochemist named Harry Steenbock discovered that ultraviolet radiation increased the vitamin D content of foods. A vitamin D deficiency can cause rickets, a crippling bone disease that affected mostly children at that time. Inspired by the Wisconsin Idea, Steenbock filed for a patent, declined a licensing offer of $1 million from the Quaker Oats Company (quite a lot of money in those days) and instead rounded up a group of likeminded friends to found the Wisconsin Alumni Research Foundation (WARF) in order to use the proceeds from the patent to support the university. It was licensed to Quaker Oats and a pharmaceutical firm. The irradiation process is especially effective in milk. It essentially eliminated childhood rickets worldwide within about twenty years, and it brought oodles of money to WARF.

Over the decades, the foundation has licensed thousands more patents, developed an equity fund of about $2 billion, and disbursed the better part of $1 billion to the Madison campus. (Another huge moneymaker was Warfarin, which is named for the foundation. It is the most widely prescribed oral anticoagulant in North America. One of its trade names is Coumadin.) In an average year, WARF distributes $50 to $70 million. The money can be used for pretty much anything related to its broad mission: funding research projects, building buildings, purchasing land, endowing fellowships, subsidizing visiting scholars. Vernon Barger used WARF money to bring Francis Halzen to Madison in 1971, for the visit that still has not ended.

Twenty years later, when the money crunch hit AMANDA, it so happened that the man in the strongest position to steer WARF funds was John Wiley, Bob Morse's old drinking buddy from the days of the 602 Club. They'd been graduate students together and even part-time roommates.

Wiley had earned his doctorate in solid state physics, gone off for a six-year stint at the legendary Bell Telephone Laboratories, returned to a faculty position at Madison, and proceeded to rise like cream. By 1993, he was dean of the graduate school and by virtue of that position the university's primary liaison with WARF, which is a separate, non-profit entity. The foundation's profits "are funneled through the graduate school," he says, "so the dean of the graduate school is kind of Santa Claus on campus." Like Peter Wilkness, John Lynch's boss at NSF, Wiley loves the big idea.

According to Lynch, "Smart guy like Francis goes and talks to a smart guy like Wiley, things can happen, you know?" The dean responded to AMANDA's money crunch by arranging a million-dollar line of credit through WARF. The hope was that NSF grants would eventually right the balance—and eventually they did—but in the meantime the credit line would buffer the project against the rollercoaster of federal funding.

With their newfound flexibility, the Wisconsinites did everything they could to keep the project and the Californians afloat. Francis even sent an installment from his endowed professorship, the Gregory Breit Chair in the Madison physics department, to Steve Barwick, to pay for salaries. Steve also got what he wanted to finish the last two strings. (The Swedes kicked in some money for hardware as well.) By the time AMANDA headed south for the 1993–94 drilling season, Bob and Francis owed WARF a quarter of their credit limit, and WARF had also guaranteed a $100,000 loan that Buford had obtained from Berkeley.

In the two-year break while the physicists went through this exercise, Bruce Koci went off on other adventures. During the summer after the Bucky-1 debacle, he and Bill Barber joined Lonnie Thompson on a three-month expedition to the highest icecap in the remote Kunlun Shan range in far western China, where they drilled on a wide snow dome at 20,000 feet and recovered an ice core that held a climate record reaching back three-quarters of a million years. Bruce did some non-AMANDA hot water drilling in Antarctica the following winter, and over the summer of 1993 joined Thompson on Huascarán, the highest mountain in Peru's Cordillera Blanca. Camping for more than fifty consecutive days in the 19,850-foot col between the mountain's twin summits, the drilling team recovered two long cores to bedrock that altered the course of thinking in climatology. These were the first tropical ice cores to reach back to the last ice age, 20,000 years ago.

A couple of months after descending from Huascarán, he followed what was becoming his own annual pattern by returning to Pole to take a second stab at AMANDA.

The other key personnel on the Ice that year were Bob Morse, Steve Barwick, Per Olof Hulth, and Serap Tilav, a Turkish woman who had earned her doctorate under Tom Gaisser at Bartol and was now working as a post-doc under Francis. Adam Bouchta, a first-year grad student in Stockholm, and Pat Mock, a post-doc of Steve's at Irvine, also joined in. And the Swedes provided two drillers.

They got a little more organized this year and put Pat in charge of string deployment. He, Bruce, and Bob caught one of the first Hercs south, reached Pole at the end of October, and began getting the ball rolling all around. Bob still has the hand-drawn map he made from his survey of where to drill the holes. Since compasses don't work at the pole, he made a homemade sundial by planting a wooden pole in the snow and estimating directions based on the direction of the shadow it cast under the ever-present sun and the time of day on his wristwatch.

Through a beneficial alignment of the political stars at PICO, Bruce was given clear leadership of the drilling team that year, so the drilling went well.

But that doesn't mean it was easy. The weather was miserable, unusually cold, with blizzard after blizzard. Drilling and deployment again took place in the wide open, with the one saving grace that the computers that ran the drill had to be kept warm, so the humans could retreat to a small plywood shack equipped with a space heater when it really got rough. Again there were only two drilling crews, tag-teaming twelve-hour shifts, and to save time and assure that reinforcements would always be handy, the drillers slept in a large unheated tent out by the drilling site.

Bruce recalled that it was actually colder inside the tent than out and that they kept the door open in order to let the "liquid air" out. Their meals were delivered from the galley in the dome, half a mile away. "It was pretty barbaric conditions," he said. "Of course, *we* were pretty barbaric, too. Still are."

Their first task was to assemble the drill by hand in wind chills Bruce described as "below a hundred below." "There were six or eight of us huddled around any kind of shelter we could find by the garage, trying to assemble this stuff, freezing our tushes off." (I believe this was the only time I ever heard Bruce complain about the weather.) And the hot water boilers for the drill burned diesel fuel, which doesn't want to ignite even in good weather, so it took them three or four hours simply to light them, which they had to do whenever they started a new hole.

The basic message that Bruce carried home that year was that humans are more resilient than machines. Drifting snow covered the cables and hoses snaking around on the snow, and even the large reels that held them. The machines broke down constantly. Sometimes the drillers had to lie on their backs on the snow to fix them, often without gloves. The ice in the hole lost about 8 percent of its volume as it turned to water (ice is less dense than water, the most obvious consequence being that ice cubes float), but they had to keep the holes topped up to the level of the firn layer in order to keep them from collapsing. So they had to shovel about 50,000 pounds of extra snow, by hand, into their hot water reservoirs for every hole that they drilled. Everyone helped shovel, including Per Olof.

The benefits of Bruce's new drill, and especially the larger hose, demonstrated themselves immediately. They had no problem reaching a thousand meters, and they also drilled faster and used less fuel than they had two years earlier: 4,500 gallons per hole, as opposed to 12,000 for the one successful hole on the previous attempt, which probably hadn't reached a thousand meters. And progress was steady. They began drilling the first hole on December 20 and reached a thousand meters four days later, on Christmas Eve. Then the physicists started in on deployment—a process Bruce monitored closely, since it was basically the report card on how well he had drilled his holes. He didn't get much sleep.

You might be surprised to learn that it takes about forty hours for the water in a hole surrounded by minus fifty degree ice to freeze, the main reason being that water is an insulator. Bruce often said that no matter how many times he did the calculation he couldn't convince himself that it wouldn't freeze instantly. This meant that the physicists had about two days to drop their strings of pearls into the ice before they would freeze in place forever. They had only one shift, so they worked non-stop until the job was done.

For both drilling and deployment, a tall metal arch is set in place over the hole by means of a crane. It supports a large pulley or sheave, fifteen or twenty feet off the ground (see photograph 10). The huge reel that holds either the hose for the drill or the cable that provides mechanical support and electrical connection to the optical modules (the string for the pearls) is placed on an axle, like the reel to a fishing rod, sitting on the Ice some distance away from the sheave. The hose or cable runs across the Ice from the reel and up to the sheave and then drops straight down into the hole under the pull of gravity. A powerful winch is used to lower and raise the cable or hose, by rotating the reel.

The cable has so-called breakouts at the specified interval—ten meters that year—where the pearls are attached one-by-one as the cable is lowered into the hole. The first thing that goes on the cable is a weight at the bottom end, six or eight hundred pounds, to keep it straight and pull it downward. The cable is lowered to the first breakout and stopped, the first optical module is attached mechanically and electrically, the cable is lowered ten meters to the next breakout, and so on. Owing both to the cold and the two-day time window, the entire process needs to be thought out in advance. A few of the steps are intricate and require bare hands. Bruce remarked that "the physicists suffered equally," since they had no shelter either.

Having supported quite a few science experiments in his day, Bruce had observed that at the beginning of a project the engineering always seems more difficult than the science, but that sooner or later a crossover occurs: the engineering reaches a point where the scientists can begin facing their own challenges, and these, he was humble enough to admit, are "orders of magnitude more difficult" than the engineering. Sometime in January 1993, that crossover occurred. AMANDA began producing its first live data—and confounding the scientists, not for the last time.

Along with the main electrical cable, they had run individual optical fibers to each of the detector modules on every string, so that they would be able to pipe light down to the modules from a laser on the surface. The fibers terminated in nylon balls that broadcast this light more or less evenly in all directions. By sending pulses of laser light through the different fibers they could send flashes between different modules and strings and thereby measure the optical characteristics of the ice. Not long after they'd deployed the first few strings and got them running, they realized they had a problem on their hands.

The strings on the perimeter of the array were a hundred feet from the string in the center and about sixty-eight feet from each other. (Bob Morse had scandalized the Swedes by doing his survey in feet and inches, because that's what he had on his tape measure.) That meant that the flashes should have taken about a tenth of a microsecond, one ten-millionth of a second, to travel from string to string. Instead, they were taking a full microsecond and more—more than ten times what the scientists expected. The flashes were changing shape as well. An initial short square pulse would arrive at the next string as a longer rounded pulse: it would spread out in time, an effect known as dispersion.

When something goes wrong in a physics experiment it can be hard to tell whether your instrument is acting up or you're measuring a real effect. Steve Barwick, the hardware guru, checked what he could; they decided the instrument probably was working correctly . . . and slowly it began to dawn on them that something was wrong with the ice! The light was clearly bouncing around down there—the technical term is *scattering*. Bob Morse remembers thinking things through with Per Olof and "arguing with Steve, which is always an irrational event." Within "thirty minutes it hit with a painful clarity that I remember today. It was bubbles . . . lots of bubbles. . . . We had not gone deep enough—SHIT!!!"

It was as if the light particles, the photons, were bouncing around randomly like pinballs off millions of tiny mirrors, describing what are called random walks through the ice. This meant that any kind of light, whether it was a pulse from the end of an optical fiber or a Cherenkov cone from a muon, would diffuse in all directions, rather than travel in straight lines. Demonstrating his knack for vivid, and in this case aromatic metaphors, Bob writes, "It was not like the sound of a fart crossing the room, which is what we hoped for and expected, but it was like the diffusing smell of the fart arriving much later."

He, Per Olof, and Steve did some back-of-the-envelope calculations that showed that the light was hitting a bubble every twenty or thirty centimeters. It was as if their instrument was looking at the sky through a pane of frosted glass (another of Bob's metaphors). There was plenty of light, but no information. By the time any photon reached an optical module it had lost memory of its original direction. There was no way they would be able to build a telescope—in this ice, anyway.

Although the scattering was definitely a show-stopper, the three physicists' quick analysis also revealed a "bright" side: as the average photon followed its random path through the ice, it was traveling almost ten times farther than the actual distance between strings, and the simple fact that it made it from one string to the next indicated that it was not being *absorbed* over that long a distance. This told them that intrinsically the ice was an excellent medium for neutrino detection, if only they could get away from the bubbles. And that meant drilling deeper.

After deploying four strings, they stopped near the end of January, figuring it wasn't worth the effort to deploy any more in view of the scattering.

And in spite of the bad weather, it was a long season. Bob's logbook tells him that he arrived at Pole on October 29 and left on February 21. His notes also include "a lot of wonderful and sometimes really snide comments about a host of people . . . because we had to drink that shitty, twice-frozen Bud Light beer at the Pole."

10. A Supernova of Science

THIS WAS THE FIRST TIME FRANCIS THOUGHT THE PROJECT WAS GOING TO die. (There would be three more.) They had done a fine job with the drilling, and they had managed to deploy working hardware, but they had "goofed up" by placing their instrument in shallow ice. They were also in debt, and a thought that had occurred to him and Bob the previous fall came back to haunt them. During the tense financial discussions with Buford and Steve, they had suggested that it would be "better to borrow than to fail, for any type of failure will hamper future funding." Now some could argue that they *had* failed—and a few of their rivals did just that.

This was probably the most competitive moment that the small world of neutrino astronomy will ever see. There were four neutrino telescopes in development at the time, DUMAND, Baikal, AMANDA, and NESTOR, the last led by Leo Resvanis, the Wisconsin-style physicist from Greece. NESTOR's proposed site was in the Ionian Sea, off the Peloponnesian town of Pylos, and Leo had gone a good distance out of his way to name his instrument after the mythical king of the town, who appears in the Iliad: NEutrinos from Supernovae and TeV sources, Ocean Range. Nothing much of a practical nature had transpired on NESTOR by that time, and not much more has transpired since, although Leo has come up with some creative ideas.

The most prominent of the projects, DUMAND, had just begun to well and truly flail. In mid-December, six days before Bruce and his crew had begun drilling the first hole for AMANDA-A, the DUMAND collaboration had failed miserably in an attempt to deploy the first string for DUMAND-II. After numerous mishaps, they had lost contact with the string about five hours into their complex deployment operation, having obtained only two minutes of seriously compromised data.

The way they deployed a string was to attach anchors to the top and bottom ends, lower it to the ocean floor, and release the top end from its "sacrificial anchor" by remote control. A buoy at that end would float up and hold the string vertical. Well, the anchor at the top of the first string failed to release, so what data they did get came from an arch-shaped, inclined string—which proceeded to go silent two minutes after they began listening to it. They managed to retrieve it late in January, whereupon an inspection revealed a small leak in one of the nearly sixty connectors on the "string controller." (That many connectors was asking for trouble.) In February, another piece of hardware failed: the "junction box" on the ocean floor, which transferred the electrical signals from the light detectors to the optical cable that carried the signals to shore.

Typifying their high-tech approach, they had placed video cameras three miles down on the ocean floor in order to watch the deployment from their boat. Sometime later, an AMANDA post-doc somehow got ahold of some especially damaging footage from a different deployment, showing a robot picking something up, dropping it, and stirring up all kinds of dust and muck from the ocean floor. "It just looked horrible," he said. "It kind of explained how the . . . it didn't look good."

The general picture was one of daunting objective challenges, excessively complicated hardware, and inadequate operating procedures.

Baikal by contrast had continued to make steady progress in the face of daunting political challenges.

When we last visited the lake, 1988, the Soviet Academy of Sciences had elevated the status of the project and increased its level of funding, and Christian Spiering's group from East Germany had joined the collaboration. The strengthened collaboration had then stepped back and designed a small array that they called NT-200 (Neutrino Telescope with "about" 200 optical modules; it ended up with something like 192).

Then, in November 1989, the Berlin Wall came down. Funding shrank rather than grew, and the Soviet scientific establishment, which had once been one of the greatest in history, began to unravel.

Christian writes,

> The construction of NT-200 coincided with the decay of the USSR and an economically desperate period. Members of the collaboration and even some industrial suppliers had to be supported by grants from Germany; nevertheless many highly qualified experimentalists left the collaboration and tried to survive in the private sector. Over a period of three years, a large part of the food for the winter campaigns at Lake Baikal had to be bought in Germany and transported to Siberia. Still, a nucleus of dedicated Russian physicists heroically continued to work for the project. Under these circumstances, the construction of NT-200 extended over more than five years.

He is understating his group's influence: they saved the day. The crumbling of the wall signaled the reunification of Germany. Christian's small institute in Zeuthen became part of the former West Germany's high-energy physics establishment, Deutsches Elektronen Synchrotron, or DESY ("Daisy"), and they found themselves flush with cash. They also gained access to western markets for electrical components, glass pressure spheres, and the like, which they could now transfer to Baikal. Christian himself smuggled tens of thousands of deutschemarks across the border in his pockets on his visits to the lake, and the food that his group sent to their Siberian colleagues was packed in crates labeled as scientific equipment.

The Russian government is notorious for paying its workers late. This wasn't so bad in Moscow, where paychecks generally arrived within a month of their due date, but at the University of Irkutsk, the closest collaborating institution to the lake, it sometimes took six months, by which time deflation of the ruble had rendered the checks nearly worthless. And Russian scientists weren't paid handsomely to begin with. They earned less than the engineers who worked at their sides and only 20 or 30 percent more than the people who cleaned their labs.

Those who remained in Irkutsk supplemented their income by tutoring the children of wealthy neighbors, installing security systems in the same people's homes, or working in the construction industry. Even Nikolaij Budnev, the leader of the group (he of the ingenious device for rescuing the string from the bottom of the lake), came nowhere near surviving on his university salary, which amounted to about twenty U.S. dollars a month. Zeuthen scientist Ralf Wischnewski gave him his old car, which Budnev drove as a taxi for three hours every morning and evening, before and after his academic work. Zeuthen also paid him a stipend.

Despite these hardships, in March and April of 1993, Baikal established a beachhead in neutrino astronomy by deploying the world's first three-string Cherenkov detector. (You need at least three strings to reconstruct a three-dimensional muon track, and one of them needs to be out of the plane defined by the other two. In other words, they need to be arranged in a triangle.) This was about nine months before DUMAND's disastrous deployment attempt and ten months before Bob Morse, Per Olof Hulth, and Steve Barwick realized that they had placed four strings in bubbly ice. John Learned and Francis Halzen demonstrated the collaborative nature of all good science, and especially of this field at that point in time, by sending e-mails to Christian, congratulating Baikal for "winning the three-string race."

Not long after Christian received Francis's e-mail, his group commenced a more direct collaboration with AMANDA. They employed software they had developed for Baikal to run computer simulations of its icy sibling. They refrained from officially joining AMANDA just yet, however.

In late February 1994, with Baikal's success and DUMAND's failure as a backdrop, and hot on the heels of AMANDA's second seemingly disastrous field season, the Sixth International Workshop on Neutrino Telescopes took place in Venice.

Perhaps you've noticed that four years into the project Francis Halzen still had not visited Pole. Well, more than twenty years later, he still hasn't. He's always said that it would be a farce, since there would be nothing of a practical nature for him to do (although he does add, jokingly, that he'll go when there's a good French restaurant there). Still, one would think that once or twice in almost thirty years the leader of a project might find the time for a symbolic visit at least, or had some curiosity about seeing his creation at first hand. I suspect that there's a deeper reason for Francis's reticence, and that is that he wants to send a message that the real adventure is in the science, not the derring-do of traveling and working in Antarctica.

His job—and it's been a harrowing one—has been to keep an eye on the big picture and the highest levels of the physics.

Francis was in Venice, of course. He was scheduled to give his usual theoretical talk about high-energy neutrino astronomy: sources, how many neutrinos they should produce, and so on, and he planned to include a few remarks about AMANDA, by the by. Bob Morse had been scheduled to talk about recent progress, but he was still traveling north from his long season at Pole, so Adam Bouchta, the young Swedish graduate student, took his place. It was Adam's first big conference.

This was a delicate moment, obviously. It is important to proceed deliberately in science, and the data from Pole was very fresh. Even if it had been clean and clear and just what they had expected, they might not have presented it at this meeting (well, probably they would have). It would turn out that the scientists on the Ice were correct about the bubbles, but even they admitted that their calculations were crude and that other explanations, including hardware malfunctions, had not been be ruled out.

They had not shared the bubble hypothesis with Francis. He was aware of the confusing data—it had been sent north by e-mail—but he had not spoken to Bob, Per Olof, or Steve, and neither had he found time to focus on the data himself. He claims that he did not hear the word "bubbles" before flying to Venice.

The two AMANDA talks were scheduled for the last day of the meeting. On the previous day, Francis happened to chat with Grigorii Domogatsky, the leader of the Baikal project, and Domogatsky told him about a paper about the Vostok ice core he had recently run across in an obscure Russian glaciology journal. It reported that between eight hundred and a thousand meters down—the depths of AMANDA-A—the bubbles at Vostok were about fifty microns in diameter, a twentieth of a millimeter. It seems that the American glaciologist Tony Gow had been right about the *density* of the bubbles, how many there were per unit volume, but wrong about their size; he'd figured they were only about one micron in diameter. A fifty micron bubble would scatter about fifty times more light.

This got Francis to thinking. In his hotel room that night, overlooking the Grand Canal, he worked out the random walk problem by hand on a piece of hotel stationary. This was before the days of the internet and Google. He "had no books, . . . no library, no nothing," (Neither did his colleagues on the Ice.) At about four in the morning, he came up with a formula, plugged in the diameter he'd gotten from Domogatsky and the density from Tony Gow, and "everything fell in place." He constructed a probability distribution for the arrival times of the laser pulses between strings that matched the experimental data well enough.

Now Francis is an excitable boy. The thrill of his insight left him so exhilarated that he didn't sleep at all that night—or take the next logical step and realize that it was very bad news. "I'm a theorist," he says. "I'd solved a problem!"

When he caught his first glimpse of Adam Bouchta the next morning, he ran up and told him he "understood everything. We have big bubbles in our detector!"

Adam hadn't heard about the bubbles either. He'd taken a vacation after returning from Pole, and as a grad student focused mainly on hardware, had not been privy to the conversations between the senior scientists as the data came in. Despite his wetness behind the ears, however, Adam understood the implications right away—and swiftly turned white.

"What do I do now?" he asked.

This brought Francis to his senses. He thought for a moment and advised, "Nothing. You just give your talk, pretending you didn't talk to me."

He was reasonably confident in his all-night calculation, but again, it was far too new to be shouted from the rooftops. He drilled Adam on sticking to his script and what to say if uncomfortable questions arose.

Adam remembers his talk as being "pathetically short" and "pure hardware." It had nothing to do with data. He was there to review the recent deployment and announce that it had gone very well. They had shown that they could drill to a thousand meters and that the hardware would survive refreezing. (Even the strings installed two years earlier were still sending up signals.) These simple achievements demonstrated after only one year of real effort that it would probably be easier—and definitely cheaper—to build a neutrino telescope in Antarctic ice than in the tropical ocean or even a relatively shallow Siberian lake.

The young man had already been dealing with some aggressive questioning in the halls by John Learned, who had heard through the rumor mill that something was up with AMANDA and was understandably anxious, since he would be giving a bleak status report on DUMAND later in the same session. During the question period at the end of Adam's talk, John asked a few more nasty questions, but Adam kept his poise.

"We have some fresh data, and we're looking at it," he said. "Here we're presenting the hardware and the status, and everything is working very well." John congratulated him afterward for doing a fine job.

They were off and running, but they were running scared. And in response to these manifold pressures, not unlike a star collapsing under force of gravity, they produced a supernova of fine science. They now had a small amount of data. And data, as long as it's good data, is always a good thing, especially in capable hands.

The Swedes took the lead on the "flasher data." They developed both a computer model and a physical model, based on calculations from first principles, for the propagation of laser flashes in the ice. (In physics parlance, this type of computer model is known as a "Monte Carlo," because it relies on a random number generator to accumulate statistics about how an instrument or a physical system will behave. Such a method was quite appropriate in this case, since the photons were doing random walks off the bubbles.) The models not only agreed with each other, they fit the data precisely. This put the bubble hypothesis beyond dispute and gave solid numbers for the optical parameters that the scientists on the Ice and Francis in Venice had roughed out by hand. The scattering length, the average distance a photon was traveling before bouncing off a bubble, turned out to be about ten centimeters at the top of the array, eight hundred meters down, and twenty-five centimeters at the bottom of the array, a thousand meters down: there were fewer bubbles in the deeper ice. The trend indicated that the bubbles might disappear altogether at about 1,150 meters, and their diameter also seemed to be shrinking with depth.

The models also gave an estimate for the absorption length of the laser light, and this too confirmed earlier hunches. It turned out to be about sixty meters—it was growing as their measurements got better. The ice at the South Pole appeared to be at least as pure as the ultra-pure water that the Kamiokande and IMB collaborations had used in their detectors.

This fine work demonstrates just how powerful the methods of physics can be, and it was also deemed interesting enough to find publication in *Science*. All in all, the paper concluded, "the results of this study suggest that the ice cap is indeed an ideal medium for a neutrino telescope."

✦ ✦ ✦

Meanwhile, Buford Price approached the bubble question from a more fundamental perspective, based on his background in the physics of solids. In the process, he not only helped the collaboration decide how deep they needed to place their next array, he also made an impact on the fields of glaciology and climatology.

At that time, no one in the world of glaciology understood why there was such a long transition zone between the depth at which the bubbles in a glacier began to disappear, in other words, when the air inside them began to migrate into the ice to form clathrate crystals, and the depth at which they disappeared altogether. In some ice cores this zone is as long as 800 meters. The theories of the time held that the bubbles should disappear instantly at a depth that reached the right combination of temperature and pressure.

After returning from Venice, Francis gave Buford his first lead by telling him about the paper Domogatsky had pointed out to him. (It was fortuitous that Buford read Russian.) The paper summarized a painstaking study by a man name Vladimir Lipenkov, who had spent months in a cold room with an ice core recovered at Vostok in 1980. Inspecting the core manually with a microscope and reticule, Lipenkov had catalogued the size and concentration of the air bubbles as a function of depth, down to 1,400 meters.

Looking further in the literature, Buford found laboratory studies of clathrate formation, which showed that the reason the transition doesn't occur instantly is that it takes awhile for the air molecules to *diffuse* into the ice. Thus the ice has to reach a certain *age* in addition to the right temperature and pressure. Putting together empirical data from ice cores from Greenland and Antarctica, he used the venerable diffusion equation, first developed in the mid-1800s, to produce a theory of the rate of clathrate formation that could be applied to all ice cores. It accounted convincingly for the very different transitions in the Byrd and Vostok cores. His paper presenting the theory, which again appeared in *Science*, is now recognized as one of the classics in the journal's archive.

The main factors controlling the disappearance of the bubbles are the age and temperature of the ice, as a function of depth. Bruce Koci had developed a model for the temperature profile at Pole, but since no deep core had ever been obtained there, the age profile was unknown. So Buford went ahead and developed one of his own.

In the course of all this work, he realized that every core that had ever been obtained in the polar regions showed evidence of a series of dust layers, which had been laid down all over the world during cold intervals and ice ages in the distant past, when the atmosphere had been drier and dustier than it is today. These layers would both scatter and absorb light and thus hinder muon tracking significantly.

In both his papers, Buford made specific suggestions about how deep the collaboration needed to drill the next time around. His clathrate theory predicted that the bubbles should cease to be an issue at about 1,400 meters, and his theoretical age profile predicted that the two most recent, that is, highest, dust layers would be found in the next thousand meters or so. These, he suggested, were depths to avoid.

This impressive tour de force had obvious practical value, and it was accomplished in time to be useful as well. He submitted his papers several months in advance of the next possible deployment season, 1994–95.

The third contribution came from Serap Tilav, who was the only pure cosmic ray physicist in the group at the time. You could say that Serap was "greedy on" the data. She was looking for muons from the minute she landed on the Ice in the same year that the instrument landed in the bubbles.

She had "trained her eye" beforehand so that she would know what to look for when the data began coming in. She grilled Steve Barwick on the electronics in order to understand exactly which so-called observables the instrument would produce. She developed a Monte Carlo for the instrument and sent down-going muons through it, using a Monte Carlo for air showers that she had brought to Madison from Tom Gaisser's lab at Bartol. At first, since she had nothing to go on, she "dipped" her simulated instrument in perfect ice, with no absorption and no scattering—and, of course, no bubbles.

Remember that in 1987, Gaisser, Alan Watson, and friends had installed the South Pole Air Shower Experiment, or SPASE, at Pole. This instrument was designed to detect down-going showers of secondary particles produced by cosmic rays hitting the atmosphere above the instrument. (As we've seen with the Oh-My-God particle, a single cosmic ray can produce billions of offspring, and many of them turn out to be muons.) SPASE was located on the surface of the Ice, not far from AMANDA, and it could also tell the direction of the showers, so it could be used to "tag" the ones that were headed in AMANDA's direction.

When the data began rolling in, Serap realized just as quickly as her friends looking at the flasher data that the ice was not perfect. She would see an overall rise in the amount of light hitting AMANDA when an air shower tagged by SPASE passed through, but the pattern of the light as it spread across the array in no way resembled the perfect air showers she had trained her eye to see. As with the laser pulses, the signals from the air showers lasted much too long a time. The Cherenkov light from the muon showers was bouncing around in the bubbles just as the laser light was.

"So nobody slept," she says. "We tried to think."

She couldn't solve her problem as the others had with pencil and paper. It required a lot of "Monte Carlo-ing," as she says, to begin to get a handle on it. When she returned to the real world, she kept track of the Swedes' progress on the optics, and she incorporated their work on bubbles and ice properties into her Monte Carlo, but it still didn't reproduce the patterns in the data.

As it worked out, the collaboration did not drill during the 1994–95 season. At that point, in fact, they weren't sure they'd be given the chance to drill ever again. They did some maintenance on the instrument, and Barwick installed a new flasher system that could send different colors of light through the ice. This proved key for Serap, since it showed that the shorter the wavelength of the light, that is, the more blue or violet it was, the more transparent the ice. At the shortest wavelength they tried, which was violet but still visible, the absorption length was an astounding 230 meters. This was fortuitous, because the Cherenkov light given off by muons is weighted heavily toward the ultraviolet, just beyond the range that the human eye can see.

Serap incorporated every new piece of knowledge into her Monte Carlo. She adjusted and tweaked every parameter she could think of, trying to reproduce the patterns in the data. At one point she even tried increasing the speed of light, which is a no-no, thanks to Einstein's theory of relativity. This produced some good matches with the data, but Francis reined her in. Finally, she arrived at a more sensible conclusion: she could match the patterns in the experimental data by adjusting absorption lengths in the ultraviolet range of the spectrum—beyond the range they had probed with lasers—to more than three hundred meters! This was more reasonable than violating relativity, but it was still outrageous enough to evoke strong resistance within the collaboration. It took her colleagues almost a year to accept the idea. Antarctic ice turns out to be clearer than diamond. It is the clearest natural substance known.

18. *The geographic South Pole. The stakes receding in a line into the distance are successive markers placed by the United States Geographical Survey in a ceremony that is observed every New Year's Day. Their separations mark the distance that the East Antarctic Ice Sheet moved during successive years.* (Howard Matis, IceCube/NSF)

19. *Members of the Swedish team at the geographic South Pole on Christmas Day 2008, with the gleaming new South Pole station behind them.*

Front row: Sven Lidström, Jonas Kalin, Sarah Amandusson, Per Olof Hulth (holding the USGS marker that was planted in the Ice exactly a week later), Karl-Fredrik Karlsson, Jonas Enander. Back: Jimmy Vinblad, Anders Nilsson, Fredrik Sörqvist, John Bengtsson.
(Mark Krasberg, IceCube/NSF)

20. The IceCube drill at work. Hoses carry hot water across the Ice from the heating plant (bottom center) to the Tower Operations Structure, or TOS (upper left), where the current hole is being drilled. Two IceTop tanks await burial in the trench at the upper right, two more at left center, and two more in the center. The building at the right is the Martin A. Pomerantz Observatory, "MAPO." (Forest Banks, IceCube/NSF)

21. A closer view of the five megawatt heating plant for the IceCube drill, with the TOS to its left. The large reel in the center holds the hose nearly two miles long for drilling the hole. Left of the TOS and reel is a trench containing two IceTop tanks. (Ethan Dicks, IceCube/NSF)

22. The IceCube heating plant, with the two Tower Operations Structures on the horizon. The TOS on the right is being used to drill the current hole. The one on the left stands waiting at the position of the next hole, so they can "leapfrog" over to it in order to save time. Electrical wires dangling from the "IceCube hitching post" in the foreground supply power to block heaters in the snowmobiles. (Jim Haugen, IceCube/NSF)

23. Drilling inside the TOS.
(John Jacobsen, IceCube/NSF)

24. Someone shines a light down the hole at a digital optical module (DOM) about seventeen meters below. The DOM at the left is poised to go down next.
(Ben Stock, IceCube/NSF)

25. Driller Melany Zimmerman controlling the winches during a deployment. (IceCube/NSF)

26. Distinguished cosmic ray theorist and experimentalist Tom Gaisser in the dish pit at Pole. The guy in the chef's hat is Jeff Cherwinka, principal engineer of the IceCube drill. (Darryn Schneider, IceCube/NSF)

27. A "changing of the guard" at Pole. Laura Gladstone, left, has just arrived; Karen Andeen is about to board the awaiting Herc. "The two of us girls verified more than half of the DOMs that were finally deployed in the ice," writes Karen. "DOM-testers were the last line of defense." (Jim Haugen, IceCube/NSF)

28. The surface cable for the very last string is laid out on the Ice in front of the IceCube Laboratory (ICL). (Jim Haugen, IceCube/NSF)

29. A cable drag. The surface cables from a group of strings are fed into the culvert on the right, which leads to the base of one of the towers on either side of the ICL. The cables are pulled up through the tower and into the building, where each is attached to a dedicated computer. (Jim Haugen, IceCube/NSF)

30. *An IceTop tank with the station building on the distant horizon.* (Bakhtiyr Ruzybayev, IceCube/NSF)

31. *Per Olof Hulth, third from the right, makes some remarks just before the last IceCube DOM is deployed. Albrecht Karle, center, casts an inspired look at Per Olof. Jim Haugen smiles at Albrecht's right. Vladimir Papitashvili, NSF Program Director, far left, snaps a photo. Gary Hill, far right, and Kurt Woschnagg, between Gary and Per Olof, look on.* (Jim Haugen, Ice-Cube/NSF)

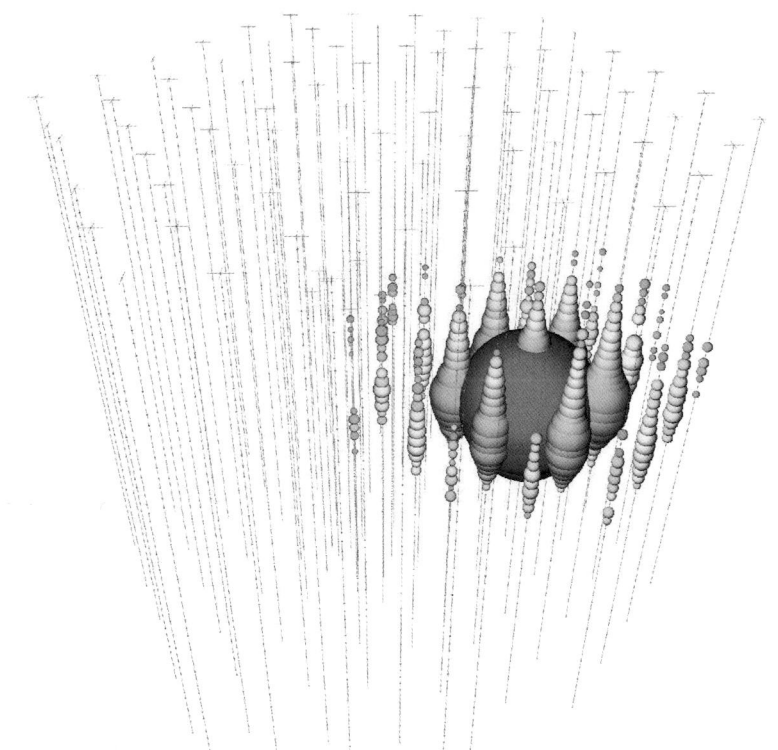

32. "Ernie," a 1.2 PeV neutrino, explodes inside IceCube. Each small dot represents a light detector. The size of the spheres indicate how many Cherenkov photons hit the corresponding detector, and their shade, darkest first, indicates when. Since the entire detector is about a kilometer on a side, Ernie would have more than filled a professional football stadium. (IceCube Collaboration)

33. Olga Botner and Francis Halzen at the banquet during the epochal collaboration meeting in the spring of 2013. (Tom Gaisser, Albrecht Karle, IceCube/NSF)

34. Francis Halzen. (Bernardo Pérez, EL PAIS [www.elpais.com])

35. An LC-130 sends a cloud of ice crystals into the air as the last "summer people" leave Pole at station closing, February 15, 2017. This left forty-eight people to enjoy the next, dark, eight and a half months at the station, alone. (Martin Wolf, IceCube/NSF)

Serap's discovery was serendipitous in two respects. Not only did it demonstrate that the ice is remarkably clear in just the part of the spectrum where most Cherenkov light is produced, it turns out that photomultipliers reach a peak in sensitivity in the ultraviolet, too. So both the hardware and the ice happen to work best at just the right wavelengths for neutrino astronomy.

So, Buford made contributions to glaciology and climatology; Serap and the Swedes uncovered new knowledge about the fundamental properties of one of the most heavily studied substances there is, ice; and the collaboration as a whole had made steps toward building a better instrument. But this wasn't astrophysics exactly; it was a bit down in the weeds. Francis provided the pièce de résistance by demonstrating that their small bubble-filled instrument was already capable of cutting-edge particle astrophysics.

Markov-type, plum pudding instruments like AMANDA were not supposed to be sensitive to the low-energy neutrinos given off by supernovae. During the good old days of collaboration with the Russians, the early pioneers had come up with UNDINE, a shell-type, Greisen design, for that side of the problem. The two Cherenkov instruments that had detected Supernova 1987a, IMB and Kamiokande, were both Greisen designs.

But back in 1993, before AMANDA got "stuck in the bubbles," Francis, his student John Jacobsen, and his former post-doc Enrique Zas had shown that Markov-type instruments *could* detect supernovae.

Jacobsen is unique even among the varied individuals who have collaborated on AMANDA and IceCube over the years. At that point, he was enrolled in both the art and physics departments at Madison. He'd been what he calls an "art nerd" in high school and then majored in physics at Madison, where he was introduced to research by none other than Ugo Camerini.

One day Camerini walked in to John's office at the campus computer center, where John had a student job, and asked him to drop by his office sometime; he had something he wanted to talk to him about. These were the early days of personal computers, and IBM was handing out its landmark PC to professors free of charge in order to promote its use in educational environments. These were new-fangled gadgets to Camerini, who was decidedly old school.

"So I showed up in his office," John recalls. "And he says, 'I have this data, and I want to shove it up the ass of this computer.' Hah, hah! He's like—you could smoke in your office back then—like chain smoking, and [this] grimy keyboard and gravelly voice. . . . And so I wound up working for him."

John's first project was to help analyze data from an early gamma ray satellite named COS-B. He then got involved in the Haleakalā telescope in Hawaii and made the acquaintance of Bob Morse.

After graduation, he worked at CERN for a year, programming again, and then returned to Madison to go to art school. Needing a way to pay the rent, he approached Morse, and Bob offered him a job doing Monte Carlos for AMANDA. This was August 1991, just before the first deployment season when they stuck Bucky-1.

After about two years, Francis got on John's case and said, "You know, you're doing a Ph.D. project, why don't you just get a Ph.D.? This is stupid. All you have to do is take a couple classes."

John considers this a good example of what he calls Francis's "criminal optimism": it ended up taking him about four years to earn his doctorate.

But they were rewarding years. Francis showed him that there are different ways of being a physicist. "The standard idea is that . . . you're going to live in this very formally-specified field where there are definitely wrong ways to do things and it's all about dotting your i's and crossing your t's," John observes. "Any sloppy thinking and basically you're out; you're going to fail, and nobody's going take you seriously.

"But there are different approaches, right? Some people take a particular corner of the room and they'll start just digging; . . . they're going to escape from jail by digging one spoonful at a time. And there are other people who are just going to think about things and write a few expressions down on a napkin or have a drink with a friend and, like, be creative! You might not be the person to fully realize whether this idea is . . . true or not, but someone's got to come up with those ideas. Someone has to be sort of the artist of physics and do that. And I realized from Francis that you could do that, that he does that, that that's actually why he does it, that this very creative kind of thinking was a deep source of pleasure and he could actually live that way."

The other collaborator on the supernova work, Enrique Zas, was brilliant mathematically. After a post-doc with Francis in Madison, he had returned home to Spain to become a professor. The way the three worked together, according to John, was that Francis "would go off—he would have an idea. He and Enrique would sit down, and they would say, 'Oh, what if we did this?' And then they would toss it over to me, and we'd do some Monte Carlo together, and then we'd write a paper. And then, you know, hundreds of people would cite it for the next decade, or three decades, or whatever." Such was the case with their supernova work.

Although supernovae produce all six varieties of neutrino and antineutrino in roughly equal measure, a Cherenkov detector sees only the electron antineutrinos, because at supernova energies the other types don't interact appreciably with water or ice. The typical energy of an electron antineutrino emitted by a supernova is about twenty million electron-volts. It will interact with a free proton to produce a positron, which produces a Cherenkov track that is only about twenty centimeters, or eight inches, long. This distance is so much shorter than the distance between the modules in AMANDA that the instrument can't possibly reconstruct such a track.

What the three physicists realized, however, was that you don't need to *track* the positrons in order to *detect* a supernova. As the ten-second neutrino flash from a supernova passes through the instrument, a large enough number of positrons should be born within ten or twenty centimeters of a large enough number of phototubes to produce an uptick in the total amount of light sensed by the array. It's hard to tell which direction the flash came from, where in the sky the event occurred, but you should be able to tell that it *did* occur if you monitor the collective behavior of the entire array: the sum of all the signals coming from every single phototube. And since the electronics of the instrument respond so rapidly, on the order of a few billionths of a second, it should be possible to reproduce the shape of the neutrino flash in exquisite detail—in great contrast to IMB and Kamiokande, which detected only a handful of neutrinos altogether.

This could provide a rigorous test of the competing theories about how a supernova explodes, each of which predicts a slightly different shape. Different features in the shape of the neutrino flash should provide evidence of the swift collapse of the iron core, the ultraluminous burst accompanying the explosive bounce-shock, the streaming out of neutrinos from the proto-neutron star that presumably revives the shock, and so on. Two interesting new possibilities also arise: The sudden cutting off of the flash would indicate that a black hole rather than a neutron star had formed at the center of the star, because a black hole would pull even the neutrinos back into itself.

And more speculatively, a brief spike at a specific point during the flash might signal the formation of an exotic quark star.

Greisen-style instruments like IMB and Kamiokande have other advantages, such as the ability to tell the rough direction of the incoming neutrinos, but it turns out that you can make the most *sensitive* basic detector much less expensively with the Markov design, since it reaches equivalent sensitivity with far fewer phototubes.

In the innocent days before they met the bubbles, Francis and his students estimated that a completed AMANDA-A or DUMAND II, comprising only two hundred or so optical modules, could detect a supernova occurring in our own galaxy with some confidence, but not quite enough. Random noise from the detectors would result in a fake signal more than once a century, but you have to do better than that, because galactic supernovae are expected to occur at about the same rate. Therefore, a small Markov instrument could not "watch" for supernovae on its own; it would produce too many false positives. It would need to be backed up by an optical telescope or an instrument like Kamiokande.

When AMANDA dropped its first four strings into the bubbles, Francis began searching desperately for "some way to make this disaster look good." "I was thinking, 'What can you do when you have a scattering length of less than half a meter and an absorption length of two hundred seventy meters?' And the answer was supernova."

Or, as Bob Morse said at the time, "If you've got lemons, make lemonade."

In 1995, Francis and his students revisited their earlier calculations, employing the astonishingly long absorption lengths they had now observed in South Pole ice, and discovered that the tiny, four-string instrument they had already built was by far the most sensitive supernova detector on the planet. Owing to the long absorption lengths at ultraviolet wavelengths, every single one of the seventy-three working optical modules in AMANDA-A (seven had failed for one reason or another) was monitoring roughly the same volume as an entire Kamiokande or IMB instrument. And the sensitivity was actually enhanced by the bubbles, since they tended to keep the light inside the array! Scaling up from the eleven neutrinos that Kamiokande had detected from Supernova 1987a, the AMANDA trio estimated that their small instrument would have detected about *twenty thousand* neutrinos from the same event. AMANDA-A was already capable of acting as a stand-alone supernova watch for the entire Milky

Way galaxy and would continue to work for a decade or more "with mini-mal maintenance or investment in manpower and operation." Thus, even their handicapped instrument represented a breakthrough.

"Suddenly our disaster had a mission," Francis recalls.

Christian Spiering was following these developments closely from Zeuthen, Germany. Baikal was still going strong—the collaboration had reached an-other milestone by reconstructing the first-ever up-going muons in the summer of 1994—but the long-term prospects did not look good. There were serious obstacles to working in Russia, and the shallowness of the lake made for additional hurdles. He was casting about for other opportuni-ties, and he was a strong player. Not only was his group now associated with DESY, one of the preeminent particle physics enterprises in Europe, they also had an impressive track record with Baikal. All three of Baikal's com-petitors, DUMAND, AMANDA, and NESTOR, were courting him. In July 1994, he and his Zeuthen colleagues came to a decision. In the full knowledge that AMANDA was mired in bubbles, they decided to join anyway—and retain their membership in Baikal. Just as they had given a shot in the arm to the Russians, they gave another to the Swedes and Americans, who were preparing a second grant proposal to NSF: a stron-ger collaboration made for a stronger proposal.

The first thing the Zeuthen group did—before they officially joined the collaboration—was to implement the new supernova idea. This required a relatively simple data acquisition system on the surface of the Ice, a so-called hardware trigger, that would count all the photons that the array picked up in rolling time-slices of about ten seconds, roughly the length of a supernova's neutrino flash, and send an alert if the count reached a cer-tain threshold. The Germans built the trigger, paid for it, and installed it in AMANDA-A during the 1995–96 Antarctic season—before Francis and his students submitted their updated paper to *Physical Review*. (Francis likes to joke that if the paper had been rejected he could have told the editors that the idea was already working in the ice.) Between AMANDA and IceCube, the collaboration has been monitoring the cosmos for a nearby supernova now for more than twenty years.

This engenders a certain anxiety, because what if a supernova were to occur while the instrument was down or being repaired? In fact, who knows? They may already have missed one. AMANDA was taken down frequently during the austral summer when it was being added to or updated, and IceCube also had a lot of down time during the years it was being installed. Blood pressures are dropping since the larger instrument was completed and has become increasingly stable, but as a sign of how critical this issue remains, "detector uptime" is displayed in large letters on the first page of the On Ice Report that is sent north every week. It hovers mercifully in the range of 99.8 percent.

Since the neutrinos from a supernova arrive a few hours before the photons do, it would be ideal if a neutrino detector were capable of telling optical telescopes where to look. And although IceCube can't tell the direction of supernova neutrinos very well, the community has come up with a creative solution. At this moment, IceCube and several other neutrino detectors, including Kamiokande's successor, Super-K, are wired in to the SuperNova Early Warning System, or SNEWS, which will broadcast an automated e-mail message the moment one or more of the instruments senses what it thinks is a supernova neutrino flash. Since the instruments are located in different spots on the globe, the general direction of the event can be inferred through triangulation: the flash will pass through the different instruments at slightly different times. SNEWS will then alert a network of earthbound and space-based optical telescopes of the impending fireworks and give them a rough idea where to look.

During a discussion of supernovae at an IceCube collaboration meeting in 2010, Francis pointed out that only at that meeting had he begun to appreciate that if a supernova were to go off at a distance of about one kiloparsec, IceCube would be so overwhelmed with signal that it would probably miss the event. A certain register in the supernova trigger would saturate after about twenty seconds, the instrument would crash, and all record of the neutrino flash would be lost.

A kiloparsec (kpc) is a thousand parsecs, which is about 3,200 light years—one-fiftieth the distance to Supernova 1987a. A so-called stellar nursery, a region of active star formation, lies at about that distance in the constellation Cygnus, and supernovae are expected to occur more frequently in such nurseries. Francis was implying that the instrument ought to be fixed, obviously, in case this possibility happens to bear fruit. That night I had dinner with Per Olof Hulth, who explained that actually there was no need to fix the instrument, for if a star were to explode that nearby, the human race and most of life on Earth would be extinguished. I checked this with Francis, who did a short calculation and assured me in one of his typically pithy e-mails: "No, 1 kpc is safe. To kill the dinosaurs you need 10 pc"—about a hundred times closer than the Cygnus nursery. When he was presented with Francis's calculation, Per Olof did "not disagree." It is comforting that the collaboration has gone ahead and fixed the instrument. . . . And such is the logic that neutrino astronomers occasionally entertain.

11. Doubling Down

Francis is fond of pointing out that the collaboration didn't have a speakers committee in those days. Now, with more than two hundred members and several significant accomplishments to its credit, the IceCube collaboration is asked to present at one conference or another roughly once a week, all over the world. The speakers committee has rules about what can be said and who is authorized to say it, and there is some competition about who will travel to the more important meetings or exotic locales. In the days of the bubbles, however, Francis was the only member of the AMANDA collaboration who was willing to make a fool of himself in public: "I have no pride. I'm a theorist. What did I care?" Although he also admits that he wasn't just giving a scientific talk, he was also campaigning for more money. The funding agencies take the views of the community into account, so it's important to maintain one's visibility and credibility.

With an apparent failure to his credit, however, he was having trouble in the credibility department. He encountered aggressive criticism almost every time he spoke. He observes that "there's an infinite possibility of being cynical, right?" It can be surprising how little an expert in the exact same field sometimes understands about a competing project even when he tries. Details can be subtle, and scientists are just as susceptible to hearing what they want to hear as anyone else is.

John Learned of DUMAND and Leo Resvanis of the Greek project, NESTOR, were especially vocal. "AMANDA will never work!" Resvanis would yell from the back of the room. (Leo continued in this vein all the way through IceCube.)

Physicists being a witty lot, there were humorous moments as well. About a month after Francis had worked out the random walk problem in Venice, David Nygren, a by-then legendary experimentalist at Lawrence Berkeley Laboratory, a Department of Energy facility perched high on a hill above the Berkeley university campus, convened a meeting on neutrino astronomy in Arcadia, California, near Caltech's Jet Propulsion Laboratory. He and a group from JPL were thinking of building a kilometer-scale instrument, which they called Km3, and wanted to take stock of the field. AMANDA's bubbles were common knowledge by then, and overhead projectors were the lecture-giving technology of the day. The Monte Carlo expert from DUMAND gave a talk, and near the end he put a blank transparency on the projector and said, "This is what AMANDA sees." Francis was sitting nearby. He reached over, switched off the projector, and responded, "And this is what DUMAND sees!"

The thing is that Francis actually believed his own propaganda: "It's not like I was trying to convince myself or deceive the world. I believed one hundred percent what I was saying." Dave Nygren remembers having breakfast with him in Berkeley's wonderful old Women's Faculty Club at about this time. "It's amazing," Dave remembers. "He announced that the scattering length was ten centimeters but the attenuation length was two hundred meters or something, and he declared victory! You know, I thought, 'What *is* this guy talking about?' ... But he persevered"—and convinced Dave to join AMANDA a year or so later. Criminal optimism can be a good thing.

It is a measure of the professionalism of John Lynch at the National Science Foundation that twenty years after the disaster in the bubbles, his friends from Madison still believed they had almost lost funding, when in fact it was never in doubt. John says the foundation was "not dissuaded." "We're used to initial failures. Any funding agency is used to initial failures. They better be or they'd never get anything done."

Thanks to its huge programs in the polar regions, NSF has extensive connections in the glaciology community. Several eminent glaciologists assured John that AMANDA-A simply wasn't deep enough. They were surprised that there were bubbles at a thousand meters, but not shocked.

"I don't remember thinking, 'Oh, Christ, it's failed, it's never gonna work,'" he says. "I knew Peter [Wilkness] wasn't gonna cut my budget because there were bubbles in ice. I knew that."

But Bob Morse believes John isn't giving himself enough credit. "He was a lot more out on the edge than a lot of the other people at the NSF. And when he used the word 'we' he was being charitable to the rest of that group . . . because they're pretty goddam conservative. . . . [And] if it had been anyone with less courage and what you'd call sense of daring and risk-taking than Wilkness, it might have turned out differently. So, we had the combination of Lynch and Wilkness, and it worked magic for us."

Near the end of 1995, NSF and the Wisconsin Alumni Research Fund doubled down. The foundation increased its support for AMANDA by the better part of $1 million a year, funding a second four-year proposal to the tune of $6.4 million, and WARF, at the urging of the ever-enthusiastic John Wiley, added another million.

✦ ✦ ✦

Bruce Koci had "had the good sense to stay home" for the previous two years. Somewhere in there, the Polar Ice Coring Office moved back to Nebraska, but he and his wife chose to stay in Alaska. He shifted from full-time employment to consulting with PICO, mainly as an advocate for AMANDA, and from that time forth split his time between the South Pole project and his expeditions with Lonnie Thompson. AMANDA became his central interest, mainly because it offered a challenge and a chance to learn. He and Thompson had more-or-less perfected the art of high-altitude ice core drilling by then, but Bruce was still the best troubleshooter in the game, so his friend continued to request his support in the field. And Bruce, for his part, had a hard time turning down these requests, since they took him to spectacular locations in the Himalaya, Africa, and the high Andes.

✦ ✦ ✦

The plan for AMANDA-B was to double down on the drilling, too: five strings in a pentagon and a sixth in the center, all dropped to 2,000 meters in the hope that Buford's hunches about the ice were true.

After the success of AMANDA-A, Bruce was reasonably confident that he understood the physics of the drilling, and at the same time aware that they were plunging into the unknown again. He and Bob Morse have frequently expressed gratitude to Simon Stephenson, the NSF Program Manager for Antarctic Support and Logistics, for taking "a gigantic leap of faith" in supporting them. The drill cost more than $1 million at that point (the Swedes again paid for it), and NSF's Antarctic Program provided a tremendous amount of logistical support, for which AMANDA paid not a penny, to transport the drill to Pole, install it, and use it.

The hose was the most vexing aspect, as always, and Bruce spent a good deal of time developing specifications for it. In order to send enough heat into the ice to be confident that they would be able to reach the new depth, he expanded the diameter again, to one and a half inches. He also specified the wall thickness so that enough heat would flow laterally into the surrounding water to keep it from refreezing during the drilling. (After the season, when he analyzed the results, he found that he'd nailed the heat transfer coefficient to within 3 percent.) Then he found an industrial partner to manufacture the thing. It came in 800-foot lengths and ended up being the largest synthetic hose that had ever been made.

He also "a little more than doubled" the size of the heating plant, going from less than one to about two megawatts. Counterintuitively perhaps, this increased the fuel efficiency: the new drill would burn more fuel per hour, but it would more than compensate for this by working much more rapidly. The previous drill had taken 72 hours to get to 1,000 meters and burned 4,500 gallons of fuel in the process, while the new drill would do the same job in 24 hours, with 2,500 gallons of fuel. This was steady progress: their first drill, Bucky-1, which is still acting as a radar marker in its eternal grave in the ice, burned 12,000 gallons to reach something less than 1,000 meters.

Meanwhile, the drill head was becoming quite a contraption, and the whole system was getting smarter. The head weighed in at a few hundred pounds. It was equipped with an inclinometer, calipers, and temperature and pressure sensors, and it sent the signals from these devices up to a control shack on the surface of the ice via telemetry.

There is no real way to steer the head in a hot water drill. The idea is to create just the right conditions down where the action is, right at the nozzle, to allow gravity to pull it straight down into the ice. That means spraying just the right amount of liquid heat out of the nozzle and lowering it at just the right corresponding speed to allow nature to take its course. If you lower the drill too quickly, the calipers will tell you that the hole isn't wide enough: you might not be able to lower a string through it, or, worse yet, you might freeze the head in place. Or perhaps the head will become cocked in the hole and wander off course. If you lower it too slowly or stop it altogether, the hole will get too wide.

This was still an exploratory process in those days. The physicists weren't exactly sure what the specs for the holes needed to be, but they figured its diameter ought to be constant, and they wanted it to be ridiculously straight. "They want these things plumb within a foot in eight thousand," Bruce once commented. "You couldn't measure that with any inclinometer on Earth." The inclinometer didn't tell them much at all about the verticality or straightness of the hole, incidentally; it was mainly used as a finger on the pulse of the situation below. (This *was* like building a telescope in a darkroom.) They found that as long as everything was in sync the head could be cocked in the hole by as much as three-quarters of a degree and still fall free in a straight and plumb line.

The telemetry data was fed to a computer in the control shack, which adjusted the speed of the winch and the water pressure in the hose, accordingly. Over the years, Bob Morse had recruited various undergraduates to write the software code that controlled the drill and logged the telemetry data, and it was becoming quite sophisticated.

Francis remembers wistfully that the AMANDA collaboration had no management in those days. The different participants mounted midnight raids and sneak attacks in the manner of Comanche war parties. There *was* a group ethic, to be sure, but this was academia after all: there was only so much a professor could ask a student to do. People tended to do what they were interested in doing. There were only about forty people in the collaboration, and only ten or fifteen did the key work. The way the PIs would decide what to do as an Antarctic season approached was to gather in a small lecture room, walk up to the blackboard one by one and write down how much money they had; they would add up the numbers and decide together what they could accomplish. "There were no spreadsheets, no software, no nothing. That's how this was done."

During those years, the collaboration often held meetings in Irvine or nearby Laguna Beach, and the group was small enough to fit in a regular classroom. The meetings lasted the better part of a week and generally ended with a science session on Saturday morning. Several people remember that an elderly gentleman would sometimes sit at the back of the room during these sessions, listening quietly. It was Fred Reines, who was in his late seventies by then. Now an emeritus professor, he was suffering from dementia, unfortunately. Francis remembers that Reines would approach him at the end of the science session every year, and always say the same thing: that he had started out as a theorist, too.

Christian Spiering, who was involved in Baikal as well as AMANDA, remembers presenting Baikal's detection of the first-ever up-going muons at a meeting in Irvine, probably in 1995. Addressing Reines in the back row, he said, "I am proud to present the first underwater neutrino in the presence of the man who found the first underground neutrino!" "And then I looked to him and looked into these empty eyes," Christian writes. "No reaction at all, other world." Sadly, this was the year in which Reines finally won his Nobel Prize.

There was a certain amount of jockeying for position among the PIs, and so, while Francis was the de facto leader of the collaboration—meaning he was the one who spoke to NSF and had his head on the chopping block— he was by no means the decision maker. In truth, there *was* no decision maker; the group was too anarchistic for that. Collective decisions were unanimous perforce, and important ones were rarely made before the last possible moment.

A case in point was a decision that came up several times every drilling season: where to put the next hole. Bob Morse and Bruce Koci would get to Pole early each season, grab a surveyor, and map out the hole positions based on their reading of the group mind. Usually, they would not get an actual decision, however, until the drilling of the next hole was about to begin. The tower that was used both for drilling and deployment had to be set up beforehand, obviously, and it was moved into position by a large crane. Oftentimes the only thing that would goose the collaboration into making a decision was for the crane to begin making its way out to the dark sector. Bob would send a desperate e-mail north: "Crane's on its way out; it's got the tower up there . . . heh, heh, heh . . . 'Come on guys!' . . . Because there are guys down here that want to set a tower and they want to know where the hell to put it.'"

Bob remembers the 1995–96 season as being a "continual battle between PICO and Bruce Koci."

With its return to Nebraska, PICO had reverted to its old ineffective ways—further exacerbated by the fact that Bruce was now seen as an outsider. The PICO managers wanted one of "their people" in charge, and even though Bruce had been with the organization since its inception and had even designed the drill, he didn't fit their bill.

At Pole a few years later, I asked Bob and Bruce if there was a story to tell about that season.

"You mean like . . . ," said Bob, turning to Bruce, ". . . like why you didn't kill Bill Jones?" (Not his real name.)

"That's water over the dam," Bruce responded placidly. He always downplayed interpersonal conflict. Bob had no such reservations.

"We had Bill Jones, who'd never drilled a hole in his life, coming in . . . as the chief engineer and the chief architect of this operation in the field. He showed up at PSL with an organization chart. . . . And I looked at that organizational chart and I thought, 'Shit. We've got trouble.' . . . It was, 'We're going to take this project back and we're going to run it, even if we don't know what the hell we're doing.'

"Now PICO's come a long way. . . . The relationship is much, much better. But in that first season, it was awful! And, ah—"

"So were the holes," interjected Bruce.

"Hah! Yeah. And so were the holes. The holes were terrible. They were just crooked as hell. . . . Hole one drifted off as much as six or seven meters from where it was supposed to be."

A stellar addition to the field team that year was a talented and hardworking experimentalist named Albrecht Karle, who had recently moved to a post-doctoral position at Zeuthen after earning his Ph.D. at the University of Munich. Albrecht's adviser had been Eckart Lorenz, one of the great pioneers of gamma ray astronomy, and his thesis research had been conducted on an air Cherenkov telescope high on a mountainside in the Canary Islands. This type of instrument employs the atmosphere as the Cherenkov medium, rather than water or ice, and is more reminiscent of the usual telescope in that it sits on the ground looking up.

The season got off to a slow start because of trouble with the drill head. Every time they dropped it into the water-filled hole it would short itself out after fifty or so meters. "It was the first week of December, second week, third week; it was Christmas, and we still hadn't any holes drilled," Albrecht recalls. "[The deployers were] getting really antsy. One was behaving so weird that Pat Mock [who was again heading up deployment] sent him off station. . . . People had nothing to do."

The institutional arrangement with PICO made it difficult to address the problem, as the drillers reported to NSF rather than the scientists, so the situation was ripe for finger-pointing. Steve Barwick, the senior scientist on the Ice, advised Albrecht not to step in to help, since he might be blamed if he failed, but Albrecht demonstrated his leadership skills by teaming up with an electrical engineer from PICO to find a solution. It was already January.

Once the drilling got started, Jones relegated Bruce to the background, and in this difficult position Bruce did what he could to keep the drill running and prevent a true disaster, such as freezing it in. He got very little sleep during the drilling of a hole, which might take a week, and the deployment, which came right on its heels.

Since the holes were deeper this year, a couple of the deployments bumped up against the forty-hour deadline when the water in the hole would refreeze. During one such nail-biter, Bruce was wandering around in his tennis shoes on one of his spot checks when he noticed the main cable for the optical modules skipping on its reel as it was being lowered into the hole. It hadn't been spooled tightly enough, so it jumped periodically to release the slack. He ended up standing by the reel for seven hours, shifting from foot to frozen foot, cushioning the skipping cable by hand.

Albrecht saw right away that Bruce was the only driller who was interested in "the final purpose." "He would always point out, you know, the hole is only good when the string is safe. He was really a different, but also in many ways a very unique and marvelous and amazing person."

Also in contrast to his PICO counterparts, Bruce kept thinking about the drill in the off-season. It turned out that more than the usual number of optical modules failed during refreezing that year. A few months after the season ended, he was puzzling this over in his office in Alaska, when he noticed that most of the failures occurred at two specific depths: one at the very bottom of the strings, and the other about eight hundred feet above it—the precise length of a hose segment. At the higher of the two depths, the drilling had stopped while the drillers changed reels. "The light went on," and he realized that the failures must have been caused by changes in hole diameter: When the drillers had stopped to change reels, more than the usual amount of hot water had been released at that particular depth and the hole had developed a bulge. At the bottom of the array, on the other hand, the holes had tapered off: As the head worked its way down into the ice, the plume of hot water it left in its wake would continue to melt the ice fifty or sixty meters above it. And since they had stopped and pulled the drill up from a point just below the spot where they intended to place the lowest module, the hole was narrower there. He realized that they needed to drill sixty to seventy meters below the bottom module in order to ensure a constant diameter for the entire string.

Bruce and Bob's hand-waving explanation for these failures was that irregularities in the diameter would cause uneven shear pressures to develop as the water in the hole refroze. This would produce a net force on Benthospheres that housed the phototubes, which were essentially large corks, or pistons, bobbing in the hole. The spheres would move, however slightly, and they would snap their connections to the fixed cable. In subsequent years, the drillers avoided changing reels in the middle of the array and drilled seventy or more meters below the bottom module, and the number of failures dropped dramatically.

They deployed the string that was intended to be in the center of the pentagon first, and then began working their way around the perimeter. By the time they had three strings in the ice, they got the laser flashers up and running and began looking into the quality of the ice. The burning question, of course, was whether they had gotten below the bubbles.

Ironically, it turns out to be more difficult to attach hard numbers to the optics of a relatively clear medium than a medium with strong scattering. A few students back in the real world applied crude Monte Carlos to the problem, Bob and Per Olof did some hand calculations on the Ice, and it soon became clear that the scattering dropped dramatically between one and two thousand meters. Buford was right! The bubbles were gone! Joyful e-mails flew north, and Francis called John Lynch to pass along the good news.

But it wasn't all good news, because the flasher data also told them that there was more scattering in the deep ice, indicating more dust than Buford had predicted. This meant light wouldn't travel as far as they had hoped, and there was a possibility that their array might be too sparse. They had deployed the first three strings in the shape of a triangle, between sixty and eighty meters from each other, but they now estimated that the attenuation length of the light in the ice, absorption *plus* scattering, was only twenty or thirty meters. If they persisted with their original design, they ran the risk of leaving "holes" in their detector: the light from a passing muon might not be able to travel the full distance between strings, and this would make it difficult if not impossible to reconstruct its direction.

The delays in the drilling also led to another consideration: if they kept to their original plan and continued to work their way around the perimeter of the pentagon and then ran out of time or the drill broke down, they would end up with a weirdly shaped, open-sided array that would also complicate reconstruction.

On the fly, then, as Bob Morse sent e-mails north pleading for a decision and as the crane lumbered out to set the tower for what would turn out to be the final hole of the season, the collaboration decided to place it in the center of the triangle. This produced a nice, solid detection volume with strings close enough together to prevent blind spots.

This four-string configuration became known as AMANDA-B4, and with it AMANDA edged ahead of Baikal—equipment-wise anyway—for about two months' time. In April, the Baikal collaboration caught up by adding a fourth string to their array. And so, by the spring of 1996, there were two teams in the exclusive four-string club of neutrino astronomy.

At the Venice conference in February, straight on the heels of the Pole season, as it always was, Buford Price announced triumphantly that that bubbles were "absent below 1,500 meters," while Grigorii Domogatsky demonstrated just how far ahead Baikal was altogether by presenting their reconstructions of the first-ever up-going muons, detected more than a year earlier with their three-string array. These results were tentative, since three short strings aren't really up to the task of identifying up-going muons unequivocally, but Domogatsky nevertheless suggested that he and his collaborators may have detected one or two actual neutrinos.

John Learned spoke on "The Many Experimental Problems Connected with the UHE [Ultra-High-Energy] Detectors," and in retrospect his title is fraught with meaning.

The Division of High Energy Physics at the Department of Energy had recently formed the Scientific Assessment Group for Experiments in Non-Accelerator Physics, aka SAGENAP ("sage nap"), to review its commitments in particle astrophysics. And in addition to evoking an image of a group of wise men dozing off, the committee's name indicates just what the high-energy physics community thought of this fledgling field. The lion's share of the money in DOE's high-energy physics budget was and still is spent at the accelerators, and most of its physicists worked at them then and still do; thus everything else was designated condescendingly as "non-accelerator physics." Admittedly, several particle astrophysicists, including Tom Gaisser and Steve Barwick, sat on the committee, and its reviews were helpful, but the SAGENAP exercise still highlights the fact that neutrino astronomers, air shower scientists, and other cosmic ray physicists have lived off the crumbs from the rich man's table ever since the accelerator era began. Some, and especially Learned, rail about this, but Francis believes it's a blessing in disguise. Lower budgets tend to keep you out of the spotlight, and until recently, most cosmic ray physicists wouldn't have known what to do with $1 billion anyway.

SAGENAP held its first-ever review meeting in Washington, D.C., a few days before the Venice conference. Learned stopped off on his way to Europe to deliver a status report on DUMAND, and it didn't go well. The committee would not release its report until spring, but he must have seen the writing on the wall.

We last looked in on the project during the disastrous deployment attempt in 1993, when the buoy at the top of their one and only string had failed to release from its sacrificial anchor and the arched string had stopped giving signals about two minutes later. A junction box on the sea floor had failed two months after that. Sometime in 1995 the collaboration had attempted to repair it with an undersea robot—and failed again.

"The overall impression was that of a collaboration encountering a never-ending series of technical problems," the SAGENAP committee wrote. "As each was understood and fixed another seemed to arise." The committee noted "a general lack of rigid quality control, and a freewheeling style toward new ideas and improvements."

"Essentially every member of the SAGENAP group expressed the view that the present DUMAND management had displayed a lack of detailed planning and careful attention to quality assurance deemed essential for the success of a project like DUMAND. This fatal failure was laid squarely on the project leadership, whose managerial skills and commitment to the project were strongly deprecated. Many reviewers said that the DUMAND 'Project Director' should have been a strong Engineer/Manager of the NASA project variety."

The leader of the project was Learned. Many felt that the reviewers went overboard in their criticism of John and that some of their comments were unnecessarily personal. They hurt him deeply.

And then came the judgment.

"Further support of the DUMAND project is not recommended. . . . The project should be terminated, and its operations phased out in an orderly manner."

Thus the pioneering project in high-energy neutrino astronomy died, and its leading visionary, John Learned, was essentially banished. He has participated in—but not led—several important "non-accelerator" projects in the meantime, but they have all been in the low-energy arena.

Nevertheless, Learned is still the pied piper of neutrino astronomy. "His heart and soul are with neutrinos, and he's been remarkably steady in that love for over thirty years," says Bob Morse. John probably has the broadest view of non-accelerator neutrino physics of anyone alive, and he continues to keep his finger on the pulse of neutrino astronomy. Twenty years later, he is a friend to IceCube and welcomed into their closed collaboration meetings. Francis Halzen says John "was really the shining light in that collaboration, and then, you know, he got blamed."

With DUMAND's demise, neutrino astronomy became a two-horse race. (Although, how serious could it be with the Germans playing on both sides?) In April and May 1996, just after it was installed, the new four-string array in Siberia detected the first-ever "gold-plated," that is, irrefutable, up-going neutrinos. One could say that this made it the world's first working neutrino telescope, although it was far too small to make an astrophysical discovery: the three neutrinos in Baikal's 1996 data sample pretty much had to have been atmospheric neutrinos, born on the far side of the planet rather than in the cosmos beyond.

12. Glory Days

The children have built something out of an orange crate, something preposterous and ascendant.

—JOHN CHEEVER

BOB MORSE REMEMBERS THESE AS THE GLORY DAYS OF AMANDA: IT "WAS very frustrating at times and very demanding and caused you a lot of pain and anguish, but, you know, it was also a really wonderful, happy time in, I think, many people's lives. . . . It's hard to be bored when you're scared witless." With all the hard work they did year-round and the adventures and good times they shared at Pole, the collaboration coalesced into a kind of family. Many of those who traveled to the Ice for the 1996–97 season went on to play key roles in the crucial developments of the next several years, and a good percentage are still with the project two decades later.

Their group of forty was small for the task at hand; they were pushing the boundaries on every front. Their anarchistic "war party" approach was probably the only way they could have done what they needed to do.

While drilling and deployment were the most arduous tasks, they faced another challenge in the need to invent an electrical technology that could sense the minute impulses from single photons as they hit detectors more than a mile deep in the ice, carry them to the surface, and determine when the impacts occurred to within several billionths of a second. All the detectors in the instrument—in IceCube there would be thousands—also needed to be calibrated to at least the same timing accuracy relative to each other. This was applied physics again. The invention of an adequate technology would take about five years, as they came up with new gadgets to test in the ice pretty much annually.

Per Olof had tested the first new idea on the final string of AMANDA-B4, the one that had been dropped on the fly into the middle of the triangle.

So far, they had been relying on the standard cable used in physics labs in the real world, a so-called coaxial cable, to deliver the high voltage that drove the photomultipliers and to bring their signals back to the surface. But coaxials are thick and inflexible, so this limited the number of optical modules they could put on a string. At the bottom of string four, as it was called, Per Olof arranged to add an extra six optical modules, driven by "twisted pairs": two thin strands of insulated copper wire twisted into a spiral or double-helix. Twisted pairs are thinner and more flexible than coaxials, and it was obvious the minute they were tested that they also improved signal quality. They never used coaxials again.

But twisted pairs were not a magic bullet. When a photon hits the active surface of a photomultiplier tube it produces an asymmetrical pulse that rises to a peak in about two billionths of a second (nanoseconds) and decays back to zero in roughly another four nanoseconds. The twisted pairs were better than coaxials at carrying these pulses to the surface, but they still distorted them significantly. In a dispersive effect analogous to the one that distorted the laser flashes in the bubbly ice of AMANDA-A, the original six-nanosecond pulse from a phototube would be stretched out to about three hundred nanoseconds. This made it difficult to detect two photons arriving within, say, ten nanoseconds of each other, which often occurred, and it also made it hard to pin down the moment when a photon hit a detector, which, again, they needed to know to within a few nanoseconds.

In the off-season after AMANDA-B4, Albrecht Karle came up with an option that promised to eliminate dispersion altogether. His idea was to use a fiber-optic rather than an electrical cable to carry the pulse to the surface. A light emitting diode would be attached directly to the electrical output of the photomultiplier. It would produce a light pulse that had the same shape as the electrical pulse, and it would be coupled right there to a fiber-optic that ran to the surface, where the light pulse would be converted back into an electrical signal. Since virtually no dispersion takes place in a fiber-optic, the shape of the original pulse would be preserved.

Meanwhile, with some help from Buford Price, Francis Halzen's criminal optimism had worked some magic on David Nygren of Lawrence Berkeley Laboratory. Dave joined AMANDA in 1996.

He had been watching neutrino astronomy from a distance ever since his grad school days at the University of Washington, when John Learned was also a student there. He followed DUMAND "as a kind of spectacle," he says. "That's too strong a word or too derogatory, but it was very heroic."

Dave is a tall, rangy man, now approaching eighty, with wispy, often shaggy, silver-white hair, who could be mistaken for a perpetual grad student in his button-down shirt and jeans. Although he's a quiet presence at collaboration meetings, Dave can change the thinking in a room with one or two sentences. He has made a career of inventing what he calls "disruptive technologies" or game-changers: he likes to pull rabbits from hats. And he's something of a legend. In the mid-seventies, he revolutionized particle physics with an invention he named the time projection chamber.

One of the leading problems of the day was the difficulty of observing the many particles that are produced simultaneously in the collisions of very high-energy particles in accelerator beams. The state of the art was to do various tricks with two-dimensional images, similar to photographic plates, but these techniques were limited to about ten emerging particles. The time projection chamber made it possible to analyze as many as ten thousand particle tracks at the same time—a disruptive technology indeed, and it remains state of the art a quarter of a century later. Pretty much solely on the basis of this invention, he was elected to the National Academy of Sciences in 2000. Among other things, he has also invented an imaging technology for mammograms that reduces the risk of breast cancer by reducing the amount of X-rays needed to produce a good image by a factor of three or four.

Dave was reintroduced to neutrino astronomy by a charming and highly regarded, somewhat off-beat particle physicist by the name of Kwan-wu Lai. The Km3 concept, which was the focus of the meeting Dave eventually convened in Arcadia, California, was originally Kwan's idea. The plan was to build a kilometer-scale neutrino telescope in a deep trench off the California coast, near Santa Barbara.

During a sabbatical at the Jet Propulsion Laboratory in 1993 or so, Kwan gathered a group to look into the Km3 concept and invited Dave

Nygren to join. For a while there, Dave was flying down from Berkeley once a week. His interest was in developing technologies basic to the entire concept of large Cherenkov telescopes, rather than any specific instrument. At one of the group's weekly meetings, he suggested taking a "phase neutral" approach, that is, developing technologies applicable to both water and ice. His desire to acquaint himself with all the technologies then being employed in DUMAND and AMANDA was part of his motive for convening the Arcadia meeting in the spring of 1994.

A few months after the meeting, he experienced a moment of serendipity as he sat in on what he calls a show-and-tell session for a master's thesis by an electrical engineering student at Berkeley named Stuart Kleinfelder. Kleinfelder had "designed a chip that could swallow a waveform with high frequency components and capture it at high speed, but then read it out more slowly," Dave explains. "This is not the kind of chip that you would use for video, because something's happening all the time. But in something like neutrino astronomy, not much of anything is happening most of the time. So if something interesting happens you can capture it and then process it at low speeds. . . . This saves a lot of power." Saving power is an important consideration when you're running thousands of modules in a remote location, a mile and more out of reach.

Since Kleinfelder's chip would turn the pulse from a phototube into a string of numbers that could be sent to the surface digitally, it would eliminate the dispersion problem altogether. And it would turn out that digital signal processing had manifold other advantages that even Dave Nygren didn't appreciate at the time.

"So the whole thing started with the random occurrence of seeing a chip that Stuart Kleinfelder had designed for no purpose other than getting a master's thesis," he says. "I saw this as a truly disruptive technology that could and should be explored. . . . It was the hope that I saw to really get into this business, so to speak, and do it right."

At Dave's urging, the Km3 group at the Jet Propulsion Laboratory managed to swing down $100,000 for the design and construction of two "digital optical modules," or DOMs, incorporating a phototube, Kleinfelder's chip, and an on-board "computer-on-a-chip."

By the summer of 1996, Shannon Jackson, an undergraduate on a summer research fellowship at JPL, was assembling the DOMs. That same summer, Dave asked Jerry Przybylski, a crack engineer at his own employer, Lawrence Berkeley Laboratory, if he would consider a trip to the

South Pole the following winter in order to install them. Jerry managed to persuade his wife to let him go by promising her a vacation in New Zealand after his deployment, and hopped on board. Thus the DOM became at least the fifth creature in the increasingly complex zoology of the AMANDA optical module.

The Km3 group petered out the following year without spawning a working collaboration.

Another area in urgent need of improvement was deployment: the physicists were suffering out there. It was far too stressful and taking far too long a time to drop their strings of pearls into the ice.

Albrecht Karle left his first deployment season, AMANDA-B4, in the firm conviction that something had to be done about the so-called breakout scheme. Each string consisted of two cables, essentially, one for the electronics and the other a strong metal cable that acted as a sort of backbone to prevent the string from stretching. The problem was that over the length of a string the electrical connectors and the mechanical attachments to the metal cable inevitably "walked away" from with each other. As the deployment team attached each successive module, the mechanical and electrical connections would get farther and farther apart, and picking up the slack became a time-consuming, seat-of-the-pants affair that led to cold fingers, since some of the connections required bare hands, and long deployments, which threatened the viability of a string. It was impossible to account for the differential stretching of the two cables ahead of time. There had to be some way of adjusting for the mismatch in the heat of battle.

During his first off-season, Albrecht devised a breakout scheme that solved the match-up problem altogether. Bob Morse refers to it as "the German whips and chains." The basic idea is to attach chains to the metal cable at the desired distance between modules and pick up the slack with the electrical cable by attaching each module to whichever link on the chain happens to match up with the electrical connections. The key to this arrangement is an item known as a chain clutch sling, a handy little gadget that Albrecht ran across in a cable and rigging store that happened to be located next door to his favorite movie theater in Berlin.

Albrecht was willing to take on almost anything in those days: "As you're younger, you have this unlimited amount of confidence," he says. Thus, when Pat Mock announced at the spring collaboration meeting that he would be stepping down as deployment manager, Albrecht agreed to take his place. One of his conditions was to ask the principal investigators to sign off on an organizational chart that gave him, a lowly post-doc, the authority to make tactical decisions on the Ice. He wanted to ensure "that there was no risk that more senior people would put pressure on me at the pole, possibly by e-mail. . . . To me it symbolized the transition towards organization."

He got his mandate, but any organization it may have instilled was symbolic at best.

It's not clear what set Steve Barwick off that year. Perhaps Wisconsin finally stood up to his annual attempt to wring extra money out of them or NSF didn't give him as much as he wanted either. In any event, sometime that summer, he went on strike. He stopped answering phone calls and e-mails and refused to do any work on the project.

This may have been the worst instance of it, but in fact Steve went on strike about once a year. The previous year at Pole, for example, when the drill kept shorting out and Albrecht helped fix it, Steve had retreated to his room to read books for a week. It wasn't anger exactly, it was a sort of panic that set in whenever the pressure built.

This strike presented more than the usual challenge, since he'd placed himself in the strategic position of integrating the string cables and transporting them to Port Hueneme for their sea voyage to Antarctica. Cable integration was usually carried out in the warehouses of a company on the outskirts of San Diego, and Steve usually acted as straw boss and sent a few of his students down to help. This year, he held his students back, too. Bob Morse had to put together a group of replacements to fly to San Diego and get the job done.

That was the first forest fire of the 1996–97 campaign.

PICO's problematic head driller of the previous year had been replaced by a member of his crew named Tim Macovicka, who got on well with Bruce Koci and Albrecht. So the drilling went well, but that was about it.

Albrecht remembers that "everything that could possibly go wrong did go wrong." His own season started out badly the very day he reached Pole, when he had trouble adjusting to the altitude. The station doctor decided (incorrectly, in Albrecht's view) that he was showing signs of high-altitude pulmonary edema—a potentially fatal condition that can be resolved only by dropping in altitude—and sent him back to McMurdo for a few days.

Despite Macovicka's improved leadership, it took a while to wake the drill up from its winter hibernation, so the first hole wasn't completed until mid-December. And so, the first deployment began. The German whips and chains notwithstanding, this was still a madcap affair. There was only one shift, made up of whatever students and post-docs happened to be on the Ice; there were no engineers or trained technical staff; it was carried out in the wide open; and it was all hands on deck until the job was done.

They had attached all the optical modules to the first string, one by tedious one, and were lowering it to its final resting place at the bottom of the hole, when about a thousand meters down, the pressure sensors began telling them that the string might be stuck: the readings weren't changing even though the winch was clearly lowering it into the hole. They pulled it up a few meters and lowered it back down, and it stuck again. After a second try, Albrecht consulted Bruce and they decided to pull it out altogether. For whatever reason, it seemed that the hole had necked down at that depth.

The deployment team had no choice but to keep working, even though they'd been awake for about thirty hours. Several were falling asleep in place, and one or two were so exhausted that they had to be sent back to their berths. They figured out a way to "un-deploy" on the fly, pulling the dripping, ice-covered modules and cables out of the hole, somehow scraping the ice from the electrical components, removing the modules with bare hands, sticking them back into their cushioned boxes, and stowing them in a Jamesway where they could thaw and dry out.

The deployers got some sleep while the drillers reamed out the hole, and they finally deployed the first string on Christmas Day, thus missing the traditional South Pole Christmas banquet. They ate their holiday meal sitting in the Jamesway with their optical modules.

"So that was my very profound initial experience for how string installation can go way different from what you expect," Albrecht remembers. "And how . . . this whole operation . . . it's tough."

John Jacobsen, the artist/physicist who had helped demonstrate AMANDA's sensitivity as a supernova detector, had recently completed his doctoral thesis and was now employed as a post-doc in Madison. He made his first trip to Pole that season and remembers his initial impressions vividly.

When their LC-130 taxied to a stop, John and his fellow passengers stepped down onto the snow to find no one to greet them or tell them where to go. He wandered around in the whiteness for a while, eventually found his way down the snowy ramp into the dome, looked up at the crap-sicles hanging from the ceiling, and said to himself, "You've got to be fuck-ing kidding me. I cannot believe I have to spend the next month and a half here!" Then he entered the dingy galley to find Albrecht sitting alone at a table, looking as though "he'd just survived the bubonic plague." The sec-ond deployment had ended a few hours earlier.

John was also hit hard by the altitude. After a briefing by the station manager that he hardly took in, he struggled across the Ice to his cubicle in Summer Camp, lay down on his bunk, and at the first pangs from his blad-der realized that his room was missing an essential appointment: the piss can. This is no joke. It's quite an ordeal when you're suffering from the list-lessness, headaches, and nausea of acute altitude sickness to "get into what basically amounts to a space suit sans oxygen supply every time nature calls," as he e-mailed to some friends back in the real world, and trek a hun-dred yards across the Ice to the men's room. To compound the problem, the best thing you can do to combat altitude sickness is drink lots of water. On one of his early forays to the plywood shack that housed the toilets, he had the good fortune to meet a guy who told him he could get his very own piss can in the galley. Standard issue was an empty industrial-sized tomato can.

Being an AMANDA "beaker," aka scientist, he was pulled in on deployments whenever they came around, but his main job was to write software, which he did in a small, cluttered, oblong room filled with com-puters, known affectionately as Back of Science. It was located in one of the inverse refrigerators inside the dome.

John approached the writing of computer code in the same way he ap-proached painting: as an art form. In fact, he was beginning to realize that he was more interested in that aspect of his work than the physics. His thesis had consisted mainly of Monte Carlo simulations, and he also liked to work "close to the metal," as he put it, at the lowest level of hardware: the chips and circuit boards that actually ran the AMANDA machine. He'd been writing code for about ten years and was becoming an adept.

He ended up going back to Pole about fourteen times. The following year, he spent many happy hours in Back of Science, listening to music (espe-cially David Bowie's *Outside*) and writing a program he named polechom-per, which took the information directly from AMANDA's data acquisition system, stored it on a tape drive at Pole, and also sent it by way of a NASA satellite to a bank of computers at the White Sands Missile Range in south-ern New Mexico, where a second of his programs, sableblanc, forwarded it to a computer at Lawrence Berkeley Laboratory. There it was made available to the full collaboration on the Web. Prior to polechomper and sableblanc, AMANDA's data was stored only on tapes, which had to be hand-carried north once a year.

This seems like a good moment to consider the contrast between working at the South Pole and working at an accelerator laboratory in the real world, like Fermilab or CERN. Not only would John have been living in a comfortable apartment, which he could have stayed in year-round, a bank of supercomputers would have been attached directly to his detector. The primary data from this experiment, on the other hand, needs to be sent through an extremely narrow pipeline, low-bandwidth satellites that are above the horizon for only a few hours a day, in order to reach adequate computers. And this pipeline isn't even capable of transmitting all the raw data. The AMANDA and IceCube scientists have had to invent clever ways of filtering and compressing their data at Pole—hopefully without missing anything—since there is a hard limit on how much they can accumulate in any given time period. This severe challenge, which usually goes unmentioned, is probably unique in high-energy physics.

John's code was not only effective, it was beautiful. About a year after he wrote polechomper, he received the following e-mail from Madison grad student Tyce "Ty" DeYoung, who was working to extend the code and must have been at Pole at the time.

> **Subject:** You are a god
>
> Hi John,
>
> I've been looking over polechomper—mere words cannot express how grateful I am that you write such clear code. You are like a god to me.
>
> Slightly punchy,
>
> Ty

They were an inventive and humorous bunch. The password for one of the early AMANDA Web sites was "cheapass": Center for High-Energy AstroPhysics at Southpole Station.

Jerry Przybylski arrived at the station a week or two after John did, having stopped in Pasadena on his way south from Berkeley in order to pick up the two digital optical modules at the Jet Propulsion Laboratory. He packed the DOMs, a couple of desktop computers, and some troubleshooting equipment into plastic cargo boxes, about three hundred pounds altogether, and carried them south as checked baggage.

Jerry remembers his first few minutes at Pole as being "very dramatic, sort of like checking into the planet Hoth from *Star Wars V.*" Once he had walked clear of the wings and whirling propellers of his plane, he became disoriented, because all he could see was white. He had seen the dome from the air, but from the ground it was hidden by recently plowed snow and the drifts that had been slowly burying the structure in the twenty-plus years since it had been built.

Albrecht gave him a corner in one of the Jamesways out by the AMANDA drill shack and arranged for a 10,000-pound spool of cable, one of the strings, to be parked on the Ice beside it. Jerry found a way to attach the DOMs to the cable, and in such crude conditions tried to get them working between interruptions for deployments and "cable drags": once a string was deployed, the top end was attached to a surface cable that had to be dragged across the Ice to a building on the surface known as the Martin A. Pomerantz Observatory, or MAPO, where it was attached to the computers and other electronics that ran the instrument.

Jerry describes Pole as being a "*very* unfriendly environment" for working with electronics. The dry conditions and constant wind generate an enormous amount of static electricity, so you can destroy chips and other electrical components if you don't discharge the static on your body now and again by touching a grounded piece of metal. One day, he was simply walking past one of the DOMs when a spark jumped through the air to it from one of his hands. By a small miracle it wasn't destroyed.

When he attached the DOMs to the cable, he found that they didn't work and he had some sophisticated troubleshooting to do—a task that was additionally complicated by the difficulty of communicating with Shannon Jackson and his software expert back in Pasadena: e-mail only works when the communications satellites are above the horizon. Jerry had to rewrite the so-called firmware in the computer-on-a-chip that ran the DOMs, and that meant "burning" the new firmware into new chips. He had to depressurize and unseal the glass Benthospheres that housed the electronics, replace the offending chips, reseal the spheres, and pump them back up to pressure. Then he discovered a fatal bug in the software that ran the DOMs from the surface. He solved that problem, hung around to deploy the DOMs and run tests from the surface, and escaped Pole just in time to rendezvous with his wife in New Zealand. It was a frantic experience all around—as the DOM would be from beginning to end.

He must have felt strange leaving his babies behind, since he wouldn't have any idea how they performed for about ten months. Over the winter, their data was stored on optical disks that didn't reach his hands until someone found the time to put them on a northbound Herc near the start of the following season. Although one of the DOMs died a couple of months after it was deployed (Jerry never knew if it was the one he'd zapped), they produced just enough data to keep the project alive.

Dave Nygren had been wise to send such an accomplished engineer to the Ice. Jerry "only appreciated later how much I learned along the way." By participating in deployments and cable drags, by watching the static arc across the room from his fingers, and just by sitting in MAPO watching AMANDA and the Amandroids at work, he picked up many details that he would fold into future designs.

His biggest take-home message was that dropping one of these basketball-sized pearls into the Ice was similar to sending a satellite into space: once you let it go you'd never see it again. His need to reprogram the chips taught him that the DOMs would need to be programmable from the surface after being deployed, just as satellites are designed to be reprogrammed from Earth after they are launched. This would take "some pressure off of the software team to be infinitely clairvoyant."

Several times while he was working in MAPO he watched every single optical module in AMANDA light up simultaneously, and he guessed that this was a false signal created as the entire building released a huge load of static electricity into the air. By digitizing the signals down by the phototubes themselves, far removed from the static and radio noise at the surface, he knew that the DOM would be able to measure the light signals in the deep ice far more accurately than the analog circuitry in AMANDA ever would.

His third important lesson came from watching Steve Barwick, an experienced physicist and a professor no less, fiddle for days with lasers and fiber-optics, trying to calibrate the instrument. (Steve had eventually called off his strike, as he always did.) Not only did calibration require an expert, it was time-consuming and inaccurate and the instrument couldn't take data while it was being done. An important cosmic event, such as a supernova, might be missed. Jerry and Dave Nygren resolved to find a way to make the DOMs capable of calibrating themselves automatically.

The winterovers for 1997 were Gary Hill, an Australian who had just earned his doctorate at the University of Adelaide, and Robert Schwarz, a German diploma candidate at the University of Munich. This was the first Antarctic visit for both, and for both it was a life-changing experience.

Robert is now known as "iceman." As I write, he is about to complete his twelfth winter at Pole. Amazingly, however, he is not the clear record-holder in that regard: a fellow named Johan Booth has spent the same number of winters there, as well as an additional six at the U.S. Palmer Research Station on an island off the Antarctic Peninsula. Robert does hold the record for the most time spent at Pole altogether, however, since he's spent more summers there than Booth has.

Gary had a relevant background. He'd done his graduate work on a Cherenkov detector in a lake in a volcanic crater in Australia (which was never built, because the lake was too shallow). Initially, he hesitated about "wintering," but Bob Morse won him over by offering him a two-year postdoc in Madison after he came off the Ice—and Gary turned out to be a keeper; he's still with the project. He's easy to get along with, he's calm under pressure, he's made enormous hands-on contributions at Pole for the past two decades, and he's made significant contributions to the science as well.

In those years, when the dome was still the heart of the station, the winter team was limited to twenty-eight. They met in Denver in mid-September to become acquainted and begin developing an esprit de corps. Personality profiles were taken; everyone was interviewed by both a psychologist and a psychiatrist and took a written psychological test. There were ropes courses and a fire school, since a winter fire at Pole is probably the most horrifying possibility one can imagine. Gary remembers "actually crawling around in burning buildings," searching for dummy victims and dragging them out. He loved it.

He and Robert traveled to the various institutions to get to know their new colleagues and learn about their winter tasks, and they headed south early to get oriented in what would be their austere home for the next twelve months. They were swiftly recruited to Albrecht's deployment team.

The winterovers are the elite of the summer crew, although one tends to regard them warily, wondering what could possibly possess a person to do such a thing. In fact, however, they are generally the sanest people at Pole, for the main reason that they have to be.

These two started out with the full AMANDA experience, since their first deployment involved the exhausting "un-deployment" as well. And Gary had a second memorable experience right on its heels. Always the sort to lend a hand, he decided to help the drillers detach the drill head as they raised it out of the second hole at the end of the drilling.

"No hardhat, I just wandered in and didn't think of that. They were rocking the drill back and forth to shake off ice and water, I think. As it was lifted out, the tip of the long pipe extensions must have been just clear of the hole when some bolts up higher gave way, and the bottom section of the pipe dropped straight down, hit the ground vertically, then fell over toward me, hitting me on the head in a glancing blow as I ran away. I was decked to the snow with a pretty long gash on the head and lots of blood every-where.... Anyway, in the end they shipped me off to biomed, where Hugh [the station doctor] stitched me up.

"After that, I was confined to biomed and ordered not to fall asleep, just in case something would go wrong while I was asleep and I'd just die with-out warning. I might have stayed awake for another twenty-four hours or something. Fortunately, people brought movies.... I still have a pretty cool scar on the head, and now I buzz the hair back to a number one or two sometimes. It looks pretty scary!"

So much for Albrecht's transition to organization. It was the wild, wild west.

Gary recovered, and he and Robert took over as deployment leaders once Albrecht left the Ice in mid-January. All in all, it was a successful sea-son. They managed to sink six strings into the ice and complete the epochal incarnation that became known as AMANDA-B10.

13. Night on the Ice

THERE MAY NOT BE AN OBVIOUS CHANGE IN THE WEATHER. THE DATE IS always up for grabs. Winter at the South Pole begins at a very definite moment and with a flourish, as the last LC-130 rumbles along the runway, lifts and banks for a low pass over the station, slowly shrinks to a dot, and disappears.

Most of the winterovers go outside to watch. They tend to greet this moment with relief, since it usually marks the end of a period of sustained panic. There's always a last-minute rush to get the final necessities to the station, both for surviving and for doing science, and to get the last summer residents and whatever else out. These last escapees are usually frazzled, too, since the most likely reason they've hung around has been to finish some essential and unwieldy task under a shifting but unyielding deadline.

Suddenly it is silent. There's an all-hands meeting. And the traditional kickoff to their eight or nine months of solitude is to watch a triple feature of relevant horror movies: *The Thing*, in which the scientists at a polar research station are terrorized by a shape-shifting alien—both the classic 1951 version, which takes place in the Arctic, and the 1982 remake, which takes place in Antarctica—and "Ice," an episode of *The X-Files* in which a research team on an Alaskan ice sheet is infected by an extraterrestrial parasite that induces fits of murderous rage. In Gary and Robert's year they made it a quadruple feature by watching *The Shining*, too. (The famous face shot of a crazed Jack Nicholson used to be posted here and there around the station.) That movie is now a midwinter tradition, from what I understand.

Some then commence a figurative escape to McMurdo, the idea being to cover 850 miles on either a treadmill, a rowing machine, or a stationary bicycle before the station reopens for summer. That would be an average of about three and a half miles a day.

There's a saying at Pole that what happens on ice stays on ice, and there is an honorable tradition against helping outsiders or passers-through—fingys—truly understand the life there, especially the life in winter. I encountered this tradition when I visited the place for six weeks near the end of 1999. Many of the old-timers were openly derisive of my "book research," and once they found out who I was would pointedly clam up when I entered a room. Part of the time, I worked with the "redneck beakers," the PICO drilling crew. One day I entered the Jamesway that housed the water boilers, looking quite the journalist in my spanking-clean ECW gear with my camera dangling from my neck, and one of my co-drillers, a friend I think, a guy named Jed, who had worked as a gold miner in South America, looked at me with a smile and said, "We're here seeking fortune, not fame."

In respect of that tradition we will not delve too deeply into the stories of that winter. (There are thousands of them. Gary says there's at least one every day.) It is perfectly fine and indeed noble that some things about so sublime a place should remain hidden. I can say, however, that Gary and Robert became fast friends and that the scientific and career fruits of that winter have been highlights of their lives so far.

They were an unusual pair at Pole in that neither drank alcohol. For the bar in the old dome, a dark place on the second floor of the galley building, steeped in tradition, was the beating heart of the station. It was also the only place where people were allowed to smoke. About the only ways to escape from the boredom of the place (aside from sex, if you were one of the lucky few who had that option; relationships lasting only the winter are not uncommon) were drinking and watching videos. It is probably fair to say that the U.S. government used to subsidize alcohol abuse at Pole, and probably at the other stations as well. When I was there, the beer was cheaper than it was in the real world, and I heard stories of runs on different types of liquor—with the goal of drinking all the tequila in the store, for instance—that had occurred the previous winter. There was a secret still for making moonshine somewhere in the dome, and it is said that at least one belligerent alcoholic used to crop up every winter. The previous year, in fact, "a worker was thrown into detox three times before he was finally forced to live in the medical facility, isolated from the rest of the population."

Robert and Gary were minding a grand total of five related experiments. AMANDA was their main responsibility and required the most time. The others were SPASE, GASP, and two new ones: RICE (Radio Ice Cherenkov Experiment), yet another feasibility study for detecting neutrinos with radio antennae, and VULCAN, an array of air Cherenkov telescopes that worked in conjunction with SPASE. VULCAN was the dream child of Alan Watson, the dry-witted Scot from Leeds. He tells me that he and some friends came up with the acronym one night in a bar: Very Un-Likely Cherenkov Array Name.

After getting into the swing of things with their horror movies, the winter crew took most of the next day off and then got to work battening the hatches against the approaching cold. They cleaned and closed Summer Camp, they set lines of flags across the Ice to guide them to the locations they would need to get to in the dark in blowing snow when winter deepened, they stowed the equipment and vehicles they would not need. Much of it was stored outside and would need to be dug out of the drifted snow come spring. There was a separate category labeled "DNF": Do Not Freeze, for items that had to be kept in heated buildings.

They slowly settled in to their routines . . .

The sober lifestyle of the two AMANDA winterovers wasn't the only thing that was good for their health; their commute helped, too. In contrast to the "dome rats" with desk jobs, who had no compelling reason to go outside, they had to walk back and forth to the dark sector most days, and for the most part they enjoyed it. The majority of their actual work took place inside MAPO or Back of Science, but they did have to work outside occasionally, especially on GASP and VULCAN.

One is especially aware of one's location on the planet and even to a certain extent the solar system when one lives at the South Pole. Solstices and equinoxes carry more meaning than they do elsewhere, and the equinoxes are especially vivid since they are accompanied by sunrise and sunset.

As the station closes for winter, all signs point to sunset, the autumnal equinox south of the equator, which comes only a few weeks later. Robert calls this "the time of the long shadows." What photographers know as "magic time," the hour or so after sunrise or before sunset, when colors are most alive, extends for several weeks. In a weekly report in March 2002, AMANDA winterover Katherine Rawlins wrote, "The Aeolian ripples in the snow are cast by the low sun into bold relief. It's quite beautiful, really."

The Sun circles counterclockwise, parallel to and just above the horizon, lower every day, making the complete circuit every twenty-four hours. It undergoes strange transformations as it prepares to disappear, changing shape and color and sometimes setting and reappearing several times.

Anyone who has hiked up a mountain knows that air temperatures usually drop the higher one climbs. At Pole it's other way around. The cooling effect of the oceanic ice sheet creates a temperature inversion in which the coldest, densest air lies at ground level and the higher layers become successively warmer and less dense. Optically, this creates a kind of prism at the horizon, which separates the different colors of light given off by the Sun. The denser air lower down is like the thick part of the prism, bending the light more sharply than the upper layers. As atmospheric conditions change, the qualities of the prism change with them, and this is what causes the distortions in the shape and color of the disk. The bending of the light also allows the disk to be seen when the Sun itself is actually below the horizon geometrically.

The color that is bent least is green. This leads to the so-called green flash, which appears at lower latitudes as a momentary flash of green just as the Sun disappears. At Pole the flash can occur repeatedly for hours and even days.

Darryn Schneider, who wintered for AMANDA in 2000, was watching closely as the top sliver of the disk disappeared:

> At around mid-day the sun was getting very low on the horizon. We could only see the upper limb, and this was all distorted into a flat disk shape with jagged edges. Every so often bubbles would come off the top before popping into nothing. As the sun got very low, so that it [was] nothing more than a line on the horizon, small dots of green appeared above the

sun, blinking like a traffic light, until finally the sun disappeared. Zach and I had a beer to celebrate the start of winter, only to the see the sun rise again slightly as it went past the very slight rise in the horizon. Now we know—the horizon isn't quite as flat as it looks!!

They held a sunset party that night. But the next morning, when Darryn got up early to walk to MAPO, lo and behold, the Sun was back! He checked on it all day while he worked: "I could see the shapes the sun was being distorted into quite clearly. With binoculars I could see it changing very quickly—growing taller—being sliced by dark lines—the edges all jagged—and every so often a line or bubble of green on top."

That night they gathered in the Skylab lounge, a small room at the top of a five-story tower adjacent to the dome, which afforded a 360° view of the polar plateau:

> We suddenly become sun worshipers, sitting in chairs staring at it. The green on the very upper edge is the brightest I've seen it so far—bars of it appearing every so often. Even blues are visible. With the telescope it's an amazing sight. While the air looks fairly clear, the presence of ice crystal[s] is obvious from the faint sun pillar that extends a fair distance into the sky.

This time it disappeared for good.

While the star that gives us life plays the leading role in the drama of the seasons, the Moon becomes the gentle monthly companion in winter. It rises above the horizon for two weeks and disappears for the same duration, alternately illuminating and darkening the landscape. When the Moon is full and the sky is clear, it is quite possible to walk to the dark sector without a headlamp. Owing to the whiteness of the snow, the altitude, and the unparalleled clarity of the air, night is probably brighter on the Ice than it is anywhere else on the planet.

The most affecting memories for nearly every winterover are those of the dark days when the Moon is down and the sky is at its most sublime. The stars shine brightly in the high, clear air; some twinkle back and forth between red and blue; and the Milky Way resembles an elongated, richly textured cloud stretching from horizon to horizon. The most arresting sights are the aurorae: green, violet, or all the colors in between, shimmering and shape-shifting, sometimes covering half the sky. In New Zealander Anthony Powell's superb documentary, *Antarctica: A Year on Ice*, a woman wintering at McMurdo Station describes being so overcome by the beauty of the night sky that she drops to her knees without realizing she has done so, and finally coming to only when she is released by the spell.

Any outside work is of course severely hampered by the cold. For one thing, you have to bundle up pretty well. In addition to the predictable heavy clothing, gloves, and boots, it is necessary to cover every square millimeter of skin, and everyone comes up with his or her own unique way of doing so. Robert and Gary designed special hoods made of fleece that covered their entire heads, had cutouts for their eyes and mouth, and were long enough to be rolled into sumptuous gators around their necks. They wore goggles over the eyeholes and breathed through sort of broken-off snorkels that kept the condensation outside the hoods, so it wouldn't fog up their goggles.

At one point during their winter, when the Moon happened to be down, they noticed that one of the VULCAN detectors wasn't working. They went out in the dark, detached the garbage can–shaped object from its cables, lifted it out of its stand, and carried it into MAPO to try to diagnose the problem. Couldn't find anything. Took it back out and hooked it up. Still didn't work.

Shining their lights around, they noticed the end of a cable sticking out of the snow. Gary guesses that one of them had hooked it with his foot and snapped it while walking out to the array. They decided to put new connectors on the two broken ends in order to reconnect them. It was a coaxial cable, which uses a relatively cumbersome connector that takes five or ten minutes to attach even in a warm, well-lighted lab. You need a razor blade and a crimping tool. The temperature was -70 Centigrade. After about fourteen tries and sixteen hours, they gave up and decided to run a new cable, but this wasn't straightforward either. They couldn't carry a reel outside and spool the out cable, because it would freeze and stiffen in the cold. So, in the warmth of MAPO, they cut the required length of cable and attached the right connectors to it, and Robert took one end over his shoulder and dragged it down the stairs and out an open door onto the Ice, making sure to walk in a straight line toward the detector, while Gary payed out the cable from inside the building. In the end, a job that would have taken a single person about fifteen minutes in the real world took two people about two days.

Gary wasn't bothered in the least by the lack of sunlight. He says he has probably never felt as content or mentally fit. The simple life suited him. There were no bills to pay, no traveling to do: "You get up in the morning, you check your experiments, you eat, you work, you [do some] recreation, you sleep." He'd taken his electric guitar along, and there was another guitarist and a bassist on the crew, so they trained Robert to play the drums and put a band together, which they named Fire on Ice. They would practice up in Skylab, far enough removed from the dome that they could make as much noise as they wanted. Gary had also taken a Swahili phrasebook along, so he made a desultory attempt to teach himself that rhythmic language, and Robert took it upon himself to teach a German class, which attracted about ten students. They also got way into photography. There was a darkroom in the dome where they could even develop color slides. Robert's shots of the aurorae have since become quite well known, although everyone agrees that a photograph can't possibly do justice to the actual experience of standing under an aurora on the Ice.

This was also an excellent retreat environment for thinking about physics. Gary began writing Monte Carlos to simulate the atmospheric neutrino signal from the northern hemisphere, which was expected to be the first signal that AMANDA would see once they figured out how to use it. This is a "diffuse" rather than point-like signal: it's spread all over the northern sky. He has since become one of the collaboration's main experts on diffuse neutrino signals of all kinds and on statistical inference.

Pretty much every holiday is marked with a party, generally involving plenty of booze: St. Patrick's Day, sunset, Easter, Memorial Day, Fourth of July. Midwinter's day, the solstice, June 21, is a unique holiday for the denizens of the frozen continent. The day of deepest darkness is celebrated with a festive banquet, at which a letter of greeting and commendation from the president of the United States is always read. That year they held a barbecue on Labor Day. They held "Christmas in July" on the twenty-fifth. They threw a toga party for no reason at all.

But it wasn't all fun and games. There were two relatively scary fire alarms. One occurred at two in the morning, when a scientist named J. D.

Mayfield, who was working out at MAPO, noticed some smoke in the building. (J. D. may have been what they call "free cycling." In contrast to the support personnel who report to work on a fixed schedule, the scientists can work whenever they want, and with no sun to guide them sometimes slip into a completely untethered pattern of sleeping and eating. In Darryn Schneider's year there was a fellow named Yama who *lived* out at MAPO. Darryn would carry his meals out to him. Yama became skilled at cooking feasts on a hot plate that was otherwise employed to cure epoxy and even used it to heat up the goodies for the party he hosted at sunrise.)

The two a.m. alarm launched everyone straight from their beds. The fire crew, which included Gary and Robert, hopped into a snowcat and raced out of the garage arch into a snowstorm and full-on whiteout conditions. ("Someone commented that it looked like the Batmobile coming out of the Bat Cave as these doors sprang open and *do-do-do-do-do-do-do-do* and this thing goes flying out of there," Gary recalls.) The driver, a heavy equipment mechanic, was navigating with a spotlight, and all they could see was white. They finally found the flag line and followed it the half-mile to MAPO, getting updates from J. D. by radio, who discovered by the time they arrived that the flue to the building's furnace had become clogged with ice. They cleared it and went back to bed.

Near the end of winter, they almost realized their greatest fear when one of the engines that drove the main electrical generator blew up and the entire station lost power. A valve dropped into one of the engine's cylinders, and it spewed pieces hot of metal around the power plant building. The building filled with steam, and the fire alarms went off. They couldn't get the emergency generator started, so it became a race between starting the spare and getting the main one running again. Power was down for about four hours.

The coldest temperatures arrive about a month after midwinter, in July or early August. And a certain few look forward to the first time the temperature drops below -100 Fahrenheit with great anticipation, as it affords the first opportunity to join the legendary 300 Club.

When the temperature drops below 100 below, the aspirants of this select society set the temperature in the station's sauna to 200 Fahrenheit and warm up inside wearing nothing but winter boots. Gathering their courage, they dash madly out the doors and into the cold and sprint back to the sauna. "For extra stupidity points," writes Darryn, "you could go all the way to the Geographic Pole." In Gary and Robert's day, the roundtrip was about five hundred yards, and it was getting shorter every year, because the movement of the Ice was carrying the station closer to the pole. The point of nearest approach occurred several years ago, so the trip is now getting longer. This will undoubtedly be an important consideration when the next station is built, two or three decades from now.

Gary and Robert actually did a practice run, protecting their most sensitive appendage with pairs of shorts, but on the day of the real thing Gary just couldn't bear the pain. He reports that Robert and two other "knuckleheads" did manage to touch the flag at the geographic pole and return to the sauna, screaming in agony. Over the next several days, one developed blisters on one of his hands from minor frostbite.

Evidently, it doesn't hurt if you fall in the snow, because your skin is so warm that it melts, and although everyone is afraid of frostbite to the lungs (some believe it's safer to walk than to run, so you won't breathe so hard), the worst respiratory effect seems to be a minor form of "lung burn." The "cold, dry air parches the trachea, leaving it red and engorged," and "sufferers end up coughing up blood and mucous." A wintering doctor once observed that in the aftermath of the 300 Club, the station "sounded like a TB clinic."

Round about August, most everyone is suffering from winterover syndrome, a dysfunctional mental state "characterized by varying degrees of depression; irritability and hostility; insomnia; and cognitive impairment, including difficulty in concentration and memory, absentmindedness, and the occurrence of mild hypnotic states known as 'long-eye' or the 'Antarctic stare.'"

At Pole, it's known as being toasted. "I'm toast," or, "He's toast," the saying goes. In Darryn Schneider it manifested in its "irritability and hostility" form, and he made no bones about it. "Toasty, adj.," he wrote in early August in the weekly e-mail he sent to his friends in the real world. "The state of being forced to tolerate the presence of one or more idiots all the time. Advanced state of being in Antarctica too long; as, to be completely toasty." His curmudgeonly-ness had no effect on his work, so it wasn't a problem. The touchy-feely, group-hug approach doesn't play well at Pole. People who don't get along simply avoid each other. "Problems are not swept under the rug; they are placed under it very deliberately," writes one observer. "It's the art of containment, rather than resolution, that gets Polies through the eight-month-long night."

Psychologists and social scientists have been studying the syndrome for decades. It was first observed in 1898 by the redoubtable Dr. Frederick Cook, when he still had the good name he eventually lost by faking a first ascent of Mt. McKinley in Alaska and staking a false claim to being first to reach the North Pole. An archly intelligent man in addition to being a con artist, Cook was the physician on the Belgian Antarctic Expedition of 1877–79, which became the first ever to winter in the Antarctic region when their ship, the *Belgica*, got trapped in the sea ice near the Antarctic Peninsula. Besides noting the change in behavior among his colleagues during the long polar winter, Cook also discovered that he could cure them of scurvy by feeding them freshly killed meat—seal and seal blubber, mostly, which he went out on the sea ice and hunted himself.

A good part of the scholarly interest in the syndrome arises from the obvious similarity between wintering over and spending extended periods of time in space. The Antarctic Treaty, incidentally, which governs the peaceful international use of the continent, is seen as the template for the laws that may govern the colonization of the Moon and distant planets.

Neither Gary nor Robert got toasty in the least. (Maybe it was the lack of alcohol. The output of the still was known as "toast juice.") And Robert, after twelve winters, is famous for being utterly immune to the syndrome. In the same e-mail in which Darryn Schneider flaunted his own irritability, he referred to Robert as "Super Bert." In 2008, Robert applied to be an astronaut with the European Space Agency. He made the first four cuts, which whittled an initial group of 10,000 applicants down to 192, but was eliminated there.

At McMurdo, where they must be more intellectual or something, it's known as T3 syndrome, as in "having a T3 moment." This phrase comes from a recognized physiological condition known as Polar T3 syndrome, apparently related to winterover syndrome, in which the thyroid hormone T3 is suppressed. Interestingly, aside from an increase in moodiness and a decrease in overall cognitive function, this syndrome also leads to a need for more food and an increased ability to withstand the cold. The studies also say that they get it worse at McMurdo than they do at Pole. Surprisingly, the more severe the environment, it seems, the lower the incidence of depression.

There's a marvelous depiction of T3 syndrome in Anthony Powell's documentary. In one scene, he catches himself looking awfully bedraggled, sitting with his elbows on a table, chin on one hand, staring across the room in stupefied silence, unaware that he is filming himself, while his tablemates wave and make faces at the camera. One fellow describes putting his boots on the wrong feet, taking them off, and carefully putting them back on the wrong feet again.

Whatever they call it, the people who've studied it find that it isn't caused by the darkness or cold so much as the psychological and emotional stress of being isolated from family and friends and confined to a small space for a long time. Not surprisingly, loners handle it well, while the ones who really go off the deep end are those who lose the respect and friendship of their co-workers and become completely isolated.

The upside of this dalliance with madness is that it turns out to be good for you. Studies have demonstrated a tempering effect: people who have wintered in Antarctica visit hospitals less in subsequent years than similar groups of people who have not.

One morning in late July as they walked out to MAPO, Gary and Robert thought they discerned a faint brightening on the horizon. A week or so later they were sure, when a few of the faintest stars in Orion began to disappear.

On about the eleventh of August, the three women in the Stitch and Bitch knitting club made an all-call from the Skylab lounge to announce a distinct orange hue on the horizon. (All-calls are broadcast from intercoms all through the station.) This was quite early for such a thing, so there was some suspicion that they'd been drinking, but when about fifteen people crammed into the lounge, they saw it too. By midnight that night it was extremely bright, and several people gathered on the snow near the dome to marvel at it.

Next morning it was gone. A skeptical astronomer, who must have slept through the whole thing, suggested that it might have been a mass delusion, but Gary and Robert silenced him by posting some pictures in the galley.

They began to stand under the stars with a wistfulness now, since their sublime night was coming to an end. A definite brightness began circling the horizon. Orange and red began coloring the sky. Sunrise took a couple of weeks, and the disk finally appeared three days before the equinox, thanks to the bending of rays at the horizon. They dressed Skylab up like a

South Pacific island and themselves like pirates and held a sunrise party, replete with Beach Boys tunes. It was a bit too crowded to do much dancing.

Daylight revealed massive snowdrifts around the buildings in Summer Camp and the dark sector. Many of the flag lines they'd planted in the fall barely peeked from the snow. Their final task was a conflicted one: to bulldoze and shovel out the buildings, open up Summer Camp, and clear a runway to welcome the hordes that would bring their exquisite solitude to an end. It was a joy to greet old friends and read letters from family and eat the first fresh vegetables in six months, but for the first few days a fair fraction of the winterovers sought refuge in their rooms to escape the din of the crowd or snuck their plates off to secret spots in the dome that only they knew in order to share a few last meals together.

AMANDA might have felt the same way. For in those days winter was a peaceful time for the instrument as well. No one was adding to it or monkeying around with it, so it could simply take data. Gary and Robert's main job was just to keep the instrument running, and by Gary's account it ran pretty well. In fact, that winter, unbeknownst to anyone in the collaboration, AMANDA detected its first gold-plated neutrinos.

14. The First Nus[†]

BACK IN THE REAL WORLD, RUMORS WERE CIRCULATING THAT THE AMANDA collaboration wasn't capable of picking up-going events out of their down-going haystack or, worse, that the instrument was too unstable to detect anything at all. By the time Gary and Robert left the Ice, the collaboration had supposedly possessed a working instrument for about a year, AMANDA-B4, but had no physics to show for it. Francis was becoming increasingly anxious to produce results—and privately harboring his own doubts as well.

They had a good excuse, since even they couldn't see their data for about a year after any new version of the instrument was deployed. John Jacobsen hadn't written his polechomper program yet, so there was no way to send data north in winter. The winterovers stored it on tapes at Pole, so it

---

[†] In particle physics notation the neutrino is denoted by the Greek character ν, or nu. The plural sounds like "news."

didn't become available until the austral spring when someone found a chance to carry the tapes north.

Again, research is when you don't know what you're doing. They had only a vague idea of how they were going to grapple with this inchoate mass of information. Since they didn't know exactly what an up-going muon was going to look like in their instrument and didn't want to miss any, they programmed it to cast as wide a net as possible as they squeezed the raw data down to a manageable size and saved them. The computers in MAPO stored "events" based on a loose set of criteria set by various "triggers," the most basic of which saved an event whenever a certain number of phototubes—eight or ten—lit up within a certain, short period of time. There were also coincidence triggers set up with SPASE, the air shower array on the surface, and AMANDA-A, up in the bubbles, in order to "veto" events that occurred during down-going muon air showers detected by those instruments.

The first, large and loose dataset would need to be calibrated, using the data from the laser flashers that Steve Barwick obtained every February—always in a panic as they prepared to leave the Ice at the end of summer—and the calibrated data would need to be purged of obvious noise "hits." The next step would be to filter out the air shower and obvious down-going events, leaving an enriched and significantly smaller dataset, which would be put through a sequence of more intelligent "cuts" in order to find muon tracks, reconstruct their direction, and pick the up-going events, signifying candidate neutrinos, out of the pile. The prospects were daunting, not least because the physicists knew in advance that down-going muons would outnumber the up-going by a factor of about a million.

John Jacobsen seems to have been the first to try to put the whole ball of wax together, sometime in 1995, before AMANDA-B4 was even deployed. Although he didn't get far, the effort put him right on top of things when B4 began taking data on February 19, 1996. He asked someone to carry one day's data north by hand at the end of the season, so he could begin playing with it, and he got it February 21, so whoever it was must have been under some pressure, as the last Herc of the season couldn't have left Pole long after that.

In Madison, he and Serap Tilav collaborated on what she calls "our first attempt [at the] full chain of data processing." And DUMAND's DNA was also present at the creation, since John used an animated "event viewer" he had borrowed from the Hawaiian project—a software application that produced visual displays of the events as they unfolded in time—to pick out five "interesting events" by eye. No one would have seriously argued that they represented actual neutrinos, but he and Serap had gotten the ball rolling.

The problem was also attacked by three students in Germany, the most energetic and visible of whom was a post-doc named Christopher Wiebusch. Christopher had an extraordinary amount of experience for a person of his age, having worked on three other neutrino telescopes before taking his position in Zeuthen. He had first entered the field as a diploma candidate at the Rhine-Westphalia Institute of Technology in Aachen, under the tutelage of a mysterious character by the name of Peter Bosetti, now deceased, who was another example of the sort of wild type attracted to this pioneering field.

Sometime in the 1980s, Bosetti and his then girlfriend initiated a DUMAND spinoff named JULIA (Joint Underwater Laboratory and Institute for Astroparticlephysics), after her daughter. In January 1991, Christopher got his feet wet in this business by joining them on a sea voyage in a German research vessel named the *Sonne*, aimed at deploying a short detector string in the Mediterranean Sea.

They flew to meet the boat somewhere on the coast of the Red Sea, boarded, and headed north. But just as they were entering the Suez Canal, the first Gulf War broke out and they had to turn around. They flew back to Germany to wait for the *Sonne* to round the Cape of Good Hope and enter the Atlantic, then rejoined it and attempted to deploy their string somewhere in the vicinity of the Canary Islands. The expedition was not a success. They got a first taste of the hard lessons DUMAND had learned all too well about ocean deployments, and Bosetti drank heavily and fought constantly with little Julia's mother.

Since Bosetti was not a particularly supportive mentor, Christopher worked out a complicated arrangement where he joined the DUMAND collaboration, moved to Zeuthen, wrote a thesis related to DUMAND with Christian Spiering as his adviser, and submitted the thesis to Aachen for his doctorate. At Zeuthen, he also got involved in the Baikal project, quite naturally, so he carried both its DNA and DUMAND's into AMANDA. His specialties were Monte Carlo simulation and data analysis, and he was very much into the details.

The analysis tools that Christopher brought to the table, including some of the exact software Baikal was using at the time, were folded in with

John's and Serap's. Sometime in 1996, he, John, and a Zeuthen grad student named Stephan Hundertmark embarked on an intercontinental search for up-going muons that they named MFH for "Make Francis Happy." The project did not live up to its name.

Francis's immediate concern was a SAGENAP review that was scheduled for early March 1997. He invited Christopher to come to Madison for two or three weeks in February and asked him to work with Serap on yet a fourth search for up-going neutrino candidates. (John J. was in the process of moving to California.) They found a few, but couldn't be sure that they weren't down-going muons masquerading as up-going, because about 10 percent of this million-to-one-fold background was still sneaking through their cuts. Ever the loving mother, Francis did some hand calculations to convince himself that they *were* up-going and presented them to SAGENAP, but the committee was not convinced. They pronounced them "a bit 'too hot off the press'" and suggested the collaboration strengthen its abilities on the analysis side, even if it meant bringing in new people. They were very impressed, on the other hand, with progress on the construction side, and recommended not only continued but increased funding. In that sense the review could not have gone better, though it did nothing to ease Francis's growing concern.

As was often the case, the dark horse in the race to find the first neutrino (and you can bet it was a race) was the Swedish contingent. They even managed to get a jump on the other institutions by volunteering one of their members to carry the tapes from AMANDA-B4's only year of operation, 1996, directly to Stockholm from Pole. (Of course they shared the data, but it took some pestering to get it from them.)

It happened that Adam Bouchta, the grad student who had blanched in Venice when Francis had told him about the bubbles, was in a sweet spot in his career just then, since it had finally come time for him to write his thesis. This meant he had time to focus deeply on a project of relatively wide scope, and he and his adviser, Per Olof Hulth, chose to look at the B4 data. Sometime in the spring or summer of 1997, Adam found the very first credible up-going muon tracks ever detected by AMANDA—two of them. While they would never be called gold-plated, they were the first indication that it might be possible to build a neutrino telescope in Antarctic ice.

This relieved Francis greatly, but it was far too early to broadcast the news. In fact, just what, and how little, Adam's discovery actually proved helps underscore the immense challenge of neutrino astronomy. For one thing, there is virtually no doubt that the neutrinos (if such they were) that had given birth to his up-going muons did *not* come from outer space. They had almost certainly been created in cosmic ray interactions on the far, that is, northern, side of the planet. They were atmospheric neutrinos, which are just another form of background, similar to the manmade light pollution that makes it harder to see stars in the night sky. A true telescope would be capable of distinguishing astrophysical neutrinos, which originate beyond the northern atmosphere, from the atmospheric background.

Second, Adam was flying without a net. He had picked out his events by eye and hadn't had a chance to run them through the sanity check of Monte Carlo simulation. Like the ones Francis had shown to SAGENAP, they might have been mis-reconstructed down-going muons.

A lot was known about atmospheric neutrinos, since cosmic ray physicists had been studying them ever since Menon and Reines had first detected them in 1965. Their average flux, for instance—how many passed through a square meter of the Earth's surface every second—and their energy spectrum were pretty well nailed down. This meant that they could be used as a "test beam" for figuring out how well AMANDA was working and for improving the instrument, and the best way to do that was by running Monte Carlos. Remarkably minute details can be investigated in this way, but the first and broadest check is simply to run a simulated atmospheric neutrino beam through a simulated instrument and see if your real instrument is detecting the expected number of neutrinos. Gary Hill was working on this problem at precisely this time during his spare moments under the aurorae at Pole.

The Baikal collaboration demonstrated their skill in this regard at the International Cosmic Ray Conference that took place in Durban, South Africa, in the summer of 1996. They presented the three gold-plated events they had obtained with their four-string detector and showed that they matched up well with a Monte Carlo prediction of 2.3 events. There were six Amandroids from Zeuthen among the authors of this paper, including Albrecht Karle, Christian Spiering, and Christopher Wiebusch, so it's a bit surprising that they were outpaced by the Swedes on the AMANDA side. Maybe Baikal distracted them.

Again, Adam's events were credible, but not gold-plated. Mostly they whetted the appetite—especially since, by the time he presented them to the full collaboration at a meeting in Berkeley in the fall of 1997, the data from their new ten-string array, AMANDA-B10, the largest in the world, was soon to be released.

"In my mind, '98 was when it really happened, and it was messy, as I guess these things usually are," writes John Jacobsen. "There was a lot of back and forth discussion about how to archive, filter and distribute the fairly large B10 data set—it was more data than we'd dealt with before as a collaboration."

The "back and forth" overheated at times, as the emotions aroused by the possibility of a discovery brought out the aggression in several of the harsher members of the collaboration. One of the Swedes, for example, publicly abused John in collaboration-wide e-mails while they were hashing out the details of how to run the data through the long processing chain.

The data came north in the form of 246 digital tapes the size of cigarette packs, comprising about 500 gigabytes of information all told—a lot for those days. Once tempers cooled, it was decided that the Swedes would supply the algorithms and code for the first level of filtering and John, who had taken a half-time position at Lawrence Berkeley Laboratory (so he could spend the rest of his time making art), would load the data and software onto a massively parallel array of Cray T3E supercomputers at the National Energy Research Scientific Computing Center at the Lab—one of the most powerful such arrays in the world. (AMANDA and IceCube have been near the cutting edge in the crunching of large datasets ever since. Many of the students and post-docs who enter industry after working on the project take jobs writing code.)

There was a convenient "tape robot" connected to the Crays, which could be used to upload a few tapes at a time automatically, but each tape took several hours to read, and the entire collection took somewhere around a month. Two grad students at UC Berkeley, along with a post-doc named Kurt Woschnagg, platooned on that tedious task.

Kurt is a tall, strong, upbeat individual, with pretty much the perfect background for working in AMANDA. He was born in Sweden to German and Austrian parents, and everyone in Sweden learned English in grade school, so he spoke all three of the major languages in the collaboration. He had obtained his doctorate at Uppsala under a woman named Olga Botner, who was married to a member of the collaboration named Allan Hallgren.

(Olga would join in the middle of 1998.) By the time Kurt finished graduate school, he was tired of the constrained and competitive atmosphere at CERN, where he had done his research, and was thinking of leaving physics altogether; but just then an unusual scholarship opened up, which paid for students from Uppsala or Stockholm to work in Berkeley. The scholarships had nothing to do with AMANDA, all he needed was a host, and Buford Price was more than happy to have someone work for him for free. Kurt arrived in California in April 1996. During his first conversation with his new boss, he recalls, Buford "immediately started talking to me, hah, hah, about the properties of the ice. And so . . . I got roped into that, and that has lasted for many years." Kurt went to Pole in his very first year to help deploy AMANDA-B10 and has been one of the mainstays on and off the Ice ever since. He still works at Berkeley.

Once the tapes had been read into the Crays, it took about two and a half months of processor time to filter the data. (Since the processors worked in parallel, the elapsed time was some fraction of that.) The end result was an enriched dataset of one hundred gigabytes that was small enough for the larger workstations in most physics departments to work with.

Again one is astonished at how much this small group of people, about sixty in all, was accomplishing at that point in time. Over the 1997–98 Antarctic season they made great progress on the Ice.

This was when I learned about the project. I had met Bruce Koci on the top of the Bolivian mountain in the summer of 1997. He went from there to Tibet and then to Pole, where he proceeded to demonstrate that he had the drilling for AMANDA pretty much completely figured out. Despite major conflicts with PICO, he managed to drill three holes to 2,400 meters, 500 meters deeper than they had ever drilled before and as deep as they would need for IceCube. With that accomplishment, one could say that the drilling shifted from research to development, and the first step in that transition, as Bruce saw it, was to wean AMANDA off the habit of relying on him to make heroic efforts in the field every season. He began training other people to use the drill, while he shifted his focus to the broad aspects of the design. He and Bob Morse, his main partner in that effort, knew that a larger and more streamlined "production" drill would be needed if they were ever to build a kilometer-scale instrument.

Bruce and Bob were not working in a vacuum. Francis had been building the case for a larger instrument at various conferences around the world for about three years, and he and John Jacobsen had already come up with a name. ("I am convinced that he came up with the name, and he's convinced I came up with the name, so we'll never know," says Francis inscrutably. "So the first thing I did was go into Google and type in 'IceCube.' And what came up, of course, was an actor and a rapper who had already used the name. . . . And so then the question was whether he was going to sue us, and I said, 'I hope so, because then I'm really famous,' and he never did.")

Based on the bare hint of feasibility that had been provided by AMANDA-B4, John Lynch at NSF agreed to support an IceCube Neutrino Detector Workshop to take place in March 1998 in conjunction with an AMANDA collaboration meeting at UC Irvine. As I mentioned in the introduction, I was excluded from the collaboration meeting but allowed into the open workshop. Thus, unknowingly, I attended the very first meeting ever dedicated to IceCube.

In one of our conversations on the periphery of the collaboration meeting, Francis told me that the most important scientific contribution was made by Serap Tilav, who had actually left Madison a year earlier for Pasadena and Caltech, to work on another project that stretches the definition of a telescope. LIGO, the Laser Interferometer Gravitational-Wave Observatory, was and is a visionary attempt to detect the gravitational waves predicted by Einstein's general theory of relativity. Serap, who had been around since the early days, later told me that she "felt like a mother raising a child called AMANDA." She couldn't leave it behind; she kept working on it in Pasadena. In a similar manner to the way she had used muon air showers tagged by SPASE, the South Pole Air Shower Experiment, to determine that the deep ice in AMANDA was incredibly clear, she now used muon showers to demonstrate that there were some problematic dust layers at different depths in the instrument. This was relatively bad news, but not a showstopper, and it again took a while for the rest of the collaboration to come around. As before, she was eventually proven correct.

She also eventually realized that she could not leave her child behind. She returned to IceCube in 2001 and is still there.

At the 1998 IceCube workshop, Ariel Goobar, an assistant professor from Stockholm, presented the two up-going events that Adam Bouchta had found and showed that they dovetailed well with a Monte Carlo prediction of three. Goobar did point out that they couldn't be sure the events weren't fakes, but be that as it may, AMANDA was upping its game.

You may recall Francis telling me in the bar of our hotel in Irvine that I had come too late, the best part of the story was already over. About eight months later, he would happily eat those words.

A week or two after the workshop, John Jacobsen fired the starting gun in the race to find the first gold-plated events by releasing the enriched AMANDA-B10 dataset to the participating institutions.

At the next collaboration meeting, which took place in Zeuthen in July, the competition was intense enough that even I noticed it. The home team, for example, led by Christopher Wiebusch and Stephan Hundertmark, placed themselves in direct competition with the Swedes by presenting an analysis of the B4 data with their own Monte Carlos, in which they discovered a third event to complement Adam's two.

This was before personal computers were commonplace, but this particular demographic was at least ten years ahead of the rest of the world in terms of access to the internet and addiction to computer screens. It was crucially important for the host institution to set up a room with a dozen or so workstations, and exactly where that room was and what the passwords were was always one of the first announcements at a meeting. At Zeuthen, at different times, I saw different young scientists leave the meeting room in the midst of a talk, run to the computer room, log in to their home institution's network, run a competing analysis, make up a few transparencies, and race back to the meeting room in time to ambush the speaker. Christopher covered the most miles.

The three strongest institutions, Madison, Zeuthen, and Sweden, were developing many tools in triplicate, ostensibly as checks on each other, but also, obviously, to position themselves in the race. Each was writing its own Monte Carlos and developing its own set of cuts and reconstruction algorithms, although they did discuss their methods openly for the most part and share many of them. Gary Hill tried to inject some sanity by arguing for separating the tasks and assigning them to different institutions, but that went nowhere.

There was much discussion of the B10 data: how to calibrate it, strain out noise and false signals, and conduct basic tests to make sure it was sound. And this was the first time the collaboration as a whole ever addressed the question of double-blindness: it would be dangerous to use the entire year's data to develop their methods, as this posed the risk of tuning the methods to a specific dataset and finding things that weren't actually there. They made the major decision to split the data into odd and even days and keep the even days blind. They would develop their methods with just the odd days, freeze them, and then "open the box" to see if they got similar results with the even days.

Christian Spiering, the head of the Zeuthen group, had decided to use this opportunity to convene another workshop "as a step towards closer interaction between the present four projects": Baikal, NESTOR, AMANDA, and a new one, ANTARES, which had been started by a French group that had splintered off from NESTOR and come up with an equally strained acronym: Astronomy with a Neutrino Telescope and Abyss environmental RESearch. Christian was being perhaps unduly optimistic in calling the four projects "present," however. NESTOR and ANTARES didn't really exist—neither had any strings in the water—and Baikal, sadly, was twisting in the wind. In March, after ten years of work in increasingly difficult circumstances, the Russians had finally fulfilled their dream of building NT-200, their eight-string detector, but future prospects looked grim. Their country was in such dire economic straits that the government had essentially stopped supporting basic research, and Zeuthen, in consequence, had stopped infusing cash. NT-200 would limp along for another decade or so, plagued by reliability issues, but it would never grow. AMANDA had become the clear front-runner.

The Baikal collaboration could still do science with their instrument, and several of its members attended the workshop. To save money, they had traveled by train and bus, some all the way from Irkutsk, 4,000 miles away. All but the dapper Leonid Bezrukov, who hailed from Moscow and showed up in a fine tailored suit, arrived looking disheveled. (Bezrukov, who had been Baikal's lead experimentalist, referred to himself as an administrator and said he was no longer attached to the project. The Soviet Academy of Sciences had barely enough money to keep the lights on.) The others wore what was then the universal garb of the physicist: blue jeans and t-shirts or flannel work shirts—in their cases, threadbare. In spite of their manifest exhaustion, they presented many excellent papers, in English.

Out of respect for his pioneering achievements with DUMAND, the irrepressible John Learned was accorded the honor of presenting the introductory overview of the field and one of the future-related talks, near the end of the workshop, on his latest quixotic vision: a gigantic amoeba-like instrument made up of two concentric spheres of light detectors, buried either deep underground or deep in the ocean, which he called The Neutrino Eye. He wore a Jerry Garcia tie decorated with fish, to symbolize his faith in water as opposed to ice, and he bore a certain resemblance to the iconic guitarist, with his eyeglasses, his beard, his rounded frame, and his slightly zany smile.

John is not only the pied piper of neutrino astronomy, he is also the intelligent Forrest Gump. He was now collaborating on the Super-Kamiokande experiment, aka Super-K, the successor to Kamiokande, ten times the size, which had been completed in 1996. And he seemed to be on a sort of world tour with Yoji Totsuka, the leader of the project—who gave the most anticipated talk of the workshop. For, one month earlier, Super-K had announced the physics discovery of the decade: neutrino oscillation.

This is the wraithlike, shapeshifting process that Bruno Pontecorvo had first proposed four decades earlier, in which a neutrino changes flavor as it flies through space: a muon neutrino will transform into an electron or tau neutrino and then into a third flavor and so on. Since oscillation will only occur if the neutrino has mass, it violates the standard model of particle physics, which holds that the particle shouldn't weigh anything. This discovery was the first and is still the only glimpse into physics beyond the standard model and is a large part of the reason the neutrino is one of the major focuses of particle physics today.

The Super-K discovery also provided the first clue to a puzzle that had been vexing the physics community ever since Ray Davis and John Bahcall had first encountered it in the 1960s.

Recall that in the mid-fifties, after having failed to detect "reactor" antineutrinos at the Savannah River nuclear plant, Davis had decided to train his sights on the Sun. In 1958, he learned about some new theoretical estimates for neutrino emission by the Sun that seemed to put solar neutrinos within his reach. He proceeded to build what seems to have been his fourth thousand-gallon chlorine-based detector, just short of half a mile deep in a limestone mine near Akron, Ohio—and came up empty again. It turned out that the brightness estimates were accurate, but his method was about a thousand times less sensitive to neutrinos in the energy range given off by the Sun than he had thought.

Four years later, the young theorist John Bahcall came out with new estimates for neutrino production by the many different fusion reactions that take place in the Sun, each of which produces neutrinos of a specific energy. When Davis learned of Bahcall's work, he suggested he calculate not only production, but also the rate at which a chlorine detector would capture the neutrinos in each energy range, and about a year later, Bahcall came up with a prediction higher than any of the previous by about a factor of twenty. This was promising enough to prompt him to join Davis in proposing a larger (of course) 100,000-gallon detector. The proposal took the form of back-to-back papers, published in 1964 in *Physical Review Letters*, the first by Bahcall on the theory, the second by Davis on the experiment.

Bahcall introduced the so-called solar neutrino unit, or SNU (pronounced "snew") as a convenient measure for Davis's experiment. And as with most things neutrino, the SNU is rather small: 10^{-36} neutrino captures per target atom of chlorine per second. This was part of the reason they needed a large detector. Bahcall estimated that the Sun would emit about 7.5 SNU, which meant that a detector containing 10^{36} chlorine atoms would capture about 7.5 neutrinos every second. Davis's 100,000-gallon tank may have seemed large, but it was half a million times smaller than that; it contained "only" 2×10^{30} chlorine atoms. That meant he could expect to detect 7.5 neutrinos only every half-million seconds, which is slightly less than six days, which adds up to only about 460 neutrino captures in an entire a year. His immense challenge, therefore, demanding incredible purity in the target liquid and painstaking separation chemistry, was to sweep tiny numbers of argon atoms (the byproduct of inverse beta decay between an electron neutrino and a chlorine atom) from among the astronomical number of perchloroethylene molecules (his target liquid) in a tank about one-sixth the size of an Olympic swimming pool.

It says something about the wisdom of the physics community that the project was funded despite Davis's decade and more of null results (although, to put things in perspective, he once pointed out that the $600,00 it cost to do the experiment would have bought only about ten minutes of advertising on commercial television at the time). He installed his detector almost a mile deep in the Homestake Gold Mine in Lead, South Dakota, and revealed his first results in a talk to the American Chemical Society in September 1967.

The good news was that he had finally detected neutrinos. The bad news was that he had only detected about a third as many as Bahcall had predicted. This was a remarkable achievement, nonetheless, considering the complexity of the experiment and the exquisite sensitivity of the solar models. In subsequent years, Bahcall would point out that an agreement this close should have been a cause for rejoicing, since his estimate of the neutrino flux was proportional to the central temperature of the Sun raised to the 20th power!

But the physics community, as it will, focused on the two-thirds of the glass that were empty, rather than the one-third that was full. The tension between Bahcall's theory and Davis's experiment became known as the solar neutrino problem, although Bahcall liked to call it the solar neutrino opportunity, since it indicated to him that there was new physics in there somewhere—and it would turn out that he was correct.

The indomitable Davis continued to run his experiment, Bahcall continued to refine his model, and neither much budged. By the time of the Super-K discovery thirty years later, Bahcall's prediction was 7.6 SNU, and Davis's measurement was 2.56 SNU, which gave a ratio tantalizingly close to 3:1. It is a testament to Bahcall's intellectual integrity that he stuck to his guns: he never tried to modify his theory to match the experiment.

Bruno Pontecorvo had first suggested that neutrinos might oscillate from behind the Iron Curtain in 1958, about a decade before the solar neutrino problem reared its head. Being interested all things neutrino (and the inventor of Davis's method, after all), he also kept tabs on "the problem." Bahcall and Davis remembered him telling them at a conference in Leningrad, many years after they had published their proposal, that he had presented it at a special seminar in the Soviet Union at the time and that "he was the only person present who expressed the opinion that it would be a successful experiment."

Pontecorvo's thoughts on oscillation evolved as neutrino physics evolved. When he first came up with the idea, he suggested that the neutrino might transform into its antiparticle and back again. After the muon neutrino was discovered in 1962, he guessed that electron neutrinos might change back and forth into muon neutrinos, and since the Sun was expected to emit only electron neutrinos, he predicted—in advance of Davis's first results!—that "the flux of observable Sun neutrinos must be two times smaller than the total neutrino flux." The discovery of the tau lepton in 1975 meant that there should be three flavors of neutrino and the deficit should be precisely three, just as Davis was measuring. (By the time the electron neutrinos from the Sun travel all the way to Earth, they will oscillate back and forth enough times for there to be an equal mix of the three flavors.)

Pontecorvo did not live to see the discovery of oscillation. He died in 1993.

The Super-K collaboration made their discovery using atmospheric neutrinos. In other words, they used the atmospheric beam as an experimental beam, in contrast to AMANDA, which was using it as a test beam at the time. They did real physics with it—and impressive physics at that.

Their instrument could detect only electron and muon neutrinos, effectively speaking, and by chance, the atmospheric beam is made up predominantly of those two flavors. Without oscillation, the instrument should have detected the same number of each flavor coming up through the Earth as it did coming down from the sky, but it detected fewer muon neutrinos coming up than down. Since the up-going particles traveled farther to reach the detector—all the way through the planet in some cases—those neutrinos had more of a chance to oscillate into tau neutrinos, which were essentially invisible to the detector. The electron neutrinos did not oscillate significantly over that distance.

There was some complex logic involved in cementing their case, partially because it was tricky to work with the mixed atmospheric beam. (Manmade accelerator beams, which are pure and well-characterized, are often better suited to the pursuit of pure physics goals. It is sometimes said that cosmic rays are like illegitimate children: you can never be sure who their parents were.) But the collaboration covered all the bases. One smoking gun was that the fraction of muon neutrinos that disappeared depended upon the direction they came from: if they came straight up, which meant they had traversed the entire diameter of the Earth, more disappeared than if they came from closer to the horizon.

It seemed very likely that oscillation explained the solar neutrino problem; however, its simple discovery was only circumstantial evidence, not quantitative proof. It would take yet another large underground detector, the Sudbury Neutrino Observatory (SNO), to solve the puzzle quantitatively. This instrument consisted, as usual, of a 1,000-ton spherical tank of ultraclean water, buried more than a mile deep in a zinc mine in Sudbury, Ontario. What made it unique, however, was that the water was heavy water, in which the two hydrogen atoms in the water molecule are replaced by a heavy hydrogen isotope named deuterium. The water, which was worth several hundred million dollars, was on loan from the Canadian government's atomic energy corporation.

SNO could detect electron neutrinos coming from the Sun in the usual way, employing Cherenkov radiation. And the presence of the deuterium allowed it to detect the total flux of all three flavors simultaneously. In 2001, the SNO collaboration announced the solution to the problem that Davis and Bahcall had been working on for more than three decades: the flux of electron neutrinos matched Davis's measurement, and the flux of all three flavors together matched Bahcall's prediction.

Bahcall remarked at the time that he felt the way "prisoners that are sentenced for life do when a DNA test proves they're not guilty." "For thirty-three years, people have called into question my calculations on the Sun. . . . I feel like dancing!"

The following year, Davis shared half the Nobel Prize in Physics with Masatoshi Koshiba, who had been leader of the Kamiokande experiment when it detected the neutrinos from Supernova 1987a. The other half went to Riccardo Giacconi, for his pioneering contributions to X-ray astronomy. The prize is limited to three people, and Giacconi certainly deserved his share, but many still believe it was a travesty that John Bahcall was left out.

Davis was eighty-eight when he received the prize. Owing to his failing health, his son delivered his Nobel lecture for him. "The collision between solar neutrino experiments and the standard solar model has ended in a spectacular way," said the son for the father. "Nothing was wrong with the experiments or the theory; something was wrong with the neutrinos."

At the conference banquet in Zeuthen, a barbecue in the institute's courtyard, I sat across from John Learned, who was very much in his element, flush with the success of Super-K and friends with almost everyone there. He knew many of the Russians from the days before Baikal sprang from DUMAND, and he and Totsuka seemed to have a genuine affection for each other, beyond the dictates of protocol. Totsuka-san, a warm, large-hearted man and by all accounts an inspiring and effective leader, was clearly enjoying a high point in life.

Francis Halzen points out that the Super-K discovery was a big blow to the accelerator community and that it had a huge impact sociologically by prompting them to focus on the neutrino. Not only had a relatively inexpensive "non-accelerator" experiment made a momentous discovery in pure particle physics, it had revealed the first glimpse of physics beyond the standard model, a quest the accelerator scientists had joined roughly twenty years earlier.

It took almost another twenty "for the discovery of neutrino oscillations, which shows that neutrinos have mass," to result in a Nobel Prize. In 2015, the physics prize was awarded jointly to a younger member of the Super-K collaboration, Takaaki Kajita, and Arthur McDonald of the Sudbury Neutrino Observatory. By that time, unfortunately, Yoji Totsuka had died of cancer. McDonald, too, was a "second generation" member of his experiment. The visionary who had conceived of SNO and gotten it into the ground was Herb Chen of UC Irvine, a protégé of Fred Reines who had also died young of cancer. Experimental particle physics, especially neutrino physics, takes uncommon persistence.

Speaking of which, these were the tenth and eleventh Nobel Prizes connected to Bruno Pontecorvo's prescient insights.

15. The Peacock and Eva Events

FRANCIS KNEW THAT THEY NEEDED TO GET GOLD-PLATED EVENTS OUT OF AMANDA-B10 if they wanted to keep the momentum going for IceCube. His group set their sights on that goal as soon as John Jacobsen sent the filtered data around.

Wisconsin was the strongest group in the collaboration, owing in part to the immense resources of the institution and in part to Francis's way of attracting excellent students. (Inside the collaboration, Madison is known as The Evil Empire.) At least five people were working with the data: Albrecht Karle had moved the previous year from his post-doc in Zeuthen to an assistant scientist position at Madison. Gary Hill had taken up the post-doc Bob Morse had offered him after his winter. There were at least two grad students involved: Rellen Hardtke and Ty DeYoung (mentioned previously for his worship of John Jacobsen's code writing abilities). And Morse, who couldn't advise graduate students, since he wasn't a professor, had a knack for recruiting smart undergraduates, in this case a young man named Phil Romenesko. The most important strength of this group of five was that they worked well together.

As the collaboration returned to their separate homes from the Zeuthen meeting, they were still waiting for one last piece to the puzzle, the so-called string geometry: where the strings were located in relation to each other. This was a persistent question with AMANDA, owing to the madcap way in which they sometimes chose where to drill the next hole. Kurt Woschnagg had presented a geometry at the meeting, based on the timing of laser flashes broadcast between modules, but it conflicted with the admittedly slapdash surveys that had been done as the holes were being drilled, so he had to go back to the drawing board. He came up with a new geometry within a few weeks of the Zeuthen meeting, and the ground was laid for looking at the B10 data in earnest. Madison was ready to pounce.

Dennis Peacock, the head of Antarctic Sciences at the National Science Foundation, was scheduled to visit the campus on August 26, and Francis wanted to have something to show him. (AMANDA's original champions, Peter Wilkness and John Lynch, had left the foundation by this time.) The Madison group, mainly Albrecht and Ty, had developed a reconstruction algorithm and some primitive cuts and had tested them on about four and a half days' data, but when they scanned through the resulting events by eye they found nothing of interest. Phil Romenesko had then crunched through another thirty days, but by the day before Peacock's visit that group of events still had not been scanned.

Rellen was studying for an exam, and Ty was out of town, so Francis asked Gary to look through Phil's sample, and Phil managed to join him. At eleven in the evening on the night before Peacock's visit, Gary sent the following e-mail to the rest of the Madison group:

Subject: The "moment of truth" arrives . . .

Gidday all, well we spent about 5 hours last night scanning through 564 leftover events from the 30 days data reconstructed with the new geometry . . . in fact we went over them at least three times . . . just to be sure.

. . . what did we find? Well, I think you should each experience your own "moment of truth" (we certainly did!), so check them out . . .

Have fun . . . Gary and Phil . . .

P. S. Although we really think everyone should look through all the events, we did condense them slightly into the following file: /d13/ghill/data/mot .rc.f2k . . . in case you want a subset to show the NSF chap . . .

Francis did not check his e-mail the next morning. Peacock came into his office, and they spent "the usual half-hour talking about this and that and about IceCube and the future, all the big programs like you talk about with big bureaucrats. And then, you know, we said, now we're going to show you the status of the data analysis, and . . . I sit down with him at Rellen's work station . . ."

Rellen *had* checked her e-mail. She ran the two beautiful events that Gary and Phil had found across her computer screen.

Francis recalled that moment for me during a collaboration meeting the following November: "You cannot imagine how exciting this was. . . . Rellen shows these events, and I see them, and . . . you know, you look at these events and you know that this is something—this is like cocaine, right? I mean, this is what you have been waiting for ten years for. . . .

"The events we found with B4 . . . I mean you would bet your wallet, but not, you know, the head of your son. . . . On these I would have bet the head of my son.

"You know my first reaction? This is in really bad taste. I thought they were playing a joke on me. Seriously. But Rellen is not the type, right? . . . None of them. . . . They're not as serious as they look, but they wouldn't do that. I think they're not confident enough to play a joke like that on me. So I didn't say anything. And it was only afterwards, I talked to Albrecht, I realized. . . . I was floating on air for weeks."

And he laughed. To this day, Francis insists that this was the most exciting moment in the life of this project.

Sometime in there, he also observed that "theory is just a sideshow, really," and admitted that AMANDA's moment of truth may have been more exciting for him than for the experimentalists in the collaboration, since he had never experienced the direct contact with hard reality that one does in an experiment. Some call it "touching the mystery." Albert Einstein told friends that something in him snapped and he experienced heart palpitations for several days when he discovered that his general theory of relativity explained a previously inscrutable anomaly in Mercury's orbit about the Sun.

The more convincing of the two events was named the Peacock event (see photograph 14). As Francis wrote in an essay that was included in *Best American Science Writing 2000,*

> Hardtke's screen showed a faint blue line streaking diagonally across columns of black dots. Most of the dots, each of which represented a

photomultiplier sunk in the ice, were small and black. But a few, clustered along the line, were blue or green or red, and two, near the beginning of the line, were bright orange and very large. At five in the morning on October 12, 1997, the diagram told us, a neutrino—one of nature's smallest and most elusive elementary particles—had entered the earth in the middle of the Pacific Ocean, between Midway Island and the Aleutians, hurtled straight through the planet, and collided head-on with a proton on the underside of the Antarctic ice. Two kilometers beneath the surface, our grid of photomultipliers had picked up a subatomic spark from that collision as it flew upward through the ice and flared past them for about a microsecond.

Gary was first to "leak" the news. That afternoon, he e-mailed Robert Schwarz, who was about to witness his second sunrise in a row at Pole:

> Hi Bert, well I wanted you to know asap—looks like we found real upgoing neutrino events in the B10 data we took last winter . . . so it works! . . . Francis was really worried that we were never going to see any good events, and what he was going to tell NSF today. Well [he] saw them for the first time during the showing of the events to NSF—so he is very excited, and the NSF chap was very impressed . . .

About half an hour later, Albrecht sent an e-mail with a link to an image of the event to the entire collaboration—and was met, strangely enough, with resonant silence. There were only two responses, one from John Jacobsen, which was somewhat skeptical (Gary disabused him of that), and the other from the normally skeptical Christian Spiering, who went the other way, writing "Boah—das sieht ja wirklich irre aus," which Albrecht translated for the rest of the Madison group as "Woooow—that looks really crazy [incredibly good]."

On that very day, amazingly, Fred Reines died. Francis received the news by e-mail from two of the great man's Irvine protégés and forwarded it to the rest of the collaboration.

Two weeks later, on September 11, the collaboration received an e-mail under the title "Have a look at this" from Eva Dahlberg, a grad student in Stockholm. It directed them to an image of a near dead-vertical muon that had zoomed up through the detector, right next to one of the strings, lighting up each in succession, at 97 percent of the speed of light. That would be 25 percent faster than light itself could have traveled through the ice, given its index of refraction.

Eva was also met with silence. Only Christian responded. He allowed himself one moment of ecstasy: "Oh oh oh oh! Eva, this is the absolute hammer," and swiftly reverted to his habitual skepticism: ". . . or could it be cross-talk?"

Cross-talk was an irritating byproduct of AMANDA's analog electronics in which the electrical pulse produced when light hit an optical module lower down on a string sometimes "talked" to the higher modules on the same string as it passed by them on its way to the surface. This caused them to send out secondary pulses that were hard to distinguish from the real thing. Gary Hill observes that "cross-talk would come to dominate our lives later, but this event was surely not cross-talk." The fake, cross-talk signals moved up the string more slowly than muons did, while Eva's event traveled at just the expected speed. The fake pulses also had a different shape than those produced by the light emitted by a muon.

The watershed moment for the collaboration as a whole came on the first day of the next collaboration meeting, which took place in Madison and began on a Sunday, November 1. Everyone was anxious to see the latest results, so they scheduled the B10 analysis session for Sunday afternoon.

By now, the Madison group had processed sixty-one days' worth of data (all odd days, the even days remained blind). They presented the results in three back-to-back talks.

In those days, the resident sceptics in the collaboration were Steve Barwick and Ralf Wischnewski, Christian Spiering's right-hand man in Zeuthen. Christopher Wiebusch was also quite vocal, but his motivation seemed to be less about skepticism than about demonstrating how smart he was. The dynamic during science talks was that these three would sit, or even stand, in separate corners of the room and hector the speaker— especially if he or she had good news to report. Although the haranguing could be prolonged and tedious at times, most of the members of the collaboration agreed that it was valuable. Skepticism is the very basis of good science, after all. Steve was especially sharp.

Albrecht gave the first talk. He presented a loose set of automated cuts that netted nineteen events, which he had then scanned by eye to identify nine high-quality neutrino candidates. Steve and Ralf groused a bit as

Albrecht spoke: it isn't kosher to cherry-pick data by eye, because it allows you to find what you want to find.

Ty walked to the lectern next, and as he began to speak a rapt silence fell upon the room. Even Steve and Ralf were rendered speechless. Ty presented an entirely automated analysis with a stricter set of cuts than Albrecht's, yielding the same nine events. The Peacock event cropped up in the first thirty days and the Eva event in the second thirty-one. Since his approach was run entirely by computer—no biased human eyes involved—it was the first indication that they had a working instrument in their hands.

There was some nitpicking, but most of the questions were respectful and complementary, pointing out small concordances, for example, that supported Ty's conclusions.

Gary closed by presenting a Monte Carlo predicting that the instrument should have seen, on average, thirteen events in sixty days: reasonable agreement. It was a complete package, and the three talks were notable for their clarity and cohesion. I have noticed over the years that lucidity and simplicity of argument are hallmarks of the presentations from Madison.

"In my mind," writes John Jacobsen, "these were the first absolutely compelling high-energy neutrino events found in AMANDA—compelling in that you could just look at them and see what they were without having to squint too hard, mathematically or otherwise. Getting our act together to turn the 246 raw 1997 data tapes into those gold-plated events required not just the larger B10 detector, but also for everyone to 'level up' both technically (software/analysis) and to work better as collaborators." In view of the intense competition and politicking that would follow, it is important to keep the truth of John's last point in mind. The Madisonians were not working in a vacuum. They had synergized tools and insights from around the collaboration. In a large science enterprise of this sort ideas are always "in the air." It is never easy to pinpoint exactly where a breakthrough comes from—although scientists themselves don't always remember that.

Over dinner in an unpretentious Italian restaurant on the evening of the presentations, Francis told me that he was not only thrilled at the new state of affairs, he was greatly relieved. "I can tell you I spent a lot of sleepless nights in the last ten years thinking . . . it was clear who was going to take the blame in Washington if this failed. There should be some fairness, right? I mean . . . I would have never gotten a cent anymore."

He then went out of his way to give credit to "the original DUMAND people." "I mean the vision they had! . . . You know, every sane experimentalist would never touch this subject as being too hard and too extravagant, too risky. They took the decision that they were actually going to do this, and that was very brave. Everybody after that is kind of a follower, right?"

Synchronistically perhaps, John Learned was passing through Madison just that week. "This is the first time that there's data that is really, I think, convincing from AMANDA that shows that it's probably going to work," he told me. "So this is a major breakthrough. There's no real science in what they've seen, but it's a technology demonstration, and . . . they're, you know, manfully trying to go ahead and get out a little bit of science from it. But the importance is that—it's like the bit about Dr. Johnson's dog; it's not that it speaks badly, but the fact that it speaks at all that's important. . . . I think some of us have been convinced that it was gonna work for a while, but there's nothing like seeing the cards on the table."

✦ ✦ ✦

You might be surprised to hear that some physicists are actually afraid of discoveries. They find them upsetting. A breakthrough changes the way they've been doing business. It knocks them out of their set pattern of plodding along and perfecting things, finding fault, being skeptical. I'm not sure that this is more prevalent in high-energy physics than it is in other fields, but I suspect that it might be. For these people are dealing with exceedingly esoteric concepts and their results are usually expressed in statistical terms. The first glimmer of a discovery usually reveals itself as a small effect that has risen ever so slightly above a sea of noise. You have to be careful not to fool yourself. After the initial excitement, a fear of exposure sets in. What if you got it wrong? You don't want to embarrass yourself in public. And this has happened many times.

There was a range of responses inside the collaboration. On one end were the Madisonians, who were quite sure that this was the real thing, and at the other were the Germans. Once Ralf and Christian got over the shock of hearing such good news, they dug in their heels and refused to accept it. At a basic level, it seemed to be a kind of existential pessimism. "It's not paranoia," suggests Francis, "it's that they feel that they don't deserve it when good things happen, so they get very excited about bad things. . . . We know things must be going well when the Germans get excited."

Ralf, in particular, seemed to revel in bad news. This is not to say that he didn't make important contributions. His primary responsibility was the data acquisition system down on the Ice, and in that role he made sure that the instrument was capable of taking data every year. (On the other hand, he was about as excitable as Steve Barwick, and at the tail end of the season he and Steve were usually at Pole together, trying to complete their respective tasks—Steve's being calibration—so the last few days before station closing tended to be excruciating for the winterovers.) But his continual harping on everything that was wrong with the instrument crossed the line into sheer negativity at times.

Competitiveness was also a factor, and there was probably an admixture of arrogance as well, since the Germans had been working on Baikal for so many years that they tended to think of everyone else in AMANDA as newcomers. Christian has also made enormous contributions over the decades, but it's funny how it has always seemed unclear exactly where his allegiance lies. While he and Ralf seemed to have no problem accepting Baikal's up-going events, which had been found with a smaller, eight-string detector, and while they were okay with the questionable events from AMANDA-B4—in fact, Christian took the lead in writing the B4 journal article—they were absolutely convinced that most of the "Madison events," as they called them, were fake. And it seemed that Christopher was incensed that another group had stolen the lead again.

Some collaborations work on the dictatorship model (Carlo Rubbia comes to mind), but AMANDA, in its anarchism, was inherently democratic. And Francis, furthermore, had no interest in dictatorship. He has an overriding belief in the power of diplomacy and getting along. This can sometimes be taken as superficiality, but it has also worked wonders in the collaboration over the decades. The harsher and more aggressive people have slowly calmed down or wandered away for one reason or another, and nowadays most everyone in IceCube says it's the friendliest collaboration they've ever worked in. Francis points out that it's the only collaboration he knows of "where everybody can have dinner with anyone else in the evening. . . . In most collaborations there are people who explicitly won't sit at the same table."

Besides, disagreement has its uses. The journal article announcing a discovery will always be more solid and fully realized if it's been vetted internally, all views have been aired, and consensus has been forged. After ten years of dogged effort, Francis knew better than most that detecting a ghostly particle in the presence of an enormous down-going background with a detector they couldn't even see was a genuinely difficult business. A sanity check would not only help, it would serve constructive political ends. Everyone realized that the Germans were bound to run their own analysis again anyway, so Francis, ever the peacemaker, beat them to the punch. He suggested they go ahead.

This "adversarial scrutiny" approach (Ty DeYoung's phrase) is common in complicated physics experiments. It's basically about making sure that results are reproducible. This ideal is not always reachable in one-of-a-kind experiments that cost millions or billions of dollars. Francis points out that the Nobel Laureate Samuel Ting, who is Chinese by heritage, always has a Chinese group do one analysis and "the rest of the world" do a second, and instructs the two groups not to talk to each other. The physicists who discovered the Higgs boson went a step further and built two entirely separate detectors managed by separate collaborations, consisting of thousands of people. This is a Machiavellian tactic, which ought to be managed benevolently, however; and Francis and Christian did not do that. They set up *competing* analyses, not separate ones.

The reality was that the Germans couldn't possibly run a separate analysis, because they were using many of the same tools as the Americans and communicating with them almost every week. The dynamic that arose when they got underway was that the Americans, mainly Ty and Gary, were put in the position of defending a claim that they thought was correct, while the Germans, mainly Christopher, were trying to tear it down. Now that emotions have cooled, Ty will admit that the scrutiny from Zeuthen helped a bit. But the truth is that it caused him, Gary, and Christopher enormous anguish for quite a few years. Only recently have Ty and Gary been willing to talk about it, and Christopher still hasn't opened up.

Probably because he was the instigator. Christopher's criticisms took on a personal tone, they were aimed at getting under the Madisonians' skin, and they continued, actually, long after this particular bone of contention had been chewed to bits. For several years, he took pot shots at Gary and Ty in every possible forum: collaboration meetings, e-mails, phone calls, and at first they responded defensively. Without going into the details, let's just say that there was regrettable behavior on both sides and that it is surprising that neither Francis nor Christian put a stop to it. Eighteen years later, they still chuckle at the pain their little management experiment caused, while the scars on their students remain just below the surface.

Once the Germans made progress and began dishing out their criticism, Gary and Ty got their dander up and began competing, too. And while this did spark a tremendous burst of creativity, it also had the ironic effect of making the two analyses only more intertwined. Ty says, "There was a lot of cross-pollination, since everyone was forced to adopt every new tool as soon as it came up or fall behind in the arms race."

By January 1999, the Madison group had run all the odd days from 1997 through their constantly improving automated procedure and found seventeen neutrino candidates, one of which they identified by eye as being fake. By adding an automated improvement that had been developed in Zeuthen, they showed that they could eliminate the fake and thus improve the method. By summer they had fifty neutrino candidates.

Astoundingly, for more than a year, the Germans insisted on ignoring the progress in Madison and continued to carp about the seventeen events that had been revealed in January. They attempted to take the high ground by choosing not to scan events visually or even use real data to develop their methods! They relied entirely on Monte Carlos, using a simulated instrument with a simulated atmospheric neutrino beam, and applying it to the real data without looking at the data in advance. This way, even though they were aware of the Madison events, they could argue (speciously) that they were keeping themselves blind and could look at both the odd and even days.

In retrospect, the Zeutheners were demonstrating their lack of experience in high-energy physics. Since the mainstays, Christian and Ralf, had come of age behind the Iron Curtain, it seems that they were unaware of the standard techniques in the field. The Madisonian way of doing things is the accepted approach in most high-energy physics experiments and in Ice-Cube today. When the collaboration has a new set of data or a new question they would like to ask of an old set, they un-blind a so-called burn sample, a small fraction of the dataset, and develop their analysis methods with that. Then they freeze the methods and run them on the rest of the data, which has been kept blind. A discovery so obtained has the highest possible statistical value, because the question has been asked only once without a previous look at the data. If the method is changed and run a second time on previously un-blinded data, any discovery will have less statistical significance. This gets into the realm of *a priori* and *ex post facto* statistical inference, a slippery subject about which the experts in IceCube still argue until the cows come home.

No one argues that Christopher is not a fine physicist. He and two young Zeuthen colleagues produced a solid competing analysis by fall. Running it on both the odd and even days, they found 116 neutrino candidates and—crucially, one would think—confirmed the nine-month-old Madison analysis that had come up with seventeen.

But Christian and Ralf continued to dig in their heels, and Ralf remained obsessed with the seventeen events. At a meeting that took place in February or March of 2000, five or six months after Christopher and his colleagues had confirmed the Madison analysis, Ralf dragged an audience through a minute examination of the seventeen and explained how each could have been fake. Some of his arguments were so over the top that even Christian didn't buy them. He was so buffaloed by the Eva event, for example, which was probably the most unassailable neutrino candidate in the sample, that his imagination failed him and he ascribed it to random noise.

Things seemed to be at an impasse.

The subtext to this little drama was that it was just one front in the ferocious political maneuvering that had commenced when it began to look as though IceCube might become a reality. This brought up even stronger emotions than the prospect of a discovery. Most of the Europeans wanted to delay the transition, realizing that since the lion's share of the funding would come from the National Science Foundation, a U.S. institution was bound to lead it. They feared that they might be marginalized.

The only two institutions that stood a realistic chance of taking the lead were Berkeley and Madison, and on the face of things Berkeley had the upper hand. For large physics projects aren't usually managed by universities, they're managed by national laboratories, and the Berkeley campus ran its own laboratory, Lawrence Berkeley, in its own backyard.

A year or two earlier, when NSF had solicited proposals for a second round of Science and Technology Centers, a program manager in the physics division named Gene Loh had encouraged Buford Price to fashion a center around AMANDA and IceCube. (Loh, an alumnus of Kenneth Greisen's Fly's Eye experiment in Utah, had more or less founded NSF's particle astrophysics program.) The idea was right up Buford's alley, since he loves big interdisciplinary ideas, and he remembers this as one of the more exciting and stimulating periods of his life. By early 1998, he had come up with a vision named DeepIce, which would combine research in astrophysics, through AMANDA and IceCube; paleoclimatology and geology, through ice cores and glacial seismology; and even microbiology: there was great interest at the time in looking for unusual life forms in the subglacial lakes under the Antarctic ice sheets, and the drilling of ice cores could provide access to them. The center would be administered by Berkeley, Lawrence Berkeley Laboratory (LBL) would play a large role, and IceCube would be the largest program in it. This was not necessarily a power grab by Buford—there was even talk of Francis moving to Berkeley and becoming principal investigator—but it *was* in the interest of his university and LBL, and he received much, shall we say, encouragement, especially from his LBL colleagues.

At about the same time, separate from DeepIce, LBL officially joined AMANDA. This brought in two heavyweights in addition to Dave Nygren, who was already contributing with his digital optical module: William "Willi" Chinowsky, an experienced particle physicist who had worked as an NSF administrator for a time, and George Smoot, who was reasonably well-known to the public—and notorious among physicists—for his participation in the Cosmic Background Explorer experiment (COBE), a satellite-based instrument that had made the most important advance in the study of the Cosmic Microwave Background Radiation since Penzias and Wilson had accidentally discovered it in 1965.

In the early nineties, COBE (as in Kobe Bryant), had demonstrated that the microwave background was not uniform across the sky, it rippled. And since it is believed to be a remnant of the Big Bang, the ripples were interpreted as being primordial fluctuations in the structure of the newborn universe, which had evolved, as the universe expanded and cooled, into stars, solar systems, galaxies, galactic clusters, super clusters, and so on. This was a huge idea, and Smoot made world headlines by breaking the rules of his collaboration and stepping into the limelight to claim the discovery as his own. At one point he told a gaggle of reporters that looking at COBE's map of the ripples was like "looking at God." The outrageous politicking he did in order to claim his place in the sun earned him enormous resentment within his collaboration. There was even talk of lawsuits.

The LBL group joined AMANDA with their eyes directed at IceCube and in the firm belief that Nygren's digital optical module was the only viable technology for the larger instrument. Smoot managed to obtain a $300,000 grant to develop the DOM, and in October 1998, a month before Albrecht, Ty, and Gary presented AMANDA's first gold-plated events in Madison, the Lab held a workshop that resulted in a decision to deploy one entirely digital string in AMANDA during the 1999–2000 season.

Meanwhile, the potentates at NSF were becoming increasingly serious about IceCube. We can be sure that Francis was keeping them primed with every new discovery in the AMANDA-B10 data, but they also had larger considerations in mind.

Near the beginning of 1997, the foundation had convened an external panel chaired by Norman Augustine, a former Undersecretary of the Army who was now chairman and chief executive officer of the Lockheed Martin Corporation, to evaluate the entire U.S. Antarctic Program. In the report they produced in April of that year, the panel had stressed the geopolitical importance of the "U.S. presence in Antarctica, particularly at the South Pole" and suggested that "this consideration, *in itself,* justifies a year-round presence at several locations, including a moderate-sized facility at the Pole. . . .

"National prestige is involved in participation in activity in Antarctica, particularly at the South Pole, much as there is in involvement in the space program. . . .

"Antarctica today is a continent generally characterized by peaceful, environmentally friendly, human activity. High among the reasons for this situation is the role played by the U.S. over many years in helping create a system of treaties and international agreements governing the nature of human conduct on the continent. The presence of the U.S. in Antarctica is a key element of the continued stability of the region."

The Antarctic Treaty had been created during the International Geophysical Year, a grand international research effort that took place at the height of the Cold War, in 1957 and 1958. It focused on the exploration of Antarctica, the upper atmosphere, and outer space. Sputnik was launched under its auspices. When the treaty came into force in 1961, it became the first arms control agreement of the Cold War.

Originally signed by the twelve countries that had a significant presence on the continent (there are now fifty-three treaty-states), the treaty turned Antarctica into an international peace, research, and conservation park. It prohibits any nation from establishing a military presence there, and it neither recognizes nor disputes the territorial claims, some of which overlap, that have been made by seven different nations. (Argentina once sent a woman who was seven months pregnant to its Antarctic base, hoping to buttress its territorial claim with the argument that an Argentine had been born there. Their archrival, Chile, then went one better by arranging for a Chilean child to be both conceived *and* born in Antarctica.) The stations on the continent are to be used for peaceful purposes, mainly scientific, and while they tend to be run by individual nations, all treaty-states have free access to every station.

The territorial claims, which are postponed as long as the treaty remains in effect, divide the continent into pie slices, with the center of the pie at the South Pole. The U.S. Amundsen-Scott Research Station is thus located at a strategic point.

At the AMANDA collaboration meeting at UC Irvine that I was barred from attending, a grad student from Stockholm named Patrik Ekström told me something interesting about the airstrip at the station: at one end, it ran in a circle around the marker for the geographic South Pole. Every time a U.S. plane landed or took off, he told me, it taxied around the circle, thereby treading on every territorial claim except Norway's. To Patrik, who was not a U.S. citizen, the planes seemed to be saying, "This place may not be ours, but it's not yours either."

The Augustine Panel's principal conclusion about the *entire U.S. presence in Antarctica* was that South Pole Station needed to be replaced "for economic, safety, and operational reasons." The following winter, Congress and the Clinton administration approved funding to the tune of $128 million, and construction began during the 1998–99 summer season. The new station would be completed ten years later, at a final cost of $174 million.

Although science was seen as a secondary objective of the U.S. Antarctic Program, the fact was that most of the people who worked on the continent saw it as their mission either to do science or support it, and considering its middle name, it was difficult for the National Science Foundation to justify an expenditure of more than $100 million simply for "presence." Thus the foundation needed a big project at Pole, and an *international* project would fit U.S. objectives particularly well. Under the "International Cooperation" subheading of their terse list of recommendations, the Augustine Panel singled out AMANDA specifically, noting that it was not only the largest science project at Pole, it was also the most international. Since IceCube stood to be much larger and more international, it fit the bill even better.

These considerations were undoubtedly in the minds of the NSF managers in the spring of 1999, when, on short notice, they invited the principal investigators of AMANDA to a meeting at a hotel near the foundation's headquarters in Arlington, Virginia, to take place on Monday, May 10, and made clear that its purpose was to discuss IceCube in practical terms.

About a week before the meeting, George Smoot did an end run around the rest of the PIs and secretly sent a white paper directly to Gene Loh. If it was anything like the proposals he later showed to the collaboration in the open, it would have made himself principal investigator and Lawrence Berkeley Laboratory the lead institution, and paid scant heed to European participation. Loh must have shared it with Francis or Buford, for on the Friday before the meeting, all of the original PIs got together and disinvited Smoot and Chinowsky.

Nevertheless, the meeting was a turning point. John Bahcall traveled down from Princeton to give an overview of the scientific possibilities: "I rarely give up time from talking about solar neutrinos, so the fact that I'm talking about high-energy neutrinos shows how excited I am about them." Francis presented the latest results, sweetened with his unquenchable optimism. Several institutions besides LBL that had either joined or petitioned to join IceCube sent representatives to demonstrate their support. (According to Francis, they'd been waiting in the wings to see if it would work. "You just watch," he told me. "Now, they'll be beating down the doors.") Christian Spiering reviewed the water-based efforts in Europe, and he and Per Olof must have been somewhat mollified to see how much the Americans planned to include them. Buford and Francis were wise enough to realize that this had to be an international effort, because in point of fact, good science always is. There had been some question as to whether Steve Barwick would show, as he had a double teaching load that semester, but he did in the end, and he was given the job of talking about logistics and budgets.

At that point Francis was still holding out hope that Steve would be able to assume the enormous responsibility of becoming project manager of IceCube, the technical and managerial lead responsible for getting the instrument built. Steve certainly had the technical chops. Unfortunately, however, he demonstrated once again that day his limitations in other areas. As was frequently the case, he froze the night before the meeting, and it was unclear until the last minute whether he would actually give his talk. Bob Morse had to hunker down with him in his hotel room to work on it, and the two walked into the meeting room several minutes late. Steve gave a stiff but competent presentation.

Despite the shenanigans in the background, the managers were impressed. They had seen internecine warfare before. The following afternoon, they held what was ostensibly a closed meeting with Buford, Bob, and Francis—into which Willi Chinowsky, George Smoot, and Doug Lowder marched unannounced. (Willi must have pulled some strings with his former colleagues at the foundation.)

And so, as they looked into the eyes of a fractured collaboration, the NSF managers officially embraced IceCube: they solicited a proposal due the following November first.

At that point no one knew exactly how much a kilometer-scale instrument might cost, but it was clear that it would be in the hundreds of millions, and this was far too much to be financed in the usual way. The managers explained that they wanted to fund the project through a special NSF-wide account dedicated to large infrastructure projects, known as the Major Research Equipment and Facilities Construction account, or MREFC. This mechanism appealed to the managers in specific subject areas, such as Dennis Peacock in Polar Programs and Gene Loh in Physics, as it would remove a large number from their individual budgets. It had the disadvantage, on the other hand, of making the project highly visible. It would actually appear as a line item in the budget that the foundation would present to Congress every year, so it could become a political football—as indeed it would.

Take it all around, of course, this was an excellent development, although it also added to the workload. At that point in that hectic year, the principal investigators were beginning to sag under the load. As he sat in the hotel bar on Monday evening, after they had made their presentations, even the indefatigable Christian complained of being overworked. And I have never seen Francis so agitated. He said that if the meeting hadn't been imminent Smoot might very well have been kicked out of the collaboration.

They were bouncing like pinballs all over Europe and the United States. They'd held a collaboration meeting in Berkeley the previous month; they'd hold another in Uppsala the next. After much hand-wringing, Christian was nearly ready to submit the journal article presenting the up-going events they had discovered with AMANDA-B4. (He sent it to *Astroparticle Physics* a week later, and in the relieved e-mail he sent out in order to share this signal moment with the rest of the collaboration, this unreligious man included a cross fashioned from x's.) They were planning to deliver a dozen or so papers at the International Cosmic Ray Conference in Salt Lake City in August. They were readying the many tons of hardware and mobilizing the manpower for the upcoming season at Pole, which looked to be their most challenging yet. And now they had a major proposal to write.

On Tuesday afternoon, as the meeting in Arlington came to a close, Gene Loh gave them a "non-negotiable" deadline of June 1, three weeks hence, for the first step in the proposal process: a letter of intent. Francis sent out an e-mail before he left town, suggesting who should be responsible for the different sections. As an example of his ability to let bygones be bygones, not only did he include George Smoot, he even suggested they use the rogue white paper that George had submitted the previous week for ideas.

Steve Barwick responded instantly and angrily with the circular argument that he didn't have time to work on the document and that if it was submitted without his input, he would withdraw his name. His closing words sounded dangerously close to his annual threat to undermine the approaching Pole season: "I have a hardware program to run and we must spend $800k during the next few weeks or you can kiss [AMANDA] goodbye for this year."

The other Californians and the Europeans had their reservations as well, so, in view of the deadline, Francis gave up and went it alone. He wrote the letter of intent himself and was the only person to sign it, and he finessed the politics by focusing only on the essence of the idea, not project management. He named no institutions, even Wisconsin, figuring this was the least damaging way to go, since it would signify that "nobody was really responsible" and that he would be seen simply as "a theorist who [was] out of control."

In late July, Buford learned that DeepIce would not be funded, for the main reason that IceCube, which was by far the largest project in it, stood a good chance of being funded on its own. He let the rest of the PIs know in an exceptionally gracious e-mail.

Now, the realpolitik at this early stage was that formal European support was not strictly required. The proposal was basically an agreement between the lead institution and NSF, so all it really needed in order to signify international interest was a list of potential partners. This meant that the Europeans didn't need to be involved in the writing of the proposal, which made the process that much less fraught.

Somehow, the principals from Madison and Berkeley agreed to write a proposal together, and the way they got that ball rolling was to sequester themselves in a hotel room that August during the International Cosmic Ray Conference in Salt Lake City, and hammer out a first draft over the course of about a week. They would go over to the meeting in the morning to rub shoulders and have breakfast with their friends, they'd hole up in the hotel room during the day, and they'd return to the meeting in the evening to meet friends again and go out to dinner. Francis asked one of the grad students from Berkeley to give his talk about the status of IceCube for him, and the student pointed out in his talk that one aspect of that status was that Francis was in a different hotel working on the proposal as he was speaking. (Somewhat more significantly, Albrecht Karle reported on the seventeen neutrino candidates they had uncovered six months earlier and mentioned that they now had a grand total of fifty. This made AMANDA the clear front-runner in what was now no longer a race.)

Madison was represented in the hotel room by Albrecht, Bob Morse, Gary Hill, and Francis; Berkeley by Buford; and LBL by Willi, George, and Dave Nygren. Their competing ambitions notwithstanding, this was a strong group, and what Willi and George brought to the table was long experience with large and expensive projects. COBE, the Cosmic Background Explorer, had run to more than half a billion dollars. George and Willi knew about things like management structures, and they weren't cowed by large numbers. They also worked at a national laboratory, and as Francis points out, the default assumption was that a project of this size would be run out of a lab. The labs are run by the Department of Energy, however, while this project was being funded by the National Science Foundation, and as Francis also points out, "in Washington everybody hates everybody else." So NSF was leaning strongly toward taking a risk on the University of Wisconsin rather than "god forbid having a DOE lab do an NSF project."

It seemed that Willi and George were happy to play along with this idea, in the confidence that all involved would see the good sense of locating the project at their laboratory in the end. Everyone knew it would be a stretch for Wisconsin—or any university—to carry off a project of this size and that NSF didn't have much experience with big projects either. Indeed, the foundation was very anxious just then about another megaproject in their pipeline, the project Serap Tilav had moved to, the Laser Interferometer Gravitational-Wave Observatory. LIGO had been initiated by three professors at two assuredly capable technical institutions, Caltech and MIT, but the professors had proven incapable of managing the project, even with significant support from their institutions. After a decade of false starts, the foundation had hired a new project manager, and the LIGO collaboration had finally demonstrated feasibility with a relatively small instrument, analogous to AMANDA, at around the time AMANDA detected the Peacock events. By 1999, the proposal for "Advanced LIGO," which would end up costing more than $1 billion, was only slightly ahead of IceCube's.

Francis was perfectly willing to admit that *he* couldn't manage the project. In fact, he had no desire to: he wanted to keep himself free to do physics. He was also wise enough to realize that the project was far too big to be run in the context of Madison's physics department or even the College of Letters and Sciences, which housed it. Shortly after the May meeting with NSF, he had broached this radical thinking to Terry Millar, the associate dean of physical sciences in the Graduate School, who was also an out-of-the-box thinker. (Terry, an energetic ex-Marine Vietnam veteran with a doctorate in mathematical logic, had been working with Francis and John Wiley, dean of the Graduate School, to keep AMANDA afloat since the early nineties.) Millar had responded by suggesting the Space Science and Engineering Center, a unique institution on the Madison campus that had considerable experience carrying out $10 or $12 million satellite projects for NASA and NOAA, the National Oceanic and Atmospheric Administration. Francis and Terry had then recruited an SSEC mechanical engineer-cum-project manager named Bob Paulos, with a view to making him project manager for IceCube.

Paulos took on the main responsibility for pulling the proposal together. He remembers George Smoot and Willi Chinowsky as being willing to go only so far in helping him out, however: it wasn't in their self-interest, necessarily, to make Madison look good. They also displayed a certain arrogance. "There was always this attitude like, 'You're out in hay country out in Wisconsin, you know, let the big kids handle this,'" says Bob. "One time, one of them . . . said something to me like, 'Why don't you guys just build the drill and let us do everything else.' They should take care of the tough stuff. Us farm boys, we could probably build a hot water drill." Evidently, the Berkeleyites didn't realize that the drill was probably the greatest technical challenge in the project.

The proposal was submitted on time in November 1999. It designated

Madison as the lead institution, Francis as principal investigator, and Bob Paulos as project manager. The project was expected to cost about $140 million to build and operate over the next five years, and the plan called for the first IceCube strings to be dropped into the Ice over the 2002–03 Antarctic season. Shortly after the proposal was received, Dennis Peacock at NSF told Francis it had passed with flying colors.

That's when the chess match really got interesting.

16. Y2K at Pole

AMANDA WAS NEVER SEEN AS MORE THAN A FEASIBILITY STUDY. IT HAD fulfilled its essential purpose by detecting the Peacock events. But it was also a testbed for technology, and the collaborators always held out the hope that they might make a real discovery with it. They had a working instrument in their hands, after all. They had opened a window on the neutrino universe, however small, so there was at least some chance that they'd be able to see something through it. This possibility was in nature's hands.

Understand that the holy grail in neutrino astronomy is a "point source" or "neutrino star," any single object that emits neutrinos—either steadily, like the usual star; periodically, like a pulsar; or just once, in a single blast, like a supernova or gamma ray burster. Bob Morse puts it best: "I think of myself as a cave man, stepping out of my cave, pointing at the sky, and saying, 'Look at those stars!' All I want is to see one point source of neutrinos up there. Then there are the folks who want to measure things. They're saying, 'That one's blue, and that one's red. . . .' It's like existence and essence."

The collaboration is always gripped by "point source fever." Whenever they produce a new tranche of data or enhance their instrument in some way, they search hopefully for point sources of any kind. Since their data is statistical in nature, this leads to many emotional ups and downs, because chance fluctuations nearly always produce random hot spots in the neutrino sky.

In 1999, for example, the Madison group thought it had found a point source among the Peacock events and their siblings in the data from AMANDA-B10. The even days in the dataset were kept blind, remember; they were working with only the odd days. At an October collaboration meeting in Philadelphia, about a month before the IceCube proposal was submitted, Ty DeYoung presented a sky-map (the neutrino version of a map of the visible stars in the night sky) with sixty up-going events and showed that six of them—10 percent!—were clustered at a particular spot in the sky. This was so exciting that more than fifteen years later Ty still remembers the exact coordinates of the spot. He petitioned the collaboration for permission to "un-blind" the even days in order to find out if the hot spot was real. "Steve [Barwick] in particular gave me a hard time at the presentation," he writes, "but no one found anything really problematic in the analysis, so we opened the box on the other half of the data, and found . . . nothing. Not a single event!" Improved statistics had washed the hot spot away. Some variation on this scenario has recurred several times a year ever since.

But even without a point source or some other surprise of Galilean magnitude, they could still do science with AMANDA, mostly in the way of setting new limits.

The millennial season at Pole, 1999–2000, was a banner season. Most everyone who was there remembers it fondly. The camaraderie of working together in the field put all their family squabbles in the background. They intended to add six more strings, which would open their window that much wider, and, more importantly, to make one of those strings entirely digital, in order to test Dave Nygren's potentially game-changing digital optical module.

Not only had they carved out a large chunk of work for themselves, the station as a whole was also running on all cylinders, as this was the first year of serious construction on the new station. There were twice as many people in and around the old dome as it was designed to support, and Y2K only added to the fun. Various "DVs," distinguished visitors, passed through, and numerous adventurers—skiers, "snow bug" riders, and even a group of skydivers and balloonists—got it into their heads to do something special at that special place as the new millennium was born. It was even more of a circus than usual.

I arrived about a week before Thanksgiving. As I adjusted to the altitude and the surreal surroundings, I was most struck by the humanity of the place. Amidst all the whirring and buzzing of three shifts a day, six days a week, I sensed that this unusually positive group of people were more than just co-workers, they, like the Amandroids, were a kind of family. A day or two before the holiday, as I walked across the snowy floor of the dome on my way toward the tunnel that led up to the Ice and the long walk to the dark sector, I came across the head cook, Sally Ayotte, shuffling from foot to foot in her Extreme Cold Weather gear, deep-frying one turkey and smoking another in different rigs fashioned from fifty-gallon oil drums (see photograph 13). She told me that Bruce Koci had procured the hickory for smoking the birds by arranging to have the wooden crates for the hot water boilers he needed to ship to the station that year made of that kind of wood.

Sally went to great effort to provide a welcoming atmosphere, or, as she put it, "show respect" for all the workers and visitors at Pole. Her galley was the heart and soul of the place. Over the years, she'd also learned some tricks about how to cook in those high, dry conditions. Thanksgiving dinner, which the two hundred of us enjoyed in three shifts, serving and cleaning up for each other, was a cozy, warm-hearted affair, and surprisingly elegant: the wine steward wore tails and a top hat above his blue jeans.

The season had gotten off to a good start. PICO had fielded a strong drilling team (see photograph 16). But it was early in the season, so Bruce was still engaged in the slow process of waking the sleeping dragon. ("It's the whole history of Antarctic drilling rolled up into one drill," he told me.) He got his thirty-two boilers fired up about a week after I arrived, and the next order of business was to melt enough snow to fill a few tanks with hot water and get it running through the hoses. (Once a hose was filled, the water had to be kept hot and flowing; otherwise it would freeze and the hose would become one of the most unwieldy pieces of detritus you can imagine.) We had to unwind and rewind the hose segments in order to check for leaks, and this gave me the chance to experience what was said to be the essence of the drilling experience: dragging more than a mile's worth of two-inch hose around on the Ice. It was a riot. About five of us would line up about twenty feet apart and drape the hose across our shoulders in the manner of Jesus draping his arms on the cross, then drag it across the Ice as its large winch-driven reel payed it out. If we weren't going in a straight line and one of us dropped the hose for whatever reason, there would be a slingshot effect and the people on either side would be dragged twenty or thirty feet perpendicular to the direction they thought they were going, bouncing on one leg as if doing the can-can. All this at minus forty degrees on an ocean of white, under perfectly still, bluebird skies.

As the weeks wore on, the reality of the place set in, although I can't say that it ever seemed less than surreal. I learned to be wary of the thrill and exhilaration—the sunlight relentlessly pouring into your eyes, the constant activity all around—you could run yourself ragged. You'd get out of bed, don your many layers, grab the backpack you needed to carry on your wanderings that day, and stumble across the Ice to the galley to find yourself at the end of a drunken and ebullient tablemate's day. Or on the night—or day; it was hard to tell the difference—that you really needed to sleep, a bucket loader would be rumbling back and forth outside your Jamesway, carrying snow to the showers to melt for water. That might be the same night that the woman on the other side of the paper-thin wall in the next cubicle invited a guy to sleep over. Or at five a.m. there would be an "all-call" over the intercom about a pair of skiers who had just been sighted on the horizon, halfway through their attempt to traverse the whole continent, and you'd rush out to the geographic pole marker to meet them. (The two I met were man-hauling surprisingly lightly laden sleds in an "unassisted" attempt. They refused even a cup of hot tea as they stood by the marker having their pictures snapped. After half an hour or so, they glided off and disappeared over the opposite horizon. They'd been out for forty-seven days and were on pace to set a speed record, but unfortunately, they had to abandon their attempt two days later, when they discovered that the fuel for their stove had leaked into their food.) The phrase "inhospitable environment" took on palpable meaning for me. I got a gut feel for how herculean a task it really was to build a world-class particle physics detector at a remote field station on a two-mile-thick ocean of ice.

Bob Morse was AMANDA's "on-ice lead," cheerfully coordinating and assigning tasks; dealing with every little emergency that came up with scheduling, transportation, or health; negotiating with NSF about "population": how many beds they could find for the incoming Amandroids, or for space on the next Herc for an essential piece of equipment. As an old hand, he knew how to pace himself. "This place is like a junkyard dog," he advised. "It keeps coming at you."

Drilling commenced during the second week of December, and we pulled the drill out of the first hole of the season at three in the afternoon on the eighteenth. I was then transferred to Albrecht's deployment team. Things got off to an inauspicious start when one of the "GAs," or general assistants, the all-purpose gofers at Pole, dropped a $6,000 pressure sensor, made by a company named Paro, down the hole as she was preparing to attach it to the bottom of the string. Kurt Woschnagg, who was monitoring progress from the control shack, called out to ask if it was connected, and Albrecht told him it was gone.

"What do you mean 'gone'?" Kurt asked.

"It's gone. The Paro's gone," Albrecht responded.

Kurt walked angrily to the hole to make sure he'd heard Albrecht correctly, but when he noticed how attractive the young woman was, his attitude changed. None of us—all male, admittedly—wanted to hurt her feelings by telling her how expensive that little gadget was. Things went smoothly after that.

I couldn't quite believe that I was threading connectors together with my bare hands in minus-thirty-degree weather, but I cannot claim to have experienced the barbaric essence of the deployment experience, because this was the first year that there was any sort of shelter around the hole. Bruce had designed a windbreak shaped like a dome with a large pie slice taken out of it, about a quarter of the circle, and it worked well in spite of the gap, because the wind at Pole almost always blows from the same direction. We could even fit a couple of space heaters in there (although that led to several melted holes in people's wind pants). Bruce made note of what seemed to be "a direct correlation between the amount of shelter that you have and the kind of weather that you have. If you don't believe that, we bought all kinds of shelters this year that we've never had before and look at the weather." It was sometimes overcast, but there weren't many storms.

We completed the deployment of the first string of the year, AMANDA's fourteenth overall, at eight a.m. the next morning. When Albrecht e-mailed the news to the north, he was congratulated by Per Olof and Francis, who was "religiously following" the progress from CERN in Geneva. It was the earliest first deployment in AMANDA's history. I left the next day.

Thus I missed the traditional Christmas morning "Race Around the

World," in which the contestants, on foot, bicycle, or skis, complete two laps around a two-kilometer loop encircling the South Pole marker—and the general mayhem that accompanied the millennium. But I kept track of what my new friends were up to once I got home.

The Amandroids celebrated the big moment alone out on the Ice. The deployment team was about halfway through its job on the third string of the season. Earlier in the day, the ubiquitous Robert Schwarz had mentioned to Carlos de Los Heros, a Spanish scientist based in Uppsala, how much he liked the Spanish tradition of eating twelve fresh grapes at New Year's, one at each strike of the town clock, and they had decided to go ahead with it. Robert procured some grapes from the galley and handed them out, and since they had no clock, Carlos stood by the metal "90 South" interstate highway sign that was posted on the drill tower and hit it twelve times with a hammer. They drank champagne, posed for a photo (see photograph 17), scrawled commemorations on the first optical module to be deployed this millennium, and got back to work. The string was deployed about four hours later, just in time for those who had any energy left to wander over to the geographic pole and join in the ceremonial placement of the new South Pole marker.

The drillers had finished their work early enough to join in the station-wide celebration, which featured a parade led by bagpiper John Wright and was filmed by WGBH, the PBS affiliate in Boston, as part of its worldwide millennium coverage. My fellow driller, James "Tater" Bret, a good-natured and comfortably round individual, told me by e-mail a couple of days later that he'd been "filmed for quite a long time."

"I went as Baby New Year and seemed to draw a lot of attention in my diaper, sash and HUGE top hat. So much for fame. As you might have surmised, New Years Eve was a blow out filled with champagne; tight, skimpy dresses; naked people; tuxedos; and a lot of good energy. The band was stellar, and should play again as soon as possible."

The DVs must have been waiting for the much-anticipated—and in the event anticlimactic—Y2K computer bug to pass, as they began streaming through only after New Year's. The first wave consisted of a group of congresspeople and NSF mucky-mucks, including the foundation's director, Rita Colwell, and Dennis Peacock of Peacock event fame. It fell to Katherine Rawlins, a grad student studying under Albrecht, who was now an assistant professor in Madison, to show them around AMANDA.

"Their visit went very very well," she e-mailed a couple of minutes after they left. "I showed an event animation, talked about neutrinos and the thrill of discovery, etc., and we briefly visited the string 17 hole. . . . I tried to pitch IceCube at appropriate moments, and they seemed very responsive and enthusiastic. Rich thought we kicked ass.

"We now return to our regularly scheduled panic and chaos . . ."

Prophetic words. The next day string 17 got stuck in the ice, about five hundred meters before it reached the bottom of the hole. It turned out that the drillers had let the drill head hang in place for about four hours at one point and the hole had necked down at around that depth.

The deployment team had attached the last optical module to the string and were dropping it to its final resting place when Kurt Woschnagg and Gary Hill in the control shack noticed that the pressure sensors, which measured how fast the string was descending, weren't registering any changes. They looked out at Robert Schwarz, who was controlling the winch, and saw that he was lowering the string as fast as he could. As in the similar experience three years earlier, the string appeared to be streaming happily into the hole. The difference was that when they reversed the winch and tried to pull it out, it wouldn't budge. The guess was that the weight at the bottom of the string had somehow punched through the narrow part of the hole and become wedged below it. As they tried to pull it up, enough tension built in the cable to break the shaft that drove the winch, which started freewheeling—and the string began dropping again. Mike Boyce, one of the coming year's winterovers, was standing near the hole. He told them all later that he had resisted his natural impulse to grab the string— which was good, since if he had he would have disappeared into the hole with it. Gary raced out and hit the emergency brake, dropping the winch to the snow, and it ground to a halt.

The PICO drillers managed to weld the broken shaft back together, but the team failed to recover the string for a second time. Finally, they went for the last resort, Bruce Koci, who was preparing to leave the Ice. "It was always like that," says Gary. "It's stuck! What are we gonna do? Where the hell's Bruce? . . . So somebody goes looking for him, . . . and [Bruce] didn't even come out. He asked a couple questions and said, 'No. Forget it. You'll never budge it.'"

With this sobering experience fresh in mind, they commenced on the most critical string of the season and perhaps all of AMANDA, the now-legendary "string 18," the first and only string in the instrument to consist entirely of digital optical modules.

Dave Nygren, the leader of the DOM project, admits that "for some reason I'm more prone to generate those circumstances where we're living very dangerously. I may not look like the type, but apparently I do. My wife says, you know, stop doing this; it causes too much chaos, heh, heh, heh."

The decision to deploy an all-digital string had been made only about a year earlier, at the fall workshop at Lawrence Berkeley Laboratory, so Dave's team was given a remarkably short time in which to redesign what was essentially a specialized and oddly shaped computer and build and test about fifty of them. The heart of the system, a large circuit board, circularly shaped so as to fit comfortably with a photomultiplier tube inside a glass Benthosphere, contained over six hundred parts. Once the boards were fabricated, they had to be shipped from Berkeley to the Physical Sciences Laboratory outside Madison, where they were integrated into the DOMs, which were then pumped up to pressure and tested, and the finished product was then shipped to Port Hueneme, California, for its long voyage south. To give some idea of the split-second timing involved, the boards didn't even reach Wisconsin until November, after the season had begun, and the DOMs didn't reach Pole until just after Christmas.

To make things more complicated, the Germans had insisted that the DOMs also function as normal AMANDA optical modules, so the collaboration wouldn't be sacrificing a string, so to speak; it would still give them normal AMANDA data. The idea was adamantly opposed by a couple of the LBL engineers, but Jerry Przybylski and Dave Nygren agreed to put a fiber-optic readout on the digital board as a compromise, figuring it was the only way they could get their new technology into the Ice.

Dave and Jerry share a philosophical belief that any disruptive technology of the sort they were envisioning will tend to have intrinsic aesthetic appeal, and Dave found the fiber-optic compromise unappealing in the extreme. In his view, the DOM "became a hybrid to satisfy the weak of spirit." He also had a great distrust of fiber-optics, believing they were much too expensive and too fragile to survive the re-freezing process. In the end he was probably right.

Under the argument that they were already building the data acquisition system for the AMANDA strings—the hardware on the surface of the Ice that translated the raw signals from the optical modules into digital information that could be used by AMANDA's computers—the Germans also insisted on building the data acquisition system for the digital string. This made Jerry Przybylski extremely nervous, but he compromised again with the stipulation that they provide a "back channel," which would allow him to communicate directly with the DOMs from a desktop computer in the event that the German hardware didn't work.

The pressure really began to build when four members of the LBL team traveled to Wisconsin in November 1999 to help assemble and test the modules. In the testing stage, a small electrical part on the main circuit board kept burning out, so they had to depressurize the DOMs, take them apart, replace the burned-out parts, reassemble the DOMs, and pump them back up to high pressure several times before finally getting them out the door.

An engineer named Jozsef Ludvig and a management-level physicist from LBL named Bob Stokstad met them on the Ice. Albrecht gave them some space in the AMANDA Jamesway and placed the reel holding what would become string 18 on the snow next to the building, and they began troubleshooting the system one last time—just as Jerry Przybylski had done with the first two DOMs four years earlier. Here, they encountered a problem that required a chip on the main circuit board to be replaced. They arranged for someone traveling south to bring a bag of the chips with him—and then had to unseal, rework, and reseal the DOMs again, at a sprint.

To make matters worse, some sort of sickness was raging through the station, and Ludvig caught it during the final heroic push. As the only engineer on the Ice, he was the one who felt most under the gun. "Before it was over, some sixty module openings and re-sealings took place at Pole," wrote Stokstad by e-mail. "Forty-two DOMs were ready for deployment, crated and ordered, two days before deployment began." The drillers would have been more than halfway through the awaiting hole at that point.

Gary Hill, who had also caught the "pole plague," now had the task of leading the deployment. That went off without a hitch, thankfully, and in the end all but two of the DOMs survived the freeze-in, purring like cats.

Unfortunately, though, it wasn't over for Ludvig. When they hooked the DOMs up to the data acquisition system that the Germans had built, sure enough, it didn't work. Now he had to troubleshoot the thing with the highly excitable Ralf Wischnewski, who had designed it, and they simply could not get it to function.

Nevertheless, string 18 was an unqualified success. Thanks to Jerry's foresight in asking for a back channel, the LBL team actually managed to do some physics with it during its very first year, and when they returned the following year and installed their own data acquisition system, they demonstrated its full potential. The performance of string 18 eventually proved persuasive enough to convince the entire collaboration, in Dave Nygren's words, to "abandon the AMANDA model and go whole hog into this fully, what I would call modern digital era."

Looking back on that season more than a decade later, with IceCube built and more than five thousand of his DOMs still purring away in the Ice, Dave recalled that "the idea of building this really modern stuff, that really gave us the information that you would wish for, was galvanizing. I mean we got string 18 from decision in November of ninety-eight to shipping to the Pole in November ninety-nine and installed in January. Fantastic!

"And, you know, there were a lot of decisions and designs—it was just an amazing era. High adrenalin, high expectations, high performance . . ."

"And a lot of work by Jozsef Ludvig," interjected Jerry Przybylski.

"Yeah. It was a little bit too much for Jozsef," Dave replied. "He began, I think, to show the effects of stress. . . . He quit on several occasions. He quit while he was at Pole, and then he came back [to the real world] and he quit again, and finally the third time he quit I said, 'Okay, Jozsef, I believe you.' . . .

"It was white-knuckle all the way. . . . And now the young people [in IceCube] say, 'Oh this works, you know, I'm gonna use this and do some science.' . . . They're just clueless how it all came to be, and that's fine."

It must have been very stressful indeed, not only for Ludvig, but for everyone else on the Ice that January. They were nearing the end of the season, so the weather was beginning to deteriorate, and they were also entering the traditional end-of-season panic, in which Steve Barwick and Ralf Wischnewski, two of the more volatile members of the collaboration, worked desperately to complete their tasks, calibrating the instrument and implementing the data acquisition system, respectively. (Although the system for the DOMs didn't work, Ralf still had to hook up the other strings that were deployed that season, as well as the fiber-optic channel from string 18.)

On string 19, the sixth and last of the season, the PICO drillers suffered the supreme defeat: they stuck the drill.

Bruce Koci had left during the drilling of string 18 in order to engage in his other life, drilling ice cores on the tops of high mountains. His clima-tologist friend Lonnie Thompson had invited him on his expedition to Mt. Kilimanjaro in Africa.

Bob Morse, who was in some sense Bruce's boss, since he managed the AMANDA side of the drilling, had pleaded with him to stay at Pole until the end of the season—and he probably could have stayed a bit longer. But in Africa Bruce told me that he'd left in order to send the message that AMANDA was leaning on him far too much for heroics; it was past time for other people to take responsibility for running the drill.

They always had a second drill head with them for just this reason, so the drillers hooked it up and managed to complete the last hole—19a, they called it—overcorrecting a bit. Morse later observed that they could have deployed a Volkswagen in that hole. The string was deployed in record time, and the winter exodus began. Gary Hill and Robert Schwarz left the Ice two days later.

Since it was now the *end* of January, the season was bound to close in a headlong rush. Barwick and four accomplices didn't have time to com-plete the calibration, and several other things went wrong at the last min-ute, so the instrument wasn't actually working on the day the last summer Amandroid—Ty DeYoung—caught the last Herc out of town.

"Bloody typical. They can't even hand over a working experiment," wrote Darryn Schneider, who was left holding the bag with his fellow winterover, Mike Boyce.

Two days later, he and Mike were out in MAPO, the AMANDA com-puter building, on the verge of getting the instrument back up and running, when they heard an all-call on the intercom announcing the imminent arrival of the very last flight, come to deliver the last few critical supplies. This took them by surprise. The decision to close the station had come down quickly.

"We go up onto the roof of MAPO to see our last physical connection to the outside world come and go," wrote Darryn. "I realize it is getting cold, as I freeze a beer can to my bottom lip! The plane lands, unloads cargo as it taxis, and stops just briefly enough to take on a bag of mail. Then it's off again. We wave like crazy, hoping they realize the significance of their leav-ing us here. After it has taken off, it banks sharply and does a low run over the station. MAPO is engulfed in the dissipating contrail from the take-off. And that is that. For a moment I almost feel—well, isolated—that the win-ter has started. Maybe I'm just tired."

✦ ✦ ✦

And so ended the construction of AMANDA. It would never grow beyond nineteen strings.

In my conversation with Dave Nygren and Jerry Przybylski more than a decade later, I asked Dave if he had seen AMANDA as just a feasibility study or as an instrument that might be able to do some physics.

"I thought it was an *in*feasibility study, quite frankly."

I observed that that was a controversial statement.

"I don't mind being quoted in anything I say, because I'm not here for career development," he responded equanimously. "This is really how I looked at it: I didn't like it. I wanted to get rid of it. . . . Let me make clear, AMANDA was an essential step, but it should not have been carried further than it was."

It also ended at the perfect time. From an historical standpoint, it could be argued that the design and construction of AMANDA, which was nothing less than the invention of neutrino astronomy, took place during the one window of time when it would have been possible. It could not have been done before the 1990s, because physics and computer technology weren't ready for it, and it would have taken much longer and probably been prohibitively expensive after 2000, largely because, just that year, the contract to manage the U.S. presence in Antarctica was awarded to the military contractor Raytheon, which proceeded to impose stifling layers of bureaucracy and control. The company's motto for the continent became "No More Frontier Attitude." Everything is regulated now.

But AMANDA *required* a frontier attitude. The anarchic approach was the quickest, the best, and probably even the only way it could have succeeded.

Furthermore, as we shall see, the ascension of George W. Bush to the U.S. presidency that very year led to a tightening of U.S. science funding. And the Great Recession that came at the end of Bush's tenure put a stranglehold on discretionary funding of all kinds. On top of that, the political discourse in the United States has become so poisoned and the possibility of compromise so diminished that it has become pretty much impossible to gather support for projects as expensive and far-out as this one.

It was difficult enough as it was.

Part IV

The Real Thing

17. Sometimes You Get What You Ask For

CONSIDER FRANCIS HALZEN'S POSITION AS AMANDA BARRELED INTO the new century. His overriding interest was in helping IceCube to succeed. Two of the senior administrators in his university, John Wiley and Terry Millar, wanted to run the project there, and Francis would have liked to have been closely enough involved to have some hand in ensuring a positive outcome. But this was a shaky proposition, because the university clearly did not have the in-house expertise to manage the project.

From a personal standpoint, it would be a tremendous headache to have any sort of operational responsibility for the construction of IceCube. Furthermore, as a theorist and a professor, he would never really be in charge of construction, even though, as principal investigator, he would be held responsible and his head and no other would be on the chopping block if it failed. Another consideration was that he wouldn't mind having a little time to do some honest-to-god physics again, which he hadn't been able to do much of for quite a while, given the distractions of AMANDA.

Everyone at Wisconsin who was involved in bringing the project to Madison remembers Francis as being extremely enthusiastic about it. He is supremely adept politically, however, and he doesn't show his hand. Jim Yeck, who would join the project three years later to lead the construction of IceCube, points out that Francis is "always playing a game of chess where he's a couple of moves ahead. He's thinking of—he's always in a position where if he made a move he can undo it or change it. So, it's not chess; it's a little more complicated than that even."

The fact is that Francis had seriously conflicting emotions. To this day, he will claim that he would have been perfectly happy if NSF had decided to run the project out of Lawrence Berkeley Laboratory and named George Smoot as PI.

"Mostly, it was scary!" he says, with a weak laugh. "Scary. It was scary. . . . You know that at some point you will be spending ten, twenty million dollars a year, flowing through your hands in a very strange way, and you're responsible for it. . . . What people don't realize is that the chances it was a failure were much larger than it was a success. It would have been George Smoot's failure." And he laughed again, more heartily this time.

Keep in mind that the project hadn't been funded yet and that once it was, the broad strokes of the management structure would be etched in stone. So this was the time of the most strenuous infighting. And Smoot wasn't the only one bucking for PI. Although it was always a bit unclear whether Steve Barwick would be in or out—he withdrew so often it was hard to keep track—it soon became clear that he, too, harbored the secret and utter fantasy of becoming PI and moving the project to Irvine.

The Europeans were not going to the lengths that George and Steve were, but it did seem that they were delaying things in order to jockey for leverage as the rules for running the new collaboration took shape— memoranda of understanding and the like—among the participating institutions. Strangely enough, the European jockeying often took the form of promoting Steve's visibility at the expense of anyone's from Madison or, god forbid, LBL. The AMANDA collaboration always had two spokespeople during those years, one from the United States and one from Europe, and the majority of the Europeans always supported Steve for the U.S. role. He and Christian Spiering were elected in late 1998, and then re-elected in mid-2001, by which time Steve's antics had taken yet another disruptive turn.

Francis wasn't particularly concerned with the political jockeying— although it did waste time and energy. For him, the most maddening and painful obstacle was the conflict that surrounded their most important job—which was science, lest we forget. By the time 2000 rolled around, the discovery of the Peacock events was more than a year old, and those first few gold-plated neutrinos had acquired roughly a hundred siblings, yet the PIs still could not agree on how or even whether to publish a paper about it. Some of this infighting was competitive in nature, and some arose from scientific disagreement, but the main undercurrent was political. For the best way to undermine IceCube in the eyes of the many committees that were reviewing it was to torpedo this paper. It is amazing how far some were willing to go to damage or even kill the project rather than lose a chance at gaining control.

The Peacock events had been discovered in the data taken in 1997 by AMANDA-B10, remember, an instrument that was about half the size of the one they had now completed. By the end of 1999, Zeuthen had nearly caught up to Madison in the "arms race" to analyze the 1997 data, and given the comprehensive report that Christopher Wiebusch and his colleagues had produced in the fall, reasonable people would agree that they had also confirmed the validity of the Madison analysis. But Christian Spiering and Ralf Wischnewski still refused to accept it.

The PIs scheduled a collaboration-wide analysis meeting in Zeuthen in March 2000, and Francis asked the two mainstays of the Madison analysis, his graduate student Ty DeYoung and his post-doc Gary Hill, to visit Zeuthen for three weeks in advance of the meeting in order to try to bring Christian around. His opinion was the crucial one, since he was the leader of the Zeuthen group, and Ralf, the second-in-command, was probably unreachable anyway. As Ty writes, "Ralf's insistence that we couldn't publish anything until [everything about AMANDA] was completely understood was way over the top, and probably would have killed any hope of doing IceCube." The one thing that stands out in Ty's mind from that visit is that either owing to his concerns about keeping the analysis blind or because he viewed it as a gimmick, Christian had never observed a neutrino event with the AMANDA event viewer before. This graphic means of watching an event unfold in time was the most intuitive way to see what was going on in the instrument. Ty and Gary sat him down and showed him a few events that had passed both their analysis and Zeuthen's, as well as a few obvious fakes that the two analyses had rejected. Christian still refused to give in—on the surface at least—but Ty believes their session together had something to do with his eventual conversion.

Two months later, at a closed meeting of the PIs during a collaboration meeting in Brussels, Buford Price engineered a small break in the logjam by suggesting they present their dual (or dueling) analyses in the form of a brief letter to *Nature*. Zeuthen still had some discrepancies in their analysis—the Monte Carlo simulations weren't agreeing closely enough with the real data quite yet—which may explain why Christian agreed only reluctantly. So Madison produced a first draft, presenting their analysis alone, and sent it to Zeuthen as a kind of ultimatum. (The backbone of this draft would have been written by Ty, since his Ph.D. thesis would become the primary documentation for the Madison analysis.) Some time later, an extended draft came back from Zeuthen with their analysis added, the Germans demonstrating both their pettiness and their continued competitiveness by designating the two analyses as "A" and "B" and assigning "A" to their own, even though Madison's had been way out in front all along.

George Smoot was in the thick of this infighting, too, of course, although at some disadvantage. He knew better than most that fame and glory would come from actually making a discovery and that you can only do that if you can analyze data. His group was very close to the data stream, of course, since they were employing the supercomputers at Lawrence Berkeley Laboratory to filter the raw data coming directly from Pole. He'd recruited John Jacobsen to his group, perhaps the most accomplished codewriter in the collaboration, and he had also tried unsuccessfully to recruit Christopher Wiebusch, the driving force behind Zeuthen's analysis. But George hadn't managed to put much of an analysis team together yet, so he resorted instead to the next best thing: vicious mudslinging at both Zeuthen and Madison. Actually, he was badmouthing virtually every aspect of AMANDA and IceCube to anyone who would listen, including outsiders, the potentates at NSF, and myself.

He had employed such tactics quite successfully once before, on the Cosmic Background Explorer, or COBE, experiment, back in the early nineties. The person who actually made the COBE discovery was physicist Ned Wright from UCLA, who had invented a computer algorithm, much like AMANDA's event viewer, for analyzing the data. When Wright presented the first glimmering of the discovery at a meeting of COBE's head scientists, George, whose group was having difficulty interpreting the data, became incensed and began impugning Wright's work. George shut off access to the portion of the COBE data that streamed through his hands and then cut off all communication with the rest of the collaboration for two months while he traveled to Antarctica to work on a different project. By the time he returned, Wright and some others had improved their methods and were now sure that his initial hunch was valid, and the collaboration had prepared three papers to announce the results. George's group had fallen further behind in their own analysis while he'd been away, of course. He declared angrily again that none of the papers were ready for publication and then worked to delay them. When his colleagues decided to go ahead and announce the results anyway, in three talks at a meeting of the American Physical Society in Washington, he broke the rules of the collaboration and published a press release about the results a few days before the talks were given, thereby attracting worldwide attention to himself and Lawrence Berkeley Laboratory. No one else on the project was named in the press release. Even NASA, which had funded and managed the project and whose scientists were involved at the highest levels, received only passing mention. At the same time, George's group had not contributed in any way to the analysis. For a while there, he was perhaps the most well-known scientist in the world, rivaling Stephen Hawking, and he soon signed a book contract about COBE for a rumored $2 million.

There was still a reasonable chance that IceCube would go to LBL, although Madison seemed to hold the advantage. Among the NSF officers, only Gene Loh, who was something of an outsider at the foundation, wanted it to go to the lab. (This explains why Smoot had sent his rogue white paper to Loh before the May 1999 IceCube meeting in Arlington.)

The foundation put together an external committee to review the project, chaired by Donald Hartill, an experimental high-energy physicist from Cornell, and scheduled what Bob Paulos calls a "soup to nuts management and technology review" for the end of June. It was held in Madison, which was still being viewed as the lead institution. Most of the AMANDA/Ice-Cube PIs were there, and several high administrators from LBL demonstrated their interest by showing up as well.

It was clear that NSF was new to this game, as they didn't even send Bob an agenda until three days beforehand. But it was also clear that the members of the Hartill committee knew their stuff. And they weren't afraid to level criticism.

One of their frank judgments was that the University of Wisconsin did not know what it was getting into. Paulos remembers that he and Francis were told "the lights were going to shine very brightly." "Every time you make a move, and every time you make a mistake, it's just gonna be an order of magnitude brighter than what you are used to. So just be prepared for that." Terry Millar recalls Beverly Hartline, a hardnosed project director–type from Los Alamos National Laboratory, telling them "UW simply is not in the game." She compared them to someone who wanted to run a marathon (Terry suspected she'd run a few), "training for a few weeks at two-three miles a day, and then deciding they were ready for prime time. They might last until mile five or six, but they will not be there at mile fifteen, to say nothing of miles twenty or twenty-six." And Francis loves to tell the story of how Hartline told him and Paulos that they'd be gone in a year or two, replaced by more experienced managers. Several years later, when she turned up at another "Hartill review," he cheerily pointed out that he and Bob were still there. (These reviews continued for years and were very helpful.)

After holding their feet to the fire for several days, the Hartill committee gave them a surprisingly hearty endorsement, and the NSF officials made it clear that they planned to approve IceCube. Of the many reviews that AMANDA and IceCube have undergone over the decades—certainly more than a hundred—this is the most memorable for Francis. His name-tag from the first Hartill review still sits on his desk in his home office.

Per Olof used to love telling the story of how they all went out to dinner at a restaurant on State Street afterward and how Francis, who is known for his epicurean tastes, chose the wine for a toast. He was so excited (not only at the positive review, he claims, but also at the news that he'd soon be fired) that he didn't notice that the wine had gone bad. None of the others had the heart to rain on his parade by pointing this out to him, so he went ahead and ordered another round.

The tide began shifting toward Madison.

NSF had put the contract for the Polar Ice Coring Office, PICO, out for bid again, and Bob Paulos had written up what he thought was a longshot proposal to bring it to the Space Science and Engineering Center. "If we had the drilling contract, then the hot water drill would be a big component of that, that would help IceCube, and, oh by the way, the other stuff might be interesting, too," he explains (that "other stuff" being tens of millions of dollars' worth of ice core drilling, mostly in the polar regions). At the end of June he found out that he'd won the lottery.

"This guy calls me out of the clear blue. . . . He introduces himself and then he says, 'Well, you probably know why I'm calling.' I go, 'I have no idea why you're calling.' He goes, 'I just wanted to let you know that you guys have won the ice coring and drilling contract.' I'm like, 'Really!' And I remember my first thought was 'Oh shit!' because I had so much going on that I couldn't imagine now having to deal with this." (He already had about one and a half jobs between IceCube and his regular SSEC responsibilities.)

"I said, 'Okay, so how's this gonna go?' And he says, 'I'm here to tell you, as of today you're authorized right now to spend a million dollars.' And I go, 'Well that's really great. Are we done for now?' And he goes, 'Yup.' And I go, 'Okay, well thanks a lot. I guess we'll be in touch.' And so I hang up the phone, and I go, 'This is really kind of interesting. . . . We won the ice coring and drilling contract; at that time that included the development of the hot water drill for IceCube—assuming IceCube got funded; but IceCube wasn't funded yet. But yet, we were tasked to start developing the drill.' If that makes any sense."

This brought things to a head.

In July, Lawrence Berkeley Laboratory made a formal request to become the lead institution for IceCube. On the first of August, Willi Chinowsky followed up with a curious and condescending e-mail to Francis.

"I make you, and IceCube, an offer you cannot refuse," he began. "Accepting some personal sacrifice, for the good of IceCube, and to lessen your burden, I offer to take charge of . . . IceCube organizational matters as interim Project Director." There were sops for Madison: Albrecht Karle would be deputy director and Bob Paulos project manager. But IceCube would be led from the lab, Smoot would be PI, and Francis was nowhere to be found. Willi also made the bizarre request that Francis be the one to share the new arrangement with the collaboration as a whole.

"Do you think that I didn't sleep at night when I received this e-mail?" Francis asks. "No. Did it upset me? Yes. . . . [But] you want the science to happen; that dominates everything, right? And so then you ask, does it have a better chance of succeeding if LBL does it?"

He wasn't sure he knew the answer to that question.

In any event, LBL's request forced everyone's hands. John Wiley and Terry Millar began lobbying energetically for the project, the NSF folks began backing Wisconsin, too, and in response, it seems that something in George Smoot snapped. On the twenty-seventh of August, he sent an e-mail to the entire collaboration, students included, under the title "Why I asked my name to be taken from the *Nature* letter draft." "There are two reasons," he began:

> First, I have been nominated for a Nobel Prize, other awards, and a promotion and as a result all my postings and publications will be reviewed carefully by the appropriate committees. It is clear that the draft AMANDA Nature paper containing essentially no science but only assertions based on theoretical prejudice and wishful thinking rather than firm documented experimental results is more likely to be an embarrassment than an asset. . . .

He pointed out that this was an easy decision, since he was involved in another project that was producing *good* science. "Second," he continued,

> I am appalled that graduate students and post docs are being shown that it is an acceptable standard to publish work that is sloppy and lacking in scientific integrity.

(The other principal investigators, by and large, held quite the opposite view. They were appalled that George would have aired such dirty laundry in front of students, since it would undermine their morale.)

He ended by asking rhetorically if AMANDA or IceCube would *ever* become assets rather than embarrassments.

Of the several responses to this missive, the most entertaining came from Per Olof Hulth, always a constructive voice, who also knew a thing or two about the Nobel Prize. As a member of the Royal Swedish Academy of Sciences, he voted every year on the winners.

> Dear George,
>
> I do not think you need to worry to sign any "sloppy" paper because you are nominated for the Nobel price [*sic*]. I do not know about the other awards which you are up for. But for the Nobel price you can for sure sign several papers with wrong results, much worse than any "sloppy" paper and still get the price for one single paper which have the real discovery! It is the discoveries or the genius ideas confirmed by experimental results which counts, not the mistakes. You can e.g. think about a very good scientist like Carlo Rubbia [. . .]
>
> Anyway, I disagree with your comments about the Nature letter. It has never been the idea that it should be anything else than the announcement that we think we have a working detector and that the forthcoming

papers with all details will come in the near future. The version which Bob now has edited is nice and fulfills this limited goal.

Looking forward to see you in Stockholm.

Cheers,

Per Olof

Ironically, George's capable and diplomatic LBL colleague, Bob Stokstad, had incorporated the many beneficial comments submitted by others into the most recent draft of the paper.

The letter was submitted to *Nature*, absent George's authorship, about two weeks after his outburst. This was more than twenty months after Gary Hill and Phil Romenesko had experienced their moment of truth late that summer night in Madison. It was published the following March. Ten years of work by more than a hundred people were boiled down to three pithy pages, under the title "Observation of high-energy neutrinos using Čerenkov detectors embedded deep in Antarctic ice."

Two months later, Gene Loh called Francis to tell him that IceCube had received formal approval and Madison would run it. Francis responded just as Bob Paulos had when he'd learned about the drilling contract: "Oh shit!," a rare four-letter word coming from him.

"Halzen," said Loh, "sometimes you get what you ask for."

18. No New Starts

EXCEPT THAT IT DIDN'T HAPPEN. THAT FORMAL APPROVAL WAS GIVEN BY the Bill Clinton administration only three months before Clinton left office. Within days of George W. Bush's inauguration in January 2001, the new administration began disseminating a series of directives exercising an unprecedented level of control over the way scientists and scientific agencies throughout the federal government would now be doing business. One of the first of these diktats was nicknamed "no-new-starts": the National Science Foundation would not fund any new capital projects in the coming year. IceCube lost its funding even before it began.

At a collaboration meeting at the University of Delaware in March, Francis told me, "We were in Clinton's budget. If Al Gore had won the election, we'd be drinking champagne right now."

When he had first discussed the idea of running IceCube out of Madison with John Wiley, the dean of the Graduate School, and Terry Millar, associate dean for physical sciences, Francis had exacted a promise that it would not be business as usual: rules would be broken if they got in the way. Not only did the two administrators live up to that promise, Wiley became considerably more capable of doing so on January first, a few weeks before Bush was inaugurated, when he became chancellor of the Madison campus. He and Millar responded to the no-new-starts diktat by loaning IceCube $4.5 million to keep it afloat while the funding quandary sorted itself out. The money came from WARF, the Wisconsin Alumni Research Fund, and Wiley backed it with a guarantee that the university would repay WARF from its general fund if the National Science Foundation didn't come through. He later said it was the best loan he ever made.

The main task during this unstable period was to design and build a "production" drill. Bob Morse and Bruce Koci had been on the case since mid-1999, actually, and they'd been having some fun with it. Both were opera buffs, and Bruce and his wife, Ann, were dedicated Wagnerians. In the time I knew them they made separate pilgrimages to New York, Seattle, and Chicago to attend the complete *Ring Cycle*. (Ann points out that Bruce was the only person she ever met who could grovel in the squalor of a drilling camp for three months, come home, change into his tuxedo and rush off to the opera, and look right at home.) Thus inspired, Bruce named the drill Wotan, after the king of the gods in the *Ring*, and Bob actually referred to it as such in the drilling section of the first IceCube proposal, which he wrote.

Wotan would need to be an entirely different creature from the AMANDA drill. The plan for IceCube was to deploy no less than eighty strings, each a mile and a half long and comprising sixty optical modules, spread out over a square kilometer. If they were to accomplish this gargantuan task in the five years called for in the proposal they would need to deploy sixteen strings per season, almost three times the level of their best season with AMANDA. Wotan's heating plant would be more than twice as powerful as AMANDA's, five megawatts versus two; the hose would have a larger diameter, two and a half inches versus one and a half; and they would "remove the current necessity of changing reels during hole drilling" by winding a single hose nearly two miles long onto a reel so large that it would take two Hercs to ship it to Pole. It would be shipped in pieces and assembled once it got there. When the hose was filled with water, Wotan's reel would weigh nearly fifty tons (see photographs 20–22).

John Wiley compares the building of IceCube to the building of the Great Pyramids or the Panama Canal.

✦✦✦

It was Bruce who came up with the idea for the "circus train." Wotan was modular, all "plug and play," and all the connecting pieces, even the heating plant and hose reel, were mounted on sleds. The reason they called it a circus train, says Morse, was that "when it comes to moving stuff around, setting up and moving quickly and getting the hell outa town, nothing beats a circus." In other words, they could move quickly from one hole to the next.

But this was also the first point of impact between the anarchism of AMANDA and the sober management that would be required to build and operate IceCube. It was a new ball game, with hundreds of millions of dollars on the table, while Bob and Bruce were quintessential start-up guys. Remember Morse saying that he loved to start new things but tended to be elsewhere by the time "the messy details" of finishing them up came around? Well this was the time for attending to details. Bob Paulos and his colleagues at the Space Science and Engineering Center imposed rigorous procedures like design reviews, testing the drill against specification before sending it to Pole, and so on. The difference in cultures is perhaps best signified by the fact that the SSEC folks didn't see the humor in naming the drill Wotan nor the sense in referring to it as a circus train.

"Well, hell, I mean, 'circus train'? What does that mean, you know, really?" asks Paulos. "That's a nice phrase, but it doesn't mean anything. . . . I never called it that. And none of the people that actually built the thing called it that."

But it *did* mean something. Speed was important. In what Morse calls their "drearily competent way," the SSEC engineers named it the Enhanced Hot Water Drill.

The SSEC folks *were* quite competent, however. Paulos brought in a superb SSEC engineer named Mark Mulligan, and this team of four proceeded to design the drill.

Bruce was still the guru and the driving force behind the new design. Morse says his "paw prints" are on every aspect of it. "What we would do was, we would just give physical reasons for Bruce's tremendous intuition, heh, heh. That was our job in life, to interface between Bruce and the ordinary world. Bruce was a genius." In part because he seemed disorganized, however, and in part because he wasn't one to argue, the SSEC folks never fully understood or appreciated his contributions. Paulos says, "He wasn't like a lot of engineers that I know and work with, right? At all."

It was an interesting dynamic. Bruce and Bob were used to inventing things on the fly—they were good friends, and they'd been doing this together for more than a decade—while the others weren't as comfortable with that but knew how to control the process and keep things from getting as helter-skelter as they'd gotten in the AMANDA days. Mulligan contributed inventions of his own as well, and there is plenty of creativity involved in making a huge, complicated machine efficient, reliable, safe, and capable of running for years. Morse may not have appreciated Mulligan's contributions as much as he should have, while Bruce, who was remarkably lacking in ego, probably straddled the divide. Bob talks about how Bruce used to react to the thick engineering reports that the SSEC engineers would produce. "He never seemed to be that interested, but there was *nothing* that was going on with the drilling that he didn't know about. . . . At some point you'd see him . . . looking at one of these reports, and he'd grab some scraps of paper—everything was always sort of pencil scrawls and pieces of paper—and every point that they'd come up with, Bruce had already made the rough, pretty accurate calculation of what they were going to do and what they would need. So it was amazing."

It is to Paulos's credit that he got them rowing in the same direction. "The lesson to be learned," he says, "at least by me, is it takes a number of good people with different approaches and attitudes to make these things really work out. . . . You put those guys together and then shake it up, and it really worked out quite well."

Meanwhile, all through what would turn out to be the long and harrowing passage to IceCube, AMANDA was very much alive. Indeed, in a scientific sense it had just been born. While the huge management and applied physics problem of designing and building IceCube would require more effort overall and eat up most of the resources, intellectual progress in actual particle astrophysics was being made on AMANDA.

Both efforts were taking place in an academic milieu, of course, and physicists were at the helms of both. New graduate students and post-docs were joining in constantly, and you don't get a Ph.D. in physics by working on instrumentation alone. You need to do some physics: figure out some new way of analyzing the data or look for neutrinos from, say, active galactic nuclei—whatever you and your adviser dream up. By the time AMANDA would eventually run its course, it would produce more than fifty doctoral dissertations and forty-nine refereed journal articles. (And, since it has been observed that a Ph.D. physicist generates about $1 million in economic wealth on average, through inventions, starting new businesses, and so on, AMANDA paid for itself several times over.)

For example, during the twenty-odd months that the Swedes had been steering clear of the competition between Zeuthen and Madison, they had been focusing on a point source that might come as a surprise: our own planet. This is an area where neutrino astronomy intersects both cosmology and particle physics.

Supersymmetry, the extension to the standard model on which the particle physics community is pinning its greatest hopes, posits a heavy sister or brother particle for each of the particles in the standard model (all of which have now been seen, thanks to the discovery of the Higgs boson). Several of the lightest of these hypothetical siblings also happen to be the most promising candidates for the unseen cold dark matter, which constitutes some 85 percent of the mass of the universe and is one of the great unsolved mysteries in cosmology.

As a class, these candidates have been given the name Weakly Interacting Massive Particle, or WIMP, a name that may sound humorous but isn't as highfalutin as it might also sound. It's basically just a description of two general qualities that any dark matter candidate must have. Like neutrinos, they must be uncharged and unaffected by the strong nuclear force; they will feel only the weak force and gravity (so they should be at least as hard to detect as neutrinos). Unlike neutrinos, they must be heavy in order to exert the strong gravitational pull, affecting the shapes of spiral galaxies, for example, that has been observed by astronomers. In supersymmetry, the lightest WIMP candidate is named the neutralino.

There shouldn't be very many neutralinos floating around in empty space. If they exist at all, however, celestial bodies like the Sun and Earth are expected to gather them into their hearts by force of gravity. Once they've been brought into close proximity with each other through this process, they are expected to collide with one another every once in a while and mutually annihilate, giving birth to other particles that subsequently decay to produce high-energy neutrinos. Thus, if supersymmetry is correct—and that's a big if—the Sun and Earth might be high-energy neutrino sources.

Recall that one of the earliest and most convincing neutrinos detected by AMANDA was the so-called Eva event, discovered by Eva Dahlberg, a doctoral candidate in Stockholm. It was a dead-vertical muon running parallel to one of the strings in AMANDA-B10. That meant it was coming from the precise direction of the center of the Earth.

Since AMANDA was shaped like a cylinder, taller than it was wide, and its axis was as perfectly vertical as Bruce Koci's drill had been capable of making it, the instrument happened to be most sensitive to neutrinos coming from this direction. As it also happened, the Swedish contingent in AMANDA included several strong WIMP and dark matter theorists, so they were in an excellent position to conduct a neutralino search.

Did Eva's neutrino come from a dark matter annihilation? Impossible to say, because neutrinos don't come with labels indicating where they were born, but it was most likely an atmospheric neutrino, generated by a cosmic ray hitting the atmosphere directly above the North Pole. The key to detecting Earth-based WIMPs would be to discern a hot spot significantly brighter than the atmospheric background in that direction. Indeed, this is the trick to finding a point source anywhere in the neutrino sky. The significance of the *Nature* letter was that it gave a first measure of the atmospheric neutrino background over the full northern hemisphere: the background that any point source in the northern sky must rise above. Owing to AMANDA's odd location, this included the center of the Earth.

According to a paper that the collaboration eventually published in 2002, AMANDA's first search for neutralinos yielded "no excess over the expected atmospheric neutrino background": they found no evidence for WIMPs. This was a disappointment, to be sure, but perhaps a not-too-surprising one, since such evidence would have been one of the most important physics discoveries in half a century. Not only would it have provided a clue to the cold dark matter, it would also have been the first evidence ever obtained in support of supersymmetry.

In spite of coming up empty, this "search" (a code word in physics jargon meaning they had searched and not found), did break new ground. By setting a new limit on the maximum brightness of a possible neutrino source at the center of the Earth, it provided ever so loose a constraint on one prediction of supersymmetry. As limits of this sort become more stringent, they either lead to a discovery or rule a theory out.

Over the next couple of years, the collaboration carried out other groundbreaking searches with AMANDA, incorporating data from the years following 1997. They searched for supernovae. They searched for a point source anywhere in the northern sky. They searched for a diffuse flux of cosmic neutrinos that might be reaching our planet from every direction, in the manner of cosmic rays. They searched for neutralinos coming from the Sun, which happens to be more a promising candidate than Earth. Fourteen years into it, they were finally doing bread-and-butter particle astrophysics.

They were also breaking ground in applied physics. The 2000–2001 Antarctic season saw what was probably the game-changing breakthrough on the digital optical module.

One of the most troubling aspects of AMANDA was the difficulty in calibrating the instrument, especially in time. In order to nail down the track of a muon traveling through their array at nearly the speed of light, they needed to have a precise understanding not only of where every optical module was, but also the precise moment when any photon emitted by that muon reached a module. Since light travels about one foot in a billionth of a second, or nanosecond, the positions of the modules had to be known to within a few feet, and the modules had to be synchronized to within a few nanoseconds. Both of these problems were messy ones, not only because the modules were so far underground, but also because they communicated with the surface through wires or fiber-optic cables of varying lengths and transmission characteristics. It was the old "telescope in a darkroom" problem.

Steve Barwick and company were taking a month or so every year to synchronize the modules, using lasers on the surface and fiber-optics, and the previous year he and four accomplices had not had time to complete the task before the end of the season. This was no reflection on their abilities or work ethic; it was a huge job, even with AMANDA's skimpy number of modules—667 to be exact. If IceCube was really going to comprise its planned 4,800 modules, it was looking as though it would be impossible to calibrate even in an entire season.

A year or two earlier, Dave Nygren had done a back-of-the-envelope calculation that told him that they should be able to program the DOMs to calibrate themselves with no human intervention at all. He admits that he was "barking up the wrong tree," however, on a way of implementing his notion. It was Bob Stokstad who actually figured out a way to do it. The basic idea was to send an electrical pulse down the mile and a half of copper wire to the module, ask the module what time it thought it received the pulse, have it wait for a specified interval, and send a pulse of the same shape back to the surface. Since this was something like sending a flash of light to a distant mirror and measuring how long it took to make the round-trip back, it allowed them to not only measure the length of the wire, but also to synchronize the clock in the module with a master clock on the surface. And since it could all be done with computers, even five thousand DOMs could be synchronized in the space of five or ten millionths of a second!

A team from Lawrence Berkeley Laboratory consisting of Jerry Przybylski, a software engineer named Chuck McParland, and a physics post-doc named Azriel Goldschmidt traveled to Pole in early 2001 to test out various new ideas on the digital string, number 18. After replacing the data acquisition system that hadn't worked the previous year with one that had been designed at LBL in the meantime, they put Stokstad's solution in place and proceeded to demonstrate that the forty-odd modules on string 18 could be synchronized to within less than five nanoseconds.

Goldschmidt also put in an improvement on the science side that ended up demonstrating, with data taken over the subsequent winter, that they could reconstruct the path of an up-going muon with just that one digital string—an impressive accomplishment. And they were already capable of sitting at a computer in the northern hemisphere and programming an individual DOM deep in the Ice more than half a world away. The digital technology was more than meeting its requirements, it was blowing its analog counterpart away.

Goldschmidt and his partners were working under a certain amount of pressure, since Francis (in what he characterizes as his "main contribution in leadership, if I ever had one") was at that time trying to herd the collaboration into a decision on which optical module to use in IceCube. He scheduled a two-day meeting for late February, before an external advisory committee chaired by the widely respected Barry Barish, the experimental physicist from Caltech whom NSF had brought in to save the Laser Interferometer Gravitational-Wave Observatory after its initial management failure. To make it easy for everyone to attend, the meeting was held in a conference room at the Chicago O'Hare airport.

The LBL team performed with Hemingway-esque grace, delivering a report on the DOM just five days beforehand. And amazingly, since this was awfully far from theoretical physics, Francis himself wrote up a preparatory technical document. (Some time later, when he saw his name on the author list for a journal article about the DOM, he told me that he had now "lost all self-respect as a theorist.") Dave Nygren led the digital side of the discussion, and Christian Spiering and Albrecht Karle presented the case for the latest version of the analog/fiber-optic module. Bob Paulos remembers Nygren as being extremely helpful: he "more-or-less represented the digital approach but was very evenhanded and open-minded about discussing the merits of either." And even though Albrecht had invented the fiber-optic method, he didn't fight either; he just wanted the best decision. "And so we selected . . . digital," he says, "and everybody pulled together. . . . There were no hard feelings on any side. . . . That was a good day for the collaboration, I would say."

The distracting personal melodrama that Steve Barwick decided to serve up that year was to drop out of IceCube and begin lobbying for an upgrade to AMANDA that would, in effect, be direct competition. The previous September, he had held what was ostensibly an IceCube planning meeting on his home turf in Irvine, which turned out to be an advertisement for his desire to become principal investigator and run the project out of his school. When it became "clear," according to Buford Price, "that nobody else outside of Steve would support him being the head, he resigned in a huff." And this was his response. At the collaboration meeting in Delaware in March 2001, Steve turned up in the role of "interloper" and gave a whiny, downbeat talk in support of his renegade idea. "Well, we're all trees in the forest, fighting for a space in the sunshine," Bob Morse observed.

Since the no-new-starts policy had thrown IceCube's funding prospects back into question, Steve's gambit caused concern among the other PIs. It was an obvious vote of no confidence, and several read his move as a ploy to undermine the larger instrument and maintain what little influence he had, while hoping for IceCube to fail.

Steve would continue to champion what he called AMANDA++ (an inside joke from the computer language C) until 2005, when the AMANDA collaboration was formally dissolved and merged into IceCube. His gambit failed miserably, since it ensured that he would have zero influence on the larger instrument, which was where the train was headed.

This was a disappointment for Francis, who had hoped Steve could play a major role. In his opinion, the experiment outgrew Steve. When he could no longer analyze the data on his personal computer or hold the budget in his head, "he lost it, . . . because, you don't have it in your head, you rely on other people, and he just couldn't do that." In the early days of AMANDA, his contributions were crucial, and IceCube would not exist without him, but he has never had a role in the larger instrument.

The general opinion, not only in AMANDA but also in the community at large, was that Steve and George Smoot's antics arose largely from their desire for the physicist's version of fame. When they gave talks at conferences and colloquia, each had the irritating habit of representing himself as principal investigator of AMANDA (or IceCube, if that was the subject) and making it seem as though his institution was leading it. This was very much in line with George's behavior on the COBE experiment.

Since COBE, George had been kicked out of a different collaboration, BOOMERANG (Balloon Observations Of Millimetric Extragalactic Radiation ANd Geophysics), for the same sort of behavior. And some years later, Steve would be kicked out of ANITA (ANtarctic Impulse Transient Array) for the same thing. (BOOMERANG flew cosmic background radiation instruments in balloons over Antarctica, and ANITA flew balloons equipped with antennae tuned to detect neutrino interactions in the ice sheet below.)

Between Smoot and Barwick, AMANDA was developing a bizarre reputation. Dave Cline remembered the collaboration as having a "bad image in the scientific world," owing to the "Irvine connection," meaning Barwick, and all the bickering, which was obvious to outsiders. (Cline distinguished AMANDA from the "brilliant success" of IceCube.)

But Francis wasn't all that concerned with Smoot and Barwick's self-promotion. He recalls sitting with Peter Gorham, the principal investigator of ANITA, in a coffee shop in New Zealand during a neutrino conference in 2008 and asking him why they kicked Steve out after he contributed so much to the experiment. When Gorham replied that he tried to steal all the credit, Francis laughed and said, "Oh, if we had thrown him out for that, he would have been so long gone." They never in fact did throw Steve out. He is still a member, though noncontributing, of IceCube.

Smoot was a different story. Dave Nygren observes matter-of-factly that when George realized he would not become PI, he "lost interest" and "kind of faded out from the scene." Aside from speaking ill of the project at every chance he got, he chose to ignore his operational responsibilities. His group was charged with developing some software for running the instrument or analyzing some data, and he had obtained a grant for doing so, but he blithely decided to use the money on some theoretical work instead. This was the last straw. At the meeting in Delaware, which George did not deign to attend, the other PIs voted to expel him. They tried to gain control of the misdirected grant money and failed.

Francis claims to have no hard feelings and to remain friends with both George and Steve. (This is Francis's shtick. He doesn't like to be the bad guy.) He tells me he's got a technique for working with his fellow physicists that helps him understand and not take it personally when they act like children: he assigns them an age, the maximum being twenty-one, signifying full maturity. He only uses the method on physicists, by the way, whom he sees as being less stable than "normal" people, based on what he claims is "a lot of evidence."

Only one member of the IceCube collaboration is assigned the age of twenty-one: Dave Nygren ("You notice that only Nygren can drink alcohol"). Per Olof Hulth was "in the Nygren category," and several others follow closely at "maybe nineteen": Olga Botner from Uppsala; Lutz Koepke from Mainz, Germany, who has taken the lead on supernova detection; and three or four professors from the United States. Francis puts himself at a modest seventeen ("eighteen is already mature"). Steve Barwick comes in at "twelve to fourteen; he certainly doesn't reach puberty." And George Smoot comes in at "thirteen, but a nice thirteen."

It would seem that the ability to play with the other children in the sandbox is not a factor in the decisions of the Nobel physics committee. For, in spite of the fact that George had burned bridges with virtually every collaboration he had ever worked in, he shared the 2006 prize with George Mather of NASA, whose respect he had quite thoroughly lost when they had worked together on COBE, the experiment for which they won the prize. Per Olof remembered being very happy at seeing the smile on Smoot's face when he greeted him in Stockholm.

And in contrast to Steve, George has found fame. He once won $1 million on the game show *Are You Smarter Than a 5th Grader?*, and he's appeared as himself in an episode of *The Big Bang Theory*. Just the other day, I was astounded to glance up at the TV in a restaurant in rural Vermont, and there was "George Smoot, Nobel Laureate," walking onto the scene in a Turbotax commercial.

✦ ✦ ✦

Francis jetted about so much during the no-new-starts period that travel began to lose its charm even for him. He believes that since IceCube had been pushed back into the funding queue, NSF wanted to keep them visible—and busy—by reviewing them again and again (although he also entertains the suspicion that the foundation was looking for an excuse to drop the project). IceCube went through something like fourteen major reviews, mostly in the United States, but in Germany, Sweden, and Belgium as well (two Belgian groups had joined the collaboration sometime in 1998 or 1999). They were even reviewed by the National Research Council, an arm of the National Academy of Sciences, which doesn't usually engage in such micromanagement. Dave Nygren says they "got so tired of it, you know, I would go to Wisconsin for some review, and I'd say, 'Okay, Francis, what's the plan?' And he'd say, 'Well, keep talking until the time runs out.' And I'd say, 'Good plan.'"

At one point Francis realized he had traveled to Europe an average of once a month for several years, "which means that you live off time basically, continuously, and I think these years are going to count double later on, when it's time to go."

This unusually young-looking man was beginning to show his age. His personal clock got so mixed up that wherever he was, at home or in some faraway hotel room, he would wake up in the middle of the night and read until he got sleepy again. This is a sensitive time psychologically, of course. One night, he was reading the book *Astronomer by Chance* by Bernard Lovell, who had worked on radar during World War II and gone on to build the world's first large radio telescope at Jodrell Bank in Cheshire, England. There were so many delays and cost overruns that at one point Lovell's employer, the University of Manchester, considered prosecuting him and sending him to jail! Francis was about $5 million in hock to his university on the night he read this. He froze and looked at his hotel ceiling in horror, thinking, "This could happen to me!" and for several weeks went around telling everybody how scared he was. He can laugh about it now, and he often does, but he also says that he still doesn't think it was funny. Lovell turned things around, incidentally, and made important discoveries that essentially launched the field of radio astronomy. He was eventually knighted.

The truth was that IceCube *was* in serious risk of failing. NSF kept giving them aggressive deadlines—acting as though they had given them money even though they hadn't—so when they'd gotten the green light at the end of the Clinton era, the Madisonians had gone into full start-up mode, hiring people and designing the drill. By mid-2001, the design was nearly complete and it was time to think about building the thing, at an expected cost in the neighborhood of $10 million (it would ultimately cost about $12 million). The $4.5 million from WARF would not last long, and it would be very expensive to stop the project, which would mean laying people off and starting again; in fact it might kill it altogether. It was crucial to keep the money flowing.

The political tensions engendered by no-new-starts came to a head over the summer, when Congress got into the act. It was precipitated by neutrinos, interestingly enough, but not IceCube.

What had happened was that the company that owned the Homestake Gold Mine in South Dakota, where Ray Davis had operated his solar neutrino detector for decades, had decided to close the mine. (Davis's group then decided to end their experiment, as they had definitely proven their point. Davis would win the Nobel Prize the following year.) The imminent shutdown prompted a group of researchers led by John Bahcall to propose that the National Science Foundation turn the mine into a major underground research laboratory. They envisioned various neutrino experiments, including a competitor to Super-Kamiokande in Japan, as well as efforts in geology and "extreme biology." The cost was estimated at $281 million. A decision needed to come quickly, because Homestake intended to shutter the mine by the end of the year, at which point it would begin filling with water. This led the two Democratic senators from South Dakota, Majority Leader Tom Daschle and Tim Johnson, who happened to sit on the appropriations panel that oversaw NSF, to take the extraordinary step of trying to earmark $10 million to keep pumping out the mine while the foundation made up its mind.

Normally, this project would have been funded through the same major facilities account that would have been funding IceCube had Bush not issued his no-new-starts diktat. IceCube was probably first in line for this sort of funding, and a couple of other thoroughly reviewed projects had been put on hold as well. Meanwhile, the Homestake enthusiasts had barely submitted a proposal—NSF hadn't reviewed it yet—but here they were elbowing their way to the head of the line.

It happened that the ranking member of the *House* Appropriations Committee, David Obey, not only hailed from Wisconsin, he had gone to law school at Madison and had great affection for his alma mater. He was being briefed on IceCube's teething problems all along by the university's congressional liaison, Rhonda Norsetter.

And so, "one fine afternoon" in mid-June, according to Terry Millar, Obey called Chancellor John Wiley, and said, "John, unless I hear from you otherwise by the end of the day, I am going to earmark fifteen million dollars to keep IceCube afloat."

Wiley's first thought was to turn the money down. He figured it would not only hurt the university's reputation in the short run, it would probably hurt financially in the long. The National Science Foundation had been keeping track of federal research expenditures at universities since the early seventies, and in that time the University of Wisconsin had been in the top five every single year, usually higher. The only other university in that class was Johns Hopkins, which was way out in front, actually, because it oversees the Applied Physics Laboratory, which conducts about 50 percent classified research. The University of Wisconsin did not allow classified research on campus. Wisconsin had achieved this record by depending on a level, peer-reviewed playing field, not by relying on earmarks and pork, and Wiley intended to keep it that way.

On the other hand, IceCube was in trouble. Wiley called Millar and Norsetter into his office to help him make up his mind and went off that afternoon to deliver the keynote speech at a gala retirement party elsewhere on campus. Millar and Norsetter spent the day going back and forth with Obey's staffers, building the argument that since IceCube was one of the most peer-reviewed projects in history and had passed every review with flying colors, sure, the university would take a hit, but it would probably survive. The two drove over to the retirement party in time to catch Wiley's "little spiel," and made their pitch to him afterward over hors d'oeuvres. They succeeded in changing his mind.

Congress finally overrode no-new-starts in the spending bill that was hammered out by conference committee in early November 2001. NSF ended up getting more money than the president had asked for, and five major research facilities were funded. IceCube got just the amount Obey had promised.

It seems that Obey's staffers had it about right. His earmark did generate some controversy, but it never got too heated. According to an article in the *Chronicle of Higher Education*, "One physicist referred to the IceCube project as 'Kosher pork'—somewhere in-between a peer-reviewed and a pork-barrel project." The name stuck.

+ + +

But it wasn't over yet. Even as the conference committee was finalizing the 2002 federal budget, the Wisconsin folks learned through back channels that NSF director Rita Colwell had failed to include IceCube in her budget for the next fiscal year, 2003, which would begin in September 2002. This was an extraordinary move, since NSF had already approved it, it had originally been included in the 2002 budget, it was a multi-year project, and it had now received initial funding. It may have been that she objected on principle to pork, kosher or otherwise, but the story in Madison was that Colwell, an environmental microbiologist, was resentful over the fact that Obey had not funded *her* pet mega-project, NEON, the National Ecological Observatory Network. According to Terry Millar, she had lobbied Obey over the summer, he had turned her down, "and the following week he hands her fifteen million for something she doesn't even want": IceCube.

Her omission also had immediate consequences, because NSF needed to act in order to release the $15 million in kosher pork for the 2002 fiscal year. This had to be done by the end of January, but all signs indicated it would not.

Now, Terry Millar is an energetic and creative guy—perhaps too energetic and creative for a university administrator, he is willing to admit. Don't forget that he toured Vietnam as a Marine. He got in touch with Colwell's predecessor as NSF director, Neal Lane, who briefed him on the political complexity of science policy at this level. In an e-mail to John Wiley summarizing the conversation, Terry mentioned that Lane was "totally sympathetic toward Dr. Colwell for the difficult position she must find herself in" and that Lane had observed that "interactions among NSF, OMB, OSTP, and Congress are fluid, chaotic, and often difficult to anticipate or manage" (referring to the Office of Management and Budget and the Office of Science and Technology Policy). Nevertheless, Lane had suggested the "prudent action" of going over Colwell's head.

NSF reports to OSTP, which is an office of the White House. Its head is the president's science adviser, at that point a physicist named John Marburger.

Terry believed that one of his responsibilities as assistant dean was to know everyone who was doing research on campus and figure out ways to synergize their efforts. ("I grow neurons and get them to fire up, and then I try to nurture them along, and those connections create new kinds of possibilities within the university.") In his energetic way he had somehow found out that Madison had an obscure connection to Marburger's office: A former Apollo astronaut by the name of Harrison "Jack" Schmitt—the second-to-last person to have walked on the Moon, in fact—occasionally co-taught an engineering course at the university, and Schmitt happened to know the woman who kept John Marburger's calendar. Terry arranged a dinner with Schmitt at a seafood restaurant outside Madison named Captain Bill's.

Over dinner, Schmitt told him that Marburger's main concern at that moment was biosecurity: only a month after the September attacks on the World Trade Center and the Pentagon, letters and packages laced with anthrax had been mailed to several media outlets and the office of Senator Tom Daschle.

"And so we said, you know, we have a lot to say about that," recalls Terry. (Thanks to the Wisconsin Alumni Research Fund, the Wisconsin Idea, and the fact that agriculture is so important to the state's economy, biology may be Madison's strongest suit.) "We'd be happy to come in and talk about it. And, by the way, we'd also like to talk about this thing called IceCube. . . . We got in through the back door."

Schmitt managed to arrange a face-to-face meeting with Marburger for a delegation consisting of John Wiley, associate dean for the biological sciences Tim Mulcahy, Rhonda Norsetter, Terry, and Francis Halzen. In a letter that Wiley sent to Marburger in advance of the meeting, he came clean about their true intention of discussing IceCube. The meeting took place in mid-February.

"And so we did the Homeland Security bit," Terry continues, "and then we brought up IceCube. And [Marburger] said, 'IceCube. What is that?' So we told him the story, and he said, 'So why isn't this? . . . I don't get it.' And so we said, 'Well, it might have been because of, you know, this little thing that Obey did' and that 'we think the NSF director—we're not sure, of course, but she couldn't have been real happy because that wasn't one of her pet projects.' . . . And so he thought for a moment, and he said, 'Well, I can't get it into the budget this year, but I'll put it into the president's budget next year. And I'll make sure NSF is put in line on this.'

"And so we thanked him profusely, and I said, 'Ah, so, ah, that would be next year, and we're at the end of our first year, so what would be your recommendation for the intervening year?'"

Since Marburger couldn't get them into the 2003 budget, they would be missing a year of funding.

"And so he looked at all of us and he sort of smiled and he said, 'Well, you seem to have solved the problem for the first year. Ah, the administration will not object if you solve it the same way for the second year.'"

He was inviting them to go for more kosher pork.

It would continue to be a struggle—science funding usually is—but this broke the back of IceCube's funding problem. However it came about, NSF had gotten "in line" by the middle of the year. The project was awarded $25 million in 2003, enough to complete the drill and get them to the point where they could begin construction.

Rita Colwell eventually made peace with IceCube as well. She later told Francis he was lucky, because she slept with a man who loved the project. Her husband, Jack, was a physical chemist.

19. The Coming of Yeck

There's the IceCube period, construction basically; there's the AMANDA period. But there was a period of five years in between, which I—it's like before Jesus Christ: it's before Jim Yeck.

—FRANCIS HALZEN

AS IF THE FUNDING OBSTACLES AND INTERNAL SQUABBLES WERE NOT enough, a more fundamental problem was also rearing its head. They were discovering that Beverly Hartline, their hardnosed interrogator on the Hartill panel, had been correct: the University of Wisconsin did not know how to manage a project of this size.

It seemed that the higher-ups in the Space Science and Engineering Center just didn't get it. This project was by far the largest ever undertaken by the university—it was in the running, in fact, for the largest ever undertaken by *any* university. It was three times the size of all the other projects in SSEC combined, yet they were treating it as just another project. They were low-balling resources, they were shifting personnel in and out—even Bob Paulos, the project manager—and they were trying to make technical decisions that Francis Halzen and Bob Morse were not willing to let them make. They were insisting that IceCube fit into their internal management structure, when, at three times the size of their entire operation, it clearly needed independent management of its own. The tail was wagging the dog.

This was especially difficult for Paulos, since it caused friction with people he had known and worked with for fifteen years. But he soldiered on nevertheless, and managed to make progress—especially with the drill. In an e-mail to two other senior administrators in early 2001, Terry Millar observed that "at this point, IceCube fails without him." But even Bob knew all along that they would eventually need to hire someone above him who had managed multi-hundred-million-dollar projects before. And as the transition from planning and designing to actually building their Great Pyramid under the Ice approached, and as the budget grew from $15 million a year to more like $40 or $50 million, their overseers at the National Science Foundation began insisting that that person be found.

Francis had met Jim Yeck when they had both sat on a review committee for the controversial underground laboratory in the gold mine where Ray Davis had housed his solar neutrino detector.

Jim is an engineer by training, with an idealistic streak. He'd joined the Peace Corps after college and gone to Thailand to work as a water resources engineer, building spillways and canals and the like. He then returned to the States and earned a master's degree in mechanical and nuclear engineering with a focus on plasma physics. ("I was interested in fusion, still this idealistic type, you know. Energy was, as it is today, a big challenge.") The Department of Energy hired him from graduate school into a project management program, and he went from there to a DOE-funded fusion project at Princeton.

His big break came at the age of twenty-nine, when he was made project manager for the construction of the $500 million Relativistic Heavy Ion Collider (RHIC) at Brookhaven National Laboratory. (He suspects that the reason they gave so young a person the job was that they were having a hard time getting people to move to eastern Long Island.) Since the project was underfunded, the challenge was "to work through that in a way that in the end we could say that we had built it successfully, on cost, on schedule . . . and we did." This introduced him to the world of nuclear and high-energy physics.

Partway through the RHIC project, it was discovered that one of Brookhaven's research reactors was leaking radioactive tritium into the only aquifer on Long Island. In the resulting public and political uproar, Jim was asked to step in, and as a person who admits to having difficulty saying no, he agreed. He stepped out of RHIC for about eighteen months to lead the lab's so-called Tritium Remediation Project.

"Let's just say that it was intense and it was interesting and difficult and *60 Minutes* and the whole deal. . . . I can tell you, when you have this boom over your head, heh, heh, when you're talking to someone from the public who's saying that their child has cancer and they're sure that it's from Brookhaven Lab and the reactor, it's a very difficult thing to deal with. . . . It was in many ways a thankless job, but I got a lot of visibility through that, and I handled it in a way that was, let's just say, satisfactory."

RHIC was nearing completion at about the time the Superconducting Super Collider, which would have been the U.S. answer to CERN's Large Hadron Collider, was canceled by Congress. Desiring to remain on the high-energy frontier, U.S. physicists shifted their attention to the Large Hadron Collider, and Jim was put in charge of the entire U.S. effort—half a billion dollars or so between work on the accelerator itself and various detectors and experiments. DOE told him he could live anywhere he wanted, so he moved back to Illinois, the state where he'd grown up, and based himself at Fermilab.

This project gave him the chance to work with the National Science Foundation for the first time and through NSF with the roughly thirty universities associated with the LHC experiments. He found that he enjoyed working with the schools, and he was also coming to believe that the national labs, all of which are run by DOE, were outgrowing their utility. The concept of such huge, centralized facilities had been born in World War II with the Manhattan Project, and they had now been around long enough to have taken on self-serving lives of their own—dragons needing to be fed. "The thought of a university proving to the world or to the funding agencies that they can handle a large project is interesting to me," he says.

But he wasn't looking for a change. He was perfectly happy at Fermilab.

One day, Francis Halzen called Jim and asked if he knew anyone at Fermilab who might be interested in becoming project director for IceCube. Jim gave him a couple of names, and Francis asked if *he'd* be interested. Owing to his "inability to say no just outright," Jim responded that he "might" and promptly forgot about it.

Francis called back the day applications were due.

"You put your application in, right?"

"I'm like, 'No.' And he's like, 'Well, you need to put it in.' "

The timing was okay, because the LHC was pretty far along and Jim was confident that the U.S. side of things would end successfully. And he still had this hard time saying no.

Everyone involved in IceCube was in for a lesson in project management.

Jim Yeck does not take on jobs that are set up to fail. One has the impression that he interviewed the Wisconsinites as much as they interviewed him. He talked to Francis, he talked to John Wiley, he talked to Bob Paulos. "Francis is very likable. You know, people like Francis," he says. "That can be very helpful in something like this." He liked the other people as well, and the project "met this need to prove that universities could do stuff."

He also liked the fundamentals. AMANDA had taken care of the basic research and development. There was institutional support to the level of the chancellor: "absolutely critical, cuz you're gonna have problems and when you have a problem you need people to pull together and provide the support that's needed to get past it, as opposed to trying to find out what went wrong and blame someone or distance themselves." NSF was by now firmly behind the project, so the construction phase would probably be funded.

And he made sure that he would be in a strong place in the management structure. This made it a bit confusing at the top, owing to the coexistence of the science and construction missions. As principal investigator, Francis held ultimate responsibility in the eyes of NSF, but Jim came in as essentially his equal in the Madison hierarchy. He reported not to Francis but to the university's vice chancellor for research, who reported to Chancellor Wiley. And this was where Francis demonstrated some management insight of his own: when Jim came on board, Francis told him that if they ever disagreed, he would defer to Jim.

This was unusual. The history of big physics projects is festooned with stories of arrogant physicist control freaks blowing it the first time around and having to cede control to capable managers. LIGO is a good example.

"Ninety-nine point nine percent of physicists always think they're the smartest person in the room and that nobody can do anything better," says Francis. "I realized that was not the case. You put people in charge that know how to do things and then let them do it. And the most important thing is to choose the right person."

Yeck joined in October 2003.

Jim was neither surprised nor fazed by the chaos he discovered when he got to Madison—and chaos it definitely was. The Antarctic season was just beginning. They were planning to ship the drill to Pole at the end of November—not to drill, just to get it down there and stowed for winter—and they were planning to drill something like twelve holes the following year. It was already late enough in the day that they'd have to airlift the parts to McMurdo rather than ship them by boat, and everyone was panic-stricken. NSF's hired consultants were calling daily to ask how close they were, had they done this or that test—Jim says they didn't know what they wanted; they were just nervous—and the folks in Madison weren't admitting even to themselves that they didn't stand a chance of meeting their deadline.

Jim is soft-spoken and might even be a little shy, but he does not lack self-assurance. A couple of weeks into his new job, he called an engineering meeting that resulted in a decision to push the schedule back a year. He then sent a letter to NSF, thanking the foundation for working with them, acknowledging that everyone had worked hard, and suggesting that it would not be in their "collective interest" to ship the drill that year. They'd ship the largest subsystem, the hose reel, and get it assembled, but that was about it.

He believes the NSF managers were "delighted to get that letter, because they knew that somebody was going to be accountable and responsible—which is me. I put it on my shoulders." Until that moment, the buck hadn't stopped anywhere. Responsibilities had been blurred.

As it worked out, they barely got the drill to Pole a year later.

Jim quickly made the project his own. At the third Hartill review, which took place only three months after he arrived, he presented a revised budget, which increased the cost for eighty strings by about 13 percent, to $270 million, and a new schedule that called for construction over six Antarctic seasons, with completion in 2009–10. Instead of drilling twelve holes in the coming season, they would aim for four.

The previous cost estimate, hammered out in the second Hartill review, had been $240 million. Since the Europeans had pledged $30 million, Jim's budget took NSF's contribution from $210 million to $240 million. (All these numbers were assumed to have a 20 percent "contingency": it was understood that they could go that much higher if something big went wrong—if they had to work an extra year on the Ice, for example.)

Jim also took them to school on the federal budget game. He was quite aware that there was a political aspect to funding and that NSF wasn't dotting the *i*'s and crossing the *t*'s. They had essentially cast their vote on his competence by hiring him, and they trusted him to make this work. He'd also been around the block with the Office of Management and Budget and Capitol Hill enough times to know that Madison could probably get $240 million for the project. They went to NSF with a request for two-fifty, and when the foundation balked, "showed blood on the floor by cutting strings": they "de-scoped" to seventy strings at a total cost of $270 million and got NSF to agree to Jim's original goal of $240 million.

Then there was the rat's nest at the Space Science and Engineering Center. Aside from the management problems previously mentioned, the various administrators, in their wisdom, had set up a strange "ghost organization" within the center to house both IceCube and the Ice Coring and Drilling Service (the new PICO). It was named the Antarctic Astronomy and Astrophysics Research Institute (A^3RI), and at least one astronomer who had nothing to do with IceCube was involved.

A decision to separate IceCube from the center had been made before Jim arrived, but nothing had been done in that direction. He concurred in the decision, of course. "SSEC really didn't give a shit about the project; they wanted the overhead. And it had the wrong culture for the project we were building, which was this 'engineers are smarter than physicists' problem." He wanted to build a culture where the physicists were in charge but were respectful of the engineers.

SSEC did have the advantage, however, of enjoying a special arrangement with the university in regard to overhead. In a normal grant, a scientist will request a certain amount of money for the cost of a project, and a significant percentage of that amount will be added to the total grant award in order to pay for the services her university provides in enabling the scientist to work. The most prosaic of these would be the cost of maintaining buildings and space, keeping the telephones ringing and the internet running, and so on. More important would be the incredible capabilities that a large and thriving institution can bring to the party: the university's Physical Sciences Laboratory was building the drill, for example. At Madison at that time, the overhead rate was 46 percent.

Jim saw A^3RI as a distraction and a breeding ground for petty academic politics. Why should IceCube be tangled up in unrelated drilling projects and unrelated science? He managed to kill it, and he and Francis proceeded to work out a deal with the deans and the chancellor whereby IceCube would become its own center in the graduate school, similar to SSEC, and similarly keep its overhead. This was something of a benign shell game, perhaps, since the university would have supported the project in any case, but it did ensure that maximal resources would be focused specifically on IceCube and that the people on the front lines would be calling the shots.

Francis makes the point that a university does not plan to "make money" on a project like this. It hopes not to lose money—and most importantly not to fail. This is the biggest risk, since the school's main goal is to enhance its reputation and visibility. Being in Antarctica and being about neutrinos and cosmology, both AMANDA and IceCube attracted a lot of attention in the media. Academically, they had intangible benefits such as attracting good graduate students and even faculty. "You don't see how many undergrads come here and get motivated by projects like this," he adds. "It's kind of like the argument, if you don't do this you won't be a great university. You cannot quantify it. It's like if you don't do fundamental research you won't be a great country."

Part and parcel of the separation from SSEC was the question of space. Where would the new IceCube center be housed? Jim saw this as an important management issue. In the current situation people were spread out, because they couldn't all fit in the SSEC building. They had offices there, in some space down the street, in the physics department; they were out in the countryside at the Physical Sciences Laboratory. This bred mistrust. "I knew from my experience that if you have a project, ya gotta have this sort of we're all in it together under one roof thing, if you can pull it off."

He worked with the university, and it took about a year. At one point the administration proposed some space in a local correctional facility—a suggestion he rejected out of hand. "I just kept saying, 'Nyet. Nyet . . .'"

They finally found an entire floor of an upscale office building not far from campus and only a couple of blocks from the mall that surrounds the state capitol building. (Madison is the capital of Wisconsin.) It was large enough to accommodate everyone on the project outside of the engineers at the Physical Sciences Laboratory, who needed to be out at PSL anyway, and it was entirely unfinished, so Jim could design the space from scratch. With the help of an architectural firm, he designed it to force interaction. There would be a lot of glass: in the conference space, in the break rooms, "all good stuff, because the project makes more progress when people are communicating."

"I wanted the space to be attractive, so the physics people would want to be here and not at physics, . . . so that they would see this as their home, and they have. And then I wanted it attractive so that the collaborators would want to have meetings here. So they would come here and spend time."

He succeeded in making it reasonably attractive, in a corporate sort of a way. When Dave Nygren first visited the place, he referred to it as the Ice-Cube Savings and Loan.

Clearly, the Madisonians got more than they'd bargained for. They had figured they could handle internal management themselves and had expected Jim to be the guy who would "take care of NSF" for them. Jim says that this meant, basically, "push back at them and stuff, so they stay out of our shorts." Not only was this obviously very different from his view of his role, he also knew it was not the way to work successfully with the foundation.

He introduced the novel idea of working *with* them, rather than *against* them—partnering, he calls it. This is what he was aiming at with his allusion to "collective interest" in the letter he wrote to delay shipment of the drill. They should also partner with Raytheon, the foundation's Antarctic contractor. But when he arrived in Madison, he encountered quite the opposite: "Short version is, they say, 'NSF has got their head up their ass, and Raytheon has their head up their ass.' And I said, 'You know what? And so does Wisconsin.'" (He has an understated, frank way of speaking that makes this sound humorous rather than threatening. Jim is tough, but he's good with people.)

Neither NSF nor Raytheon knew much about partnering either. The foundation has many of the flaws inherent to a federal bureaucracy: an "inside the beltway versus outside the beltway" mentality; the occasional short-term political appointee who is partisan or unqualified; petty bureaucrats who have "burrowed in"—the term for someone who has been appointed to a certain job under one presidential administration and shifted to an empty permanent post, usually lower in the pecking order, in whatever agency he or she can find, a month or two before the next president takes office. The director of NSF's Office of Polar Programs all through IceCube's planning and construction phase was a burrower. This is a powerful position; he was in charge of all the foundation's activities in both polar regions, and in the view of virtually everyone involved in IceCube, he was an obstructionist and grave detriment to the project, much more concerned with ducking risk and protecting his scalp than in working toward success. Considering the atmosphere at the foundation, on the other hand, such fears were understandable even if inexcusable: Robert Eisenstein, the foundation's Assistant Director for Mathematical and Physical Sciences in the early 2000s, was well-respected and perhaps too good at his job. He played a major role in getting two large physics projects into NSF's "mega-project" pipeline, IceCube and LIGO. In 2002, the same year that he got some sort of a plaque from Congress or the president for being one of the outstanding administrators in Washington, he resigned abruptly from NSF. "Sources" told *Science* magazine that "he had lost the confidence" of Director Rita Colwell. The Madison folks understood that Eisenstein had earned Colwell's enmity by bringing too much notice—and money—to physics at the expense of her field of biology.

You may have noticed that AMANDA and IceCube have dealt with many different program managers over the years. This comes in part from natural organizational change and in part because NSF, like other science agencies, employs "rotators," people who step out of academia to work at the foundation for a few years and then return to their academic posts. Willi Chinowsky was a rotator. AMANDA and IceCube have been blessed with helpful managers, by and large. Without John Lynch, AMANDA probably never would have happened, and Gene Loh was the hero who did the same for IceCube. (Loh continued to help as NSF's Program Director of Astrophysics until shortly before his death in 2006, at age seventy-two.) And, luckily, about a year before Jim Yeck came on board, a superb manager named Jack Lightbody was named project director for IceCube. Francis says his arrival "was like the Sun rising in Arlington." The great thing about Lightbody was not only that he actually knew how to manage a big project, he was also senior enough to push back against the incompetence in Polar Programs. A longtime NSF employee, he had shifted over from an executive position in the Physics Division.

Jim worked very hard to partner with the foundation and Raytheon. In their regular conference calls, he says, he noticed right off that the manager of science support at Raytheon seemed to think it made him look good to make Wisconsin look bad by pointing out their mistakes. Meanwhile, the NSF folks liked to play the potentate, sitting in the middle, adjudicating disputes. Jim called the Raytheon guy separately and said, "If you ever bring up an issue on that call where you're pointing at us, then I'm gonna beat the shit outa ya. I'm not gonna let it lie, you know, cuz we gotta solve this stuff before we talk to NSF." (Unlike Francis, he is not afraid to be the bad guy.) He brought a manager from Raytheon into the IceCube organizational chart, at the level of some of the physicists, and this man participated in IceCube's weekly meetings: "So, here's our dirty laundry, you can look at it, and, you know, be part of the team." It worked.

In the spring of 2004, the NSF board that is responsible for such decisions recommended that IceCube proceed to construction, in the recognition that the total cost to the foundation over the next seven years would be slightly more than $240 million. This released about $50 million for the current fiscal year. In the memorandum formalizing their recommendation, the board indicated that the recent hiring of "permanent Project Director (PD), James Yeck," was the most important factor driving the decision, noting that he was "widely recognized as one of the leading project managers in the world."

Meanwhile, they were sprinting toward their first IceCube season. The optimistic plan was for four strings. Jim hedged his bet and committed to only one.

20. Failure and Success

THE ICECUBE PHYSICISTS MAY HAVE HAD SOMETHING TO LEARN ABOUT project management, but they were good at picking people. Francis and Bob Morse had brought two superb additions into the team the year before Jim Yeck entered the picture: Jeff Cherwinka and Jim Haugen.

Cherwinka had worked at Madison's Physical Sciences Laboratory for almost ten years in the eighties and nineties, followed that with two years at Cornell as chief engineer on a particle physics detector, and then returned to Madison to work in private industry. When the company he worked for was put up for sale, he ran across a posting at PSL and applied, thinking it would be a return to high-energy physics. The last thing he expected was to be working on a hot water drill. He was hired as principal engineer on the drill in July 2002.

Not long after he arrived, Morse, Halzen, and Paulos began looking for someone to supervise the making of the digital optical modules, and Jeff recommended Jim Haugen, who had been his boss at his industrial job. The other three resisted at first, since they thought they needed an engineering type, but Jeff convinced them that what they really needed was a production manager.

Haugen is an astoundingly upbeat and energetic individual who refers to himself as "a lifelong Wisconsin guy." He graduated from Madison with "kind of like a double major in math/physics with a minor in electrical engineering," worked in Silicon Valley for ten years, and then lived in Thailand for about five, managing contract manufacturing facilities that could produce millions of electronic chips a day "with twenty different packaging types, to forty different customers, seven days a week, twenty-four hours a day—three million units of stuff, out the door, with the right label, with the right quantity. So, yeah, that was a whole different thing." He ended up taking over responsibility for everything that would go into the Ice, all the way from two and a half miles down to the connections at the computer farm in the new "counting house" that would be built on the surface—not just the DOMs: the cables and pressure sensors, numerous one-off devices, and so on. This would add up to about one-third of Ice-Cube's $270 million budget. He may have been overqualified for the task, and that was a good thing.

One of the many enhancements that came out of the years of research and development that had gone into AMANDA was the inclusion of a two-dimensional air shower array named IceTop, which would sit on the surface of the Ice directly above the submerged IceCube array, a mile and more below. IceTop, in turn, was an improvement upon SPASE, the South Pole Air Shower Experiment, which had been installed a couple of years before AMANDA, you will remember, and proved to be unexpectedly useful in the AMANDA effort. Since SPASE could not only detect showers of muons raining down from above but also tell their direction, and since some of those showers were headed toward AMANDA, it provided all kinds of ways of testing and calibrating the neutrino telescope below. It could be used to check AMANDA's pointing accuracy and angular resolution, for example. Even though SPASE was not optimally placed, because it was offset horizontally, it was used to "veto" a small portion of the down-going air shower background. If SPASE detected an air shower headed in AMANDA's direction, AMANDA could be programmed not to be fooled by the flood of light it would perceive shortly thereafter.

IceTop would be significantly more helpful in the way of vetoing, since it would be situated directly above IceCube and cover the entire instrument.

(As with SPASE, it would also prove more helpful than the physicists ever imagined.) It would consist of eighty-one "stations," each consisting of two large tanks of clear ice and each tank containing two submerged DOMs. The tanks would be sealed tightly against incoming light, so that the DOMs would perceive only the Cherenkov light produced by air shower particles inside the tanks (see photographs 20, 21, and 30). This added up to an extra 324 DOMs for Jim Haugen to build.

When he joined the project in March 2003, Haugen discovered that the production of the hardware was about as out of control as Yeck would discover the management situation to be when he arrived six months later. The DOM was the most sophisticated component, and Haugen knew he could not possibly allow it to evolve into the hodgepodge of zoological curiosities that had populated AMANDA. They would need to build 5,500 of them all told, accounting for extras, and the last one installed, seven or eight years hence, would need to be functionally identical to the first. That meant they had to verify and fix the design and build about 350 of them in time for their first drilling season, which was only eighteen months away. Meanwhile, the physicists didn't even have a parts list! Some guy at Lawrence Berkeley Laboratory kept the parts for the main circuit board on the shelves in his office.

As a veteran of the semiconductor industry, where there is such an obsession with making products absolutely identical that companies will go to the length of building $2 billion factories in three or four different countries that are exactly the same down to the bolts in the walls and carpets on the floors, Haugen was amazed to learn that the agreement between the different IceCube institutions called for the DOMs to be assembled in three separate locations: the Physical Sciences Laboratory in Wisconsin; Christian Spiering's institute in Zeuthen, Germany; and Stockholm University. Not only did this present a challenge to making them identical, it would add a lot of cost.

He walked into Jim Yeck's office one day and explained that he could easily set up a small plant in Thailand that could produce 5,500 identical DOMs for a labor cost of five to ten dollars apiece. He recalls that Yeck "was like, ''Kay, Haugen. Calm down. That's not how we're doing it. . . . Forget that crazy idea. . . . You're not in industry anymore. This is a big science project.'" Yeck had just come from the Large Hadron Collider, where on one experiment they were building identical detectors in *twelve* different countries. It was common, especially in Europe, for different nations and institutions to contribute in kind as well as in cash, and it had the added advantage of keeping them all engaged. Aside from the cultural value, Yeck also didn't mind the idea of moving the center of gravity ever so slightly away from Madison, even though it would cost in the end about $350 per module or an extra $2 million all told.

Haugen worked successfully to bridge the cultural gap—and not just with the Europeans; this was an industry versus academia kind of thing. "I'm a bit, ''Kay, let's roll up our sleeves, let's get 'er goin', let's . . .' And, you know, we had to just kinda take 'er a little bit slowly and back off here, and do some design verification and things like that." He flew to Zeuthen and held a meeting with Christian Spiering and Rolf Nahnhauer, the "tough German guy" who was in charge of building the DOMs there, and Per Olof Hulth from Stockholm and Allan Hallgren from Uppsala, a longtime AMANDA veteran.

He explained that together they were going to build 5,500 high-quality, identical DOMs, ship them on time every year, and meet their budget. He also introduced such radical industrial notions as bills of materials and process travelers. He deployed "the force of charm and trustworthiness, and drinking beer and being, you know, mostly right—you're never right all the time . . . and eventually people started trusting you." It didn't take long for the physicists to realize that the inherently team-based ethic of manufacturing sometimes made it more fun and even more exciting than doing science.

Haugen held the final design review for the DOM in May 2004, about fourteen months after he arrived, and they managed to get 360 of them onto a ship at Port Hueneme, headed south, in September. This was pretty much the last time the DOM raised the blood pressure of anyone involved.

Six years later, as construction was nearing its end, he asked Madison physicist Mark Krasberg to do some analytical tests to see if there were any differences between the first DOMs and the last, and Krasberg couldn't find a thing.

Pressures remained high on the drill team, however. Jeff Cherwinka and colleagues had put together a testing station at the Physical Sciences Laboratory that duplicated the tower and sheave that would be used to lower the hose and electrical cables into the Ice, and they'd dug a deep hole in the ground to approximate the drill hole.

In the early spring of 2004, when they put the whole thing together and tested it for the first time, they discovered a potentially show-stopping problem with the usual suspect, the hose. With the weight of the drill head and other assemblies that would be attached at the bottom end and the fact that the hose itself was not neutrally buoyant in water, the force of gravity acting to stretch it as it wound off the pulley would come to about 10,000 pounds by the time the full two and a half miles of it had been lowered into the drill hole.

They had searched the globe for a manufacturer for the hose and finally identified one not far from Padua, Italy. Although they had specified it for longitudinal strength, the engineer from Space Science and Engineering who had been responsible for testing this spec had dropped the ball. When they put the hose under load for the first time in Wisconsin, they could see it necking down and hear it tearing as it rolled off the pulley. It was far too late to redesign it (at a cost of about $1 million a pop), so they had to find a work-around. They ended up transferring some of the load to the metal cable that supported the electrical wires that controlled the drill head by taping the hose to the cable at intervals, using IceCube's version of duct tape, a fiberglass tape with a silicon adhesive that was affectionately known as driller's tape. They would strengthen the hose twice in subsequent years, but they'd always keep using the tape for good measure.

In the end, as I've mentioned, they barely got the drill to the pole a year after they had originally planned.

The final technology that was absolutely critical to the building of IceCube was plain old South Pole knowhow, and there was no one in the world better equipped with that than the people who had worked on AMANDA.

One huge concern was deployment. To the people who actually did that work, says Gary Hill, the prospect of a cubic-kilometer detector "was just absurd, you know, 'cuz just look at us out here on the Ice. . . . It's cold! . . . It just seemed hard to believe that things would come together to do an IceCube."

As he had with his "German whips and chains," however, Albrecht Karle came up with a relatively civilized solution. Gary remembers, sometime in 2000 or 2001, Albrecht taking him out to the large courtyard by the physics building in Madison and outlining a prospective "deployment building" in chalk.

"He says, 'Yeah, this is where the hole will be. This is where the DOMs will be stored.' DOMs being stored? You know, DOMs are in boxes, aren't they? I mean you take them out and you deploy them. 'No, no. We're gonna have the DOMs inside, ready to go, on shelves and things. And this is where the control room will be.' Control room? . . . And, you know, essentially that's basically how it ended up. We have this building, it's heated, you work inside. . . . You can basically work in a fairly easy environment in there. You don't have to dress up too much. I mean sometimes it's a bit colder near the hole and things, but, yeah. You know, these things all came through."

The deployment building was integrated into the drill tower, and the entire unit, which could be sledded around on the Ice, became known as the TOS (rhymes with the *goss* in *gossamer*), short for Tower Operations Structure.

Albrecht went south as on-ice lead for the 2004–05 season and, as he puts it, just "someone who could help." Gary led one of the two deployment shifts; Kurt Woschnagg and Darryn Schneider also worked on deployment; and they all helped in innumerable ways, simply by knowing people at the station, knowing where things were, and knowing how the station worked. There were also several experienced AMANDA drillers, including a good percentage of Swedes. Per Olof and company had been supplying drillers every year since they had joined AMANDA and would continue to do so all through IceCube.

Two critical people were missing, however: Bob Morse and Bruce Koci. Bob underwent quadruple bypass heart surgery in November, and Bruce, very sadly, had been diagnosed with non-Hodgkin's lymphoma. These two gave what help they could by telephone or e-mail from Wisconsin, but this went only so far. Their physical presence—and especially Bruce's—might have made all the difference.

When the drilling team and their many tons of equipment arrived at Pole in the fall of 2004, they came face-to-face with the fact that this drill was an entirely different beast from AMANDA's. For one thing, it was extremely complicated; it took weeks to put it together. Christmas came and went, New Year's came and went. And then, when they were finally ready to drill, they learned that NSF had failed to post an environmental assessment on the Web, so they had to wait through a comment period before it would be legal.

Since South Pole Station is a very small place, this had a negative effect on the drillers' morale. Everyone was aware of the problems. Jim Yeck remembers all this "macho/ego stuff runnin' around." (He'd gone down early to help set up the TOS and get the lay of the land and then returned north to work things from the real world.) The scuttlebutt around the station was that they wouldn't be able to drill at all that season, and the prospect of complete failure coupled with the fact that the drillers found themselves without a whole lot to do in a place that didn't offer much in the way of entertainment made for a corrosive mixture. So Jim and Jack Lightbody at NSF came up with the idea of a "pilot hole." It wouldn't be official construction, just some testing, but they could use it for deployment if it turned out to be acceptable.

The water finally began to circulate in their gigantic contraption on roughly the tenth of January, only about a month before the station would close for winter. Albrecht, whose job was to keep an eye on the entire situation, compares the experience to building a large airplane and finally starting it up, rumbling down the runway, and lifting off for the first time. You're in the air; you can't just pull over and park. "It was like, 'Oh my god! We're drilling! I guess we're drilling, yes . . .'"

It was clear from the start that something was wrong.

There is a truism in engineering that when something goes wrong, it's usually more than one thing. One problem was that the water pumps kept shutting off. On the Ice, they thought it was the computerized motor drives on the pumps, but afterward they realized it was actually the pumps themselves, which weren't spec'd for the 10,000-foot altitude. They would overheat and send a signal to the drives to shut themselves off. This had a ripple effect: when the water stopped flowing, the water boilers would overheat and shut *them*selves off. By drilling more slowly than they had planned, the drillers could limp along with this handicap. But it wasn't the only thing.

This is where one can't help but think that if Bruce Koci had been there, things would have gone very differently. There is no doubt that the man on the ground that season, Jeff Cherwinka, is an excellent engineer. He eventually solved every problem and became a master of the art. But at that point Bruce was far and away the best field engineer and troubleshooter in the game. He had an uncanny ability to sense what was happening with a drill and in the ice a mile beneath his feet, while Jeff had never drilled ice before.

Everyone seems to use the word *oscillation* to describe what was happening down where the action was, at the drill head. It would surge down into the ice, melting itself a pathway, and then rise back up by as much as forty feet. This chaotic motion made for a very poor-quality hole. What one shoots for in hot water drilling is for the water spraying out of the nozzle at the bottom end of the drill head to melt the ice at just the right speed to match the rate at which the head is being lowered. This allows it to fall in a plumb line under the force of gravity and melt out a straight, dead-vertical hole of constant diameter. It seemed that they were lowering the head too rapidly even at their handicapped rate, but it was hard to tell exactly how rapidly, because the many sensors they had put in place to measure the rate were sending back different readings. They struggled along for about a day, only reaching a depth of about a thousand meters—slower than they had managed with AMANDA—and finally put their heads together and decided to abort.

This was a relief, because the drillers were exhausted. Another problem was that their eighteen-person crew—nine people per twelve-hour shift— was too small. In subsequent years they would hire thirty.

When they pulled the drill head out of the hole, they discovered that they had damaged the so-called weight stack, which acted as ballast to pull the head straight down and give it inertia. This was a set of seventy-five-foot-long metal pipes arranged concentrically around the hose, between the nozzle at the very bottom of the hose and the drill head, which housed the "brains" of the unit: the pressure and temperature sensors, calipers for measuring the diameter of the hole, and so on. Several of the pipes had broken off and were lost somewhere in the hole. That meant it was unusable; they couldn't even ream it out.

Some months later, back in Wisconsin, they concluded that even at the slow rate they had been drilling too quickly, because the nozzle they were using didn't melt the ice as quickly as they thought it would. The following year they used a smaller nozzle, which sprayed out the water at a higher velocity and melted the ice more rapidly.

About an hour after they had powered down and begun to relax into idle mode, Gary Hill and Kurt Woschnagg were hanging out in Back of Science in the old dome, glancing occasionally at two webcams that gave them a view of the area around the hole, half a mile away. The cameras didn't give them continuous video; they displayed updated stills every ten seconds or so. It was two or three in the morning. They could see their friend Sven Lidström, a Swedish driller whom they'd worked with for years, wrapping up the last few details of putting the drill to sleep. They remember wondering what Sven was doing when he started struggling with a tangle of cables that was lying on the ice and that a couple of frames later he and the cables were gone. They soon learned that Sven had been injured.

In hot water drilling, the water in the hole is continuously circulated back up to the surface and across the Ice to the heating plant, where it is reheated and sent back down. (As I pointed out in connection to the AMANDA drill, more water goes into the hole than comes out in the end, because ice is less dense than water, so the level in the hole will drop and the hole will collapse unless it's topped up.) A submersible pump is thus lowered a certain distance into the hole and held there, in order to pump water back up to the surface for reheating.

When Sven and his colleagues had closed down the drill, they had pulled everything up by winching all the hoses and cables back onto their respective reels—except one: the electrical cable for the return pump. The pump itself was left hanging over the hole, its hose running up over the sheave or pulley on the tower, which was two stories high, and then down at a slant, under tension, to its reel, perhaps fifty or sixty feet away on the Ice. The electrical cable had been detached from the pump, so its free end was simply hanging from the sheave over the hole. The reel that held that cable was not motorized. At some point during the shutdown, the drillers had gotten tired of turning the reel by hand and begun coiling the cable onto the ground on the "uphill" side of the sheave, between it and the reel, which was locked. It was a heavy cable, at least an inch in diameter and more than 160 feet long, and this arrangement had left it, as several of the physicists later remarked, in a state of "unstable equilibrium."

Sven had been standing near the coil of cable when he noticed it was moving. The free end had begun dropping into the hole, dragging the rest of the cable with it—and accelerating, as the length of cable in the hole increased and its weight along with it. The coil, which, again, was lying on the Ice on Sven's side of the sheave, was snaking up toward the sheave. Not necessarily taking all of this in, Sven instinctively grabbed the cable at precisely the wrong moment. He must have been leaning over it as it went taut, anchored by the locked reel. Since the reel was on the ground and the sheave was about two stories above his head, the cable hit him in the chest and launched him into the air. Some witnesses said he went as high as the top of the drill tower, twenty-five or thirty feet. He landed on the metal platform that supported the Tower Operations Structure—thankfully missing one of its sharp corners—breaking several ribs and some other bones, including some vertebrae, and perforating one of his lungs, which then collapsed. A collapsed lung is a serious matter wherever it happens, and in the thin air at 10,000 feet it's worse.

Albrecht Karle, Jeff Cherwinka, and the leaders of the drilling crews were in the galley when they heard the news. This was the group that had made the decision to stop drilling, and they were now caucusing about what to do next. Sven didn't look good when they got to him about fifteen minutes after the accident. He was lying on the Ice in pain and having a hard time talking. He was trying to breathe through an oxygen mask, and to add insult to injury, the oxygen bottle was frozen.

The South Pole regulars responded with their usual competence and character, and Sven benefited from the existence of the new station as well. The medical area was now a small hospital basically, and there were two doctors in residence. He was transported in a van to "biomed," where the doctors pumped air into his collapsed lung and ventilated him. Meanwhile, BK Grant, the station manager, an old South Pole hand, got on the phone to McMurdo and called for a Herc. When the folks at McMurdo resisted, she got heavy with them: "You find me a plane right now and get the engines running!"

The doctors were still stabilizing Sven when the plane arrived at about eight in the morning, so it had to wait on the Ice for about three hours with its engines running. During this lull, the station came to a standstill. Under the deep hum of the engines, no one could work. Albrecht, who had been exhausted to begin with and doing everything he could to help for the past five or six hours, finally went to his room and broke down: "I was pretty beaten up emotionally, because of, you know, somebody nearly dying on you, who you've worked with for years."

When a van finally pulled up near one of the entrances to the station to shuttle Sven to the plane, everyone went outside in their big red parkas and lined both sides of the path to see him off. There was a second plane waiting when his Herc landed on the sea-ice runway at McMurdo, and he arrived in Christchurch, New Zealand, about twenty hours after the accident. The U.S. Antarctic Program did well. Sven would return to the drilling crew and even go on to winter for IceCube—twice, in 2007 and 2012.

One of the first things Albrecht did in the lull after the Herc was summoned was to call Jim Yeck. It must have been a Sunday, because he got

Jim on his cell phone in Illinois, on his way to church with his wife. Jim had already developed a lot of respect for BK Grant, who worked for Raytheon, and remembers that she gave him much good counsel during this difficult time. He communicated constantly with her and Jack Lightbody at NSF and brought in the leadership at the University of Wisconsin right away, too.

He issued a stop work order, at the same time warning Albrecht to let him know if the order was impeding progress toward getting back to where they could drill again—safely, of course. ("Sometimes people stop work and they sit around and like 'I dunno . . . ,'" he says.) He had learned on the Tritium Remediation Project at Brookhaven that it's easy to stop a job; the hard part is getting it started again. The most important question becomes who has the *authority* to start it, and he wanted to make sure that authority stayed in Wisconsin, with him.

He assigned one of the drilling leaders to be safety officer, but when it got down to it, the man wasn't up to the job; he thought he'd be blamed or something. "So Gary Hill, who's always been a good soldier all along the way, very sharp guy, stepped in and helped him do his job in terms of the accident report and the analysis and the photos, and produce a document that could be used as the basis for a decision to move forward with drilling again." Luckily, a consultant to NSF who was familiar with the drill—he had participated in the testing in Wisconsin—happened to be on the continent. He flew to Pole and helped reassure the foundation from the inside. But there were also the issues with the drill to resolve.

It was a great help that there wasn't a fourth bureaucracy involved, as there would have been in the AMANDA days. Albrecht, who was at the center of this effort, had been particularly adamant that IceCube have complete ownership of the drill, and Yeck and the other powers that be in Madison had separated it from the Ice Coring and Drilling Service sometime the previous year.

It took them about a week to perform the due diligence, write their reports, and, not least, repair the drill and figure out how they were going to run it this time. By then they were definitely bumping up against the end of the season. It takes some time to put a million pounds of drilling equipment to bed for the winter, and it was now several days past the date they had planned to stop drilling, the twentieth of January. But Yeck points out that "you can push it."

They chose a spot about thirty feet from the first hole and commenced. It didn't go quickly or smoothly—the water pumps were still turning themselves off and getting worse by the hour—but they did manage to get to 2,500 meters. Then the question became, how good was the hole? Should they risk dropping a string of DOMs into it and possibly getting it stuck? Albrecht, who had his doubts, asked Jeff Cherwinka, who was lying on the floor of the TOS at the time, trying to fix something, if they could possibly run the drill down the hole again in order to ream it out.

Jeff looked up at him as if he was out of his mind. "Albrecht," he said, "this drill is falling apart. And the people, we're tired. Either the drill's gonna give up or the people are gonna give up, but either way it's not gonna be long."

They had a tricky decision on their hands: What would be worse? Losing a string? Or making a controlled decision to back off, which would mean chalking up their first drilling campaign as a complete "learning experience"? Sticking their nose into it and seeing what would happen, on the other hand, getting "some facts on the ground and some real experience," in Jim Yeck's words, would be a learning experience, too; and that's generally the best way to go in experimental physics.

The hole wouldn't wait long; it was freezing in by the minute.

First, Jim decided for himself that he could live with it, it would not be the end of the world, if the string got stuck. He asked Albrecht to estimate the odds that the hole was okay, and Albrecht guessed somewhere between 60 and 70 percent. Jim then gathered a small group that included Bob Morse and Bruce Koci (who was suffering the ill effects of both radiation therapy and chemotherapy) in his office in Madison to think about the problem—without including Albrecht. He wanted to shield him: "I don't want him [feeling] like he has to justify what he's doing to a bunch of people that . . . just came from the coffee shop to tell him what to do, right? Cuz he's livin' it. He's on the battlefield."

The incomparable Bruce eased everyone's fears. Evidently, he asked only one question, how long had it taken to drill the hole. When they gave him the answer, he thought for a moment and said, "You could probably deploy a semi-truck in that hole."

Jim also made it clear to Albrecht that he did not want to tell him what to do. He was trying to support him in making the decision, and he would support him either way—even though, ultimately, it was Jim's decision to make. After the conversation with Bruce, he sent Albrecht an e-mail saying it would be okay to go ahead and deploy the string. Jim and I once tried to puzzle this somewhat subtle arrangement out, and what we came up with was that Jim took *responsibility* for the decision (again providing Albrecht with some cover), but that he allowed Albrecht to *make* it—although "allowed" might be too glib a term, since Albrecht definitely agonized over it.

John Jacobsen happened to fly in to Pole at just this juncture, since he had been named de facto leader of the team that would try to get the string up and running—troubleshooting the software and hardware—if Albrecht decided to go ahead. He remembers sitting in the galley with Kurt Woschnagg and a very concerned Albrecht, poring over the drill data, which was hard to interpret. "Finally Albrecht said, 'Okay, we're gonna go. We're going in,'" John recalls. "That was a very cool and terrifying moment."

Gary Hill, who had been promoted to deployment manager, was also concerned at first. It started off very slowly—almost an hour to attach the first module. But they picked up momentum and finally got the string deployed after about eighteen hours.

Albrecht puts it as only a physicist would: one string in the Ice was infinitely more than zero!

They invited BK Grant, the station manager, and Jerry Marty, who had managed the building of the new $174 million station, to participate in the ceremonial turning on of the very first IceCube digital optical module. It was Marty who actually flipped the switch.

There'd been some serious drama with the software. A full five years after all the trouble they'd had with the data acquisition (DAQ) software for AMANDA's digital string 18, they still didn't have working DAQ software. Jacobsen had had to cobble some code together at the last minute.

It worked on the first try, and IceCube lifted off.

Amazingly, they had met the limited goals that Jim Yeck had backed off to roughly a year earlier. In the plan they had submitted to the National Science Foundation in the spring of 2004, they had indicated that they would *attempt* to deploy four strings, but they only committed to one.

They had also deployed four IceTop stations (two tanks each), and this allowed them to detect the very first muon with IceCube on February 7, even before they left the Ice. It was a down-going muon from a cosmic ray air shower, which lit up one of the IceTop stations on the surface and then, in descending order, most of the detectors on the new string. And Dave Nygren's team soon demonstrated—from their desks in Berkeley on the other side of the world—that they could synchronize all sixty DOMs on the string to within a remarkable three nanoseconds.

As the summer folks were preparing to depart, Robert Schwarz, who was about to winter over for the fourth time on another project, walked up to Kurt Woschnagg and said, "Well, congratulations. You got one hole in a year. And how many do you want to do altogether again?"

The task did look daunting, especially to the many newcomers to Ice-Cube who had not witnessed the heroics of AMANDA. But Kurt was optimistic. He knew they could pick up the pace; they'd done it before. "If you were there," he says, "especially after having been planning this for years and years, it actually felt like a great success. The first IceCube string was in!"

21. As Quickly as It All Began . . .

JIM YECK SAYS THAT ONE OF THE THINGS THEY LEARNED THAT SEASON was humility. At the scale they would need for production drilling, this job was turning out to be more challenging and even more dangerous than they had anticipated. They also found themselves in a bit of a hole, so to speak: behind schedule. Even he was entertaining doubts as to whether they'd be able to build this thing on time and on budget—which is basically his mantra.

In his view, many of the mistakes they made the first year were the simple result of amateurism. His goal was to become professional. So it seems symbolic that right on the heels of that first difficult season, what he refers to as "IceCube Corporate"—himself, Francis, Bob Paulos, Albrecht Karle, Jim Haugen, and some others—moved from their various rabbit warrens around the Madison campus into the shiny new office space he had designed the previous year, and that the AMANDA and IceCube collaborations merged about a month after that. (The only practical effect of the merger was that Steve Barwick had to throw in the towel and join Ice-Cube or else leave altogether.)

✦✦✦

"We put an enormous amount of energy and attention into trying to gather what we could from that [first year's] experience, to learn what we could," Jim recalls. "And part of that was soliciting input directly from the people participating? So we interviewed all the drillers as they came off the Ice."

It had been demoralizing for the drillers—more than four months' work without much success—and they appreciated being listened to and participating in the decision making. When Sven Lidström recovered, he wrote a fairly critical note to Per Olof Hulth, who translated it for Jim. And Jim and his fellow managers also convened an ad hoc review committee for the drill, chaired by geologist Barclay Kamb, an associate of Hermann Engelhardt at Caltech. (Engelhardt had led the expedition in West Antarctica many years earlier on which Doug Lowder and Andy Westphal had made their unauthorized and unsuccessful attempt to run the very first feasibility study for AMANDA.)

One conclusion from these deliberations was that it was critically important to have experienced drillers on the Ice. There hadn't been enough of them the first time around. Another conclusion was that there was too much "northern hemisphere" engineering: good ideas that had the side effect of making the drill too complicated. This systemic problem was exacerbated by NSF's understandable but somewhat confused insistence on safety. The many safeguards that were suggested by the foundation's consultants also made the drill more complicated, more susceptible to problems, and thus, ironically, more dangerous.

To retain drillers from one year to the next, they did everything they could to treat them well. They gave them year-round health benefits, so that even if they went to Greenland to drill during the Arctic summer they would still consider IceCube their home. (There's an entire subculture of people who migrate seasonally between the top of the world and the bottom.) They helped the drillers find jobs in the off-season, they threw parties for them, they thanked them. Jeff Cherwinka points out that if you take care of the people, the people take care of the project, and that this attitude— which emanated, he felt, from Jim Yeck—helped build a team of committed people who wanted the project to succeed. (Jeff also observes that the people you work with at Pole become more than co-workers; you develop the sort of bonds that you would if you went on camping trips together. In February 2016, when the University of Wisconsin held a celebration to commemorate the fifth anniversary of the completion of IceCube, quite a few drillers showed up, and since this was Jeff's gang, he threw a party for them at his home afterward.)

Each of the eighteen drillers from the first season returned for the second. Realizing they had been overworked, Yeck and the other managers added twelve new people and took some pains to make their work less heroic. They provided a warm space where they could take their breaks in comfort. They backed off to three eight-hour shifts a day, rather than the two twelves that the pioneers had soldiered through on their maiden voyage.

The management approach that Jim took to upgrading the drill was to ask the Technical Director of Madison's Physical Sciences Laboratory, Farshid Feyzi, to assume the role of project director for the drill, for just one year. (Jim knew Farshid could deliver, since they had worked together on the Large Hadron Collider, for which PSL had built some large and sophisticated hardware.) Up until then, PSL had been building the drill, while IceCube Corporate had been designing and managing it.

"Let's be honest," Jim said to Farshid. "We're always in resource discussions with you about what we can get out of PSL to get this done. I'd like you to manage it as a project, the upgrade of the drill. And upgrade doesn't mean make it more complicated? It means make it less complicated." (It also meant redesigning and fabricating another million-dollar hose—which would go through yet a third iteration before the project was done.) In accordance with his general philosophy of keeping management out of the hair of the people who are actually doing the work, Jim wanted Farshid and Jeff Cherwinka (who was not IceCube Corporate; he worked out at PSL) to "own" the project and commit only to what they thought they could accomplish.

Farshid took on the job wholeheartedly. In addition to simplifying the drill and making it safer, he and Jeff also set a goal of speeding up the drilling process. To that end they made two superb innovations that allowed them to "leapfrog" from hole to hole more rapidly. They fabricated a second TOS, which they could set up in advance at the next hole, ready for the drill to be mounted on the moment the present hole was complete, and they built an independent "firn drill" for melting through the fifty-meter layer of firn, or compressed snow, at the top of the Ice, also ahead of time.

The firn drill also employed hot water, but not in the same way that the primary drill did. It didn't spray water into the firn—which didn't work well, because the water would seep out into the surrounding snow. It was something like a small immersion heater you might dip into a cup of water in order to heat it for tea. Its outer surface consisted of a helix of copper tubing with hot water circulating through it. The helix was cone-shaped with the tip pointing down, an inverted Christmas tree, and it would simply melt its way down through the firn.

Farshid went to the Ice at the beginning of the second season to help set up the improved system, and when he sensed that things were on the right track, returned to his supporting role in the north. Jim subsequently asked the leader of one of the drilling crews, Alan Elcheikh, a veteran of the Australian Antarctic program, to become permanent manager of the drill.

Alan was as good a fit on the drilling side as Jim Haugen was on the equipment side. He's very smart, but he doesn't advertise the fact. And he's patient; he's running a marathon, not a sprint. The way Yeck describes it is that if you have a disagreement with Alan "he just works you really slow, methodical, and then he ends up getting you around to his position, and you feel good about it." He's a likable person, which goes a long way pretty much anywhere, and he had just the right combination of toughness and congeniality for working with the drilling crew. They didn't feel like they were in the Navy. The vast majority of the drillers liked and respected him.

Jim delegated full authority to Alan, just as he had to Farshid. At a drilling meeting early in Alan's tenure, when they came to a point where a decision had to be made, Jim turned to him and said, "Alan, you should decide. You own the drill. It's your drill." Alan responded that that was when it first dawned on him that he was "in deep shit"—which might be Jim Yeck's management philosophy in a nutshell: give people full responsibility and at the same time give them everything they need to get the job done.

Some months before their second Antarctic season, Jim decided that a good way to proceed would be to set a minimum, "baseline" goal for each season, along with a stretch, "target" goal. For the coming season the numbers were eight and ten. As they were preparing to head south, he arranged a meeting with Chancellor John Wiley to talk about the season and invited Sven Lidström to come along. Sven's injuries had healed enough that he would be joining the drill team. In the presence of the chancellor, Jim asked Sven how many holes he thought they'd be able to drill this year. Sven responded, "Four, maybe five."

Jim said to Wiley, "Okay, so you have his, ah . . . I think we're gonna do better than that, but I promise you I'm not gonna be beating people with a stick to do better. We're gonna do it safely."

It wasn't as much of a jump going from one hole to eight as you might expect, since they had spent a significant fraction of the first season simply putting the drill together. But it still took awhile to get moving. A *National Geographic* photographer was visiting Pole that season on an assignment about the new station, and he was hoping to get some shots of the drillers in action, but whenever they were almost there something new would crop up and send them back to the drawing board. He told Jim it was like trying to film wildlife: a lot of standing around and waiting.

But when they finally did get moving, they did well. They deployed eight strings and met their baseline. The tide was turning.

This gave Jim, who is almost as many steps ahead of the game as Francis usually is, "a high level of confidence" that they would succeed. He'd set up a sound organization and put most of the right people in place, so the road ahead looked clear. They'd essentially be going to the Ice and doing the same thing every year. He was confident enough, in fact, that in June 2006 he took on a second, half-time job at Brookhaven National Laboratory, as deputy project manager for the construction of a billion-dollar facility called the National Synchrotron Light Source.

When he told the folks at Wisconsin and NSF of this intention, he says, "They crapped their pants, right? I mean, you know, like, 'This project is not out of the woods.' And I'm like, 'Okay. I'm going to do the Brookhaven thing; I think I can do both these things. If you say no, I'll just go to Brookhaven.' And then I just started doing it."

He wasn't intending to see the Brookhaven project to completion, he just wanted to "set the initial vector," as he calls it: plan the project and build the *team* that would build the facility—which is essentially what he had just done for IceCube and was why he felt comfortable backing off a bit.

"My goal in each case was to make sure that nobody knew that I was spending time on the other thing. Each project needs to know that you're just living and dying by their project, right?"

He'd already proven his capabilities to the folks in Madison, so they came around quickly. NSF's Director of Polar Programs (the burrower) dragged his feet for two or three months, but Terry Millar and John Wiley eventually brought him around.

Jim did his job at Brookhaven, hired his own replacement, and in 2009 returned full time to IceCube. My guess is that few outside the top managers ever knew he was gone.

✦ ✦ ✦

As he had anticipated, the project gained momentum. "Every season we exceeded our baseline, so everybody felt good. And when you don't feel rushed and you feel good and you're doing well, it has a very positive effect. We were able to ride that wave for the duration of the installation."

Over the 2006–07 season they deployed thirteen strings against a stretch, target goal of sixteen. They now had twenty-two strings in the Ice. Their new instrument was somewhat oddly shaped, but it was already larger than AMANDA, and the physicists began doing science with it. The following year, they *made* their target: eighteen strings. In 2008–09 they exceeded it by one: nineteen. The cumulative total was now fifty-nine strings.

Every year, they learned from their experience, tweaked their procedures, and improved their equipment. Once Jim Haugen had dialed in the process for fabricating the optical modules and cables and shipping them to Pole every year (he got to the point where he was almost a year ahead; he'd leave two or three cables and their associated optical modules on the Ice to "age like fine wine" for the winter), Yeck decided they needed his organizational skills and prodigious energy on the Ice: working with people, dealing with logistics, interfacing with Raytheon—a high-amperage version of the role that Bob Morse had played in the AMANDA days. Haugen took it on with his usual enthusiasm. His mantra became, "Feed the bottleneck. Never starve the drill."

His explanation for his mantra gives a sense of his near-maniacal zeal: "If you look at IceCube in whole, the bottleneck is always gonna be drilling these goddamn holes, right? I mean two-and-a-half kilometers deep of complicated pumps and heat and pressure and—the damn thing gonna work or not? . . . So I talked with, let's call it this instrumentation team, right? The Swedes, the Germans, the guys out at PSL . . . I would use that example often, right? You can never starve the drill. . . . All the cables gotta be there. Everything's gotta be there. You can't ever starve the drill. We can't walk over here to Alan Elcheikh and say, 'Alan, I know you've got thirty guys here to drill holes. Can you just set 'em down for a couple of days, because we're waiting on digital optical modules to fly in from McMurdo?' Can't do that. And it never happened."

Haugen didn't just organize things, he also got his hands dirty. His responsibilities now encompassed not only getting the hardware to the Ice, but also hooking up each new string to its dedicated computer in the gleaming new IceCube Laboratory, which had now been erected on the surface. This meant he led the motley group effort, resembling a tug of war, that took place several times a season: the cable drag. They would attach surface cables to the tops of five to seven strings, bunch the cables together, dig a trench to the laboratory (affectionately known as the ICL), lay the cables in it, and then snake them up through one of the two towers on either end of the laboratory. Inside, each would be attached to a computer known as a DOMhub, which communicated with all sixty digital optical modules on the string at the other end. (See photographs 28 and 29.)

Meanwhile, Alan Elcheikh and his team were perfecting the art of production drilling. Alan began using two metrics to measure drilling efficiency and monitor progress from year to year: "frequency," the time it took to pull the drill up from the bottom of one hole, transfer it to the next, and drill to the bottom of that one; and the amount of fuel consumed per hole. Fuel is a critical and expensive commodity at Pole, and it took two or more full Hercs of fuel to drill each IceCube hole. This prodigious need put them in competition with station operations, because getting enough fuel stored away in time for winter was one of the major station-wide tasks every season. Over the three seasons after Alan took charge, they got the frequency down from just over three days to just under two, which allowed them to drill 50 percent more holes in a year.

The physicists came in handy in these efforts as well. A case in point would be the way they figured out how to minimize fuel consumption.

In 2001 or so, a brilliant graduate student from India by the name of Raghunath Ganugapati joined the IceCube group in Madison. A couple of years later, he began pursuing his doctorate under Albrecht Karle. His parents had nicknamed him Newton when he was a child, after Sir Isaac Newton, because he was something of a prodigy in math. Indeed, as a physics grad student, Newt, as he was known, would sometimes enter the "integration bees" offered by Madison's math department. This is the mathematical analog of a spelling bee, in which the contestants are asked to solve increasingly difficult calculus problems in under twenty seconds with a minimum of writing. Two years in a row, Newt came in fourth overall among the 40,000-odd students at the university.

When he first joined IceCube, he began working with Bruce Koci to optimize the drilling process: to make sure the holes would be large enough to accept the IceCube strings—so they wouldn't get stuck, obviously—but not too large, because that would waste fuel. Bob Paulos, incidentally, had hired both a Los Alamos–based consulting firm and a mechanical engineering professor at Madison to work on this problem, and these experts hadn't gotten anywhere. Francis Halzen had also tried his hand at it and given up.

The perfect hole would be just wide enough for the basketball-sized optical modules to pass through, but not much wider, and it would be uniform in diameter for its entire one-and-a-half-mile length. The uniformity is the hard part, because the drill takes anywhere from twenty-four to forty hours to drill the hole and another ten to fifteen to ream it out as it's pulled up, so the ice at different depths is first melted and then begins freezing back at different times. The starting temperature of the ice also varies with depth, and then you have the heat escaping through the walls of the hose as well as the return water that you're pumping up to the top, reheating, and sending back down. This is a problem in heat transfer and phase transitions—ice changing to water and vice versa—and it's what mathematicians would call nonlinear, which basically means that it's really complicated.

Newt solved part of the problem by hand and resorted to a computer to solve it completely. When he was done, he had a software package that could direct the drill to drop and pull back up at just the right speeds, depending on depth, that would leave a uniform hole of the correct diameter when it came time to deploy a string. To test his algorithm, he asked Bruce to give him some data from the AMANDA drill, and he knew he'd gotten it right when it predicted the exact depth at which string 17 got stuck during the millennial season.

With the help of Newt's software, Alan Elcheikh brought fuel consumption down from more than 7,000 gallons per hole in 2004–05 (the first difficult year) to less than 4,000 gallons per hole in 2009–10—a far cry from the first AMANDA drilling season with Bucky-1, when it took 12,000 gallons to get to only 1,000 meters.

Newt's graduate work also highlights the fact that there was plenty of astroparticle physics going on while IceCube was being built. He wrote a two-part thesis, the first on his solution to the drilling problem, the second on a way to measure the contribution of charmed quarks to the atmospheric muon spectrum. And Newt was only one of many graduate students passing through the roughly three dozen institutions of the IceCube collaboration during these years, all of whom worked on one physics problem or another.

Unfortunately, Bruce Koci's work with Newt and his crucial input to the decision making for the first IceCube string were his last contributions to the project. His radiation and chemotherapy treatments succeeded in eradicating his cancer for about a year, but in mid-2005 it reemerged and he was told that he had a year or two at most. "All-in-all it's been a pretty good life," he wrote to me. "I have no regrets or if-only's. It will be an interesting journey." He approached the end with his usual quiet humor and disheveled wisdom. "Been reading a couple books on Buddhism (hope that is how it is spelled)," he wrote the following February. "Being no one going nowhere." In July he sounded optimistic: "CAT scan yesterday found no cats which was a pleasant but confusing surprise. . . . Maybe I am just pretty crusty and not ready to exit yet." He and his wife had been traveling a lot. He spoke of getting out on his beloved bicycle again. He died on November 13, 2006.

In his memory and in recognition of the fact that he was the inventor and father of the drill, the summer crews at Pole began calling the fifty-ton reel for the two-and-a-half-mile hose "Bruce." They had to drag it many times each year from hole to hole: "Let's get Bruce over here and get 'er rolling!"

His friend and fellow start-up enthusiast Bob Morse withdrew from the project at about the same time. Between Bob's work on AMANDA (he was principal investigator for its entire existence) and some lobbying by Francis, not to mention the all-seeing eye of his grad school classmate, John Wiley, in the background, Bob was promoted from senior scientist to full professor in the Madison physics department in 2000. He remained at Madison until 2006, the year the last AMANDA grant ended, when he made good on a long-ago promise to his wife and "sort of retired" with her to Hawaii, where she had grown up. At age sixty-eight, he took a part-time teaching position at the University of Hawaii in Honolulu, in the same department as another of his friends, John Learned.

The AMANDA and IceCube collaborations had merged by then. The final blow came at a collaboration meeting in Utrecht in the fall of 2008, when they decided to shut down the smaller instrument forever—an emotional decision for the old-timers, who generally agreed with Bob that the AMANDA days were the glory days. They turned AMANDA off during the 2008–09 Antarctic season, and Christian Spiering gave a memorable "Farewell AMANDA" talk at the spring collaboration meeting in Madison.

It's not as if the rest of the construction was all smooth sailing. There were several more injuries, though none as serious as Sven's, and issues came up every year, of course. The most serious crisis from a management standpoint revolved around the software for running the instrument.

In the interim before Jim Yeck arrived, Francis and Bob Paulos had made the decision to contract with Lawrence Berkeley Laboratory to write the code for communicating with the digital optical modules; the data acquisition, or DAQ, software; and the higher-level code that would run the instrument as a whole, which they call experiment control. This was essentially a default decision, since LBL had designed the DOM and managed to produce the first functioning DAQ software for AMANDA's digital string 18 in early 2001.

As I have mentioned, there was still no working DAQ or experiment control software four years later, in January 2005, when the first IceCube string was installed. Albrecht Karle had seen this coming and asked John Jacobsen, who was consulting with LBL but living in Chicago, to lead a group that would try to get something working on the Ice, and John had succeeded.

LBL is an extremely expensive place to do business with. By 2005, more than a dozen engineers had been working on the IceCube software for several years, at a cost in the neighborhood of $3 million per year, yet, amazingly, Jacobsen didn't even use LBL software as the basis for his seat-of-the-pants code for the first string. He modified some code that had been developed at Madison and the University of Pennsylvania for the purpose of testing the DOMs before they went into the Ice.

The main architect of that software was a research scientist at Madison by the name of Kael Hanson. Kael had done his graduate work at the University of Michigan on MACRO, a neutrino detector in a tunnel under Gran Sasso, the highest mountain in Italy's Apennine Alps. He had moved to Penn in 2000, to do a post-doc under professor Doug Cowan, who had gotten involved in AMANDA in the late nineties. So Kael was originally an AMANDA guy, and he has the adventurous approach to physics—and life, apparently—that goes with the territory. He relates that when he was down on the Ice in 2002, Bob Morse and Albrecht Karle offered him a job in Madison. He got back to Philadelphia, packed up his family (he and his vibrant wife, Gloria, have five children), and drove to Wisconsin—only to find that Bob and Albrecht had forgotten about the offer!

"It was at Pole we made all of the arrangements," he writes. "Lack of oxygen, easy to see how things got messed up."

They found him a job in a couple of weeks.

Kael had a knack for electronics and software. When IceCube came into being, he was put in charge of testing the DOMs, and this led to his becoming manager of all the various gadgets that went into the Ice.

By the end of the second, successful season, when they deployed eight strings, LBL had managed to produce some semi-functional DAQ software, but none for experiment control. Again John Jacobsen had to hack something together, and this brought the situation to a head. Francis, Albrecht, and Jim Yeck got together and decided to promote Kael to co-manager of software development, on a par with the existing manager at LBL, who soon decided to leave the project in order to work on a neutrino experiment at Daya Bay, China.

In typical Jim Yeck fashion, Kael was dropped in at the deep end of the pool. He now "owned" a project that had burned through almost $20 million with little to show for it, employing about fifteen people dispersed around the country who didn't get along very well. Although he was guaranteed all the material support he would need and Jim and Francis ran political cover for him, he found that no matter what he tried, he could *not* get the group to team up and produce working software. As the Antarctic season approached, his nerves began to fray. He was having a hard time sleeping, and his health was deteriorating.

In September, scant weeks before the season would begin, he had what he calls "a vision." He realized that no matter how much money he spent and no matter how many people he hired, the situation would not improve. "The inspiration hit almost overnight," he writes. "I can understand the Muse concept—strange how it felt like it wasn't coming from my brain, but I was just the channel." He decided to lay off all but four people: himself, John Jacobsen, a programmer in Madison named Dave Glowacki, and another at LBL named Keith Beattie. Some still remember this as Kael's "DAQ coup." The cuts were everywhere, even in Madison, but LBL felt the most pain, since it lost the most people, and, needless to say, Kael incurred a lot of resentment. Collaboration-wide, however, most people realized the deed had to be done.

Kael, John Jacobsen, and Dave Glowacki traveled south in January 2007 with what Kael describes as "something really rough" to try on their new twenty-two-string array. Dave was the DAQ guru, John worked on experiment control. The software was so rough that as the countdown to station closing began, the instrument was crashing every few minutes and it was looking as though it might go through its entire first year of being large enough to take data without taking any at all. Things got so desperate that they considered buying a few days by missing the last LC-130 to McMurdo and catching a Twin Otter north instead. This is a smaller plane that can land in colder conditions. There are usually one or two Twin Otter flights after the last 130. They are operated by a contractor out of Calgary, Alberta, named Ken Borek Air, so they tend to route north toward Canada through Punta Arenas, Chile, rather than New Zealand.

At the very last minute, John figured out a way to make the instrument restart automatically every time it crashed. He remembers that they looked at each other and said, "Don't look at the screen again. Let's go!" and ran out to catch the last 130.

He had managed to get the skeleton of the experiment control system that they still use today working pretty well. Back home, they continued to improve the software, uploading updates via satellite. By April or May the instrument was capable of surviving for several hours at a time. It got steadily better from there.

To Francis, this debacle demonstrates that if NSF *had* decided to run IceCube out of LBL rather than Madison, the instrument probably wouldn't have been built. They would have run out of money when it was about the size of AMANDA. "Because remember how the game is played, right? . . . Your budget is basically the size of your detector."

In contrast, once IceCube recovered from its first season and had the wind at its back, Jim Yeck began to work the other side of the coin, searching for ways to *save* money in order to buy back scope. Remember that in the early budget negotiations with NSF he had backed off to seventy strings from their original goal of eighty. Now he wanted them back.

"I've had enough project experience to know that one way you save money is you spend money, depending on what you spend it on, right? . . . Like you spend it on the drill. You put more holes in the Ice per year, you have one less season, you save millions of dollars. Some people think that the way you save money is you don't spend any. That's not true in a project, because schedule will most likely be the biggest decider in what the cost is. If you drag it out, you're spending more money. So think schedule, think spending money to make schedule. That's your starting point."

Over the 2009–10 Antarctic summer, which would have been their last according to the baseline schedule that Jim had laid out in 2004, they deployed an astounding twenty strings against a target of twelve and brought the total in the Ice to seventy-nine. And they still had money left for one more season.

This allowed them to dream. Not only could they add more strings, they could expand their scientific reach.

Even in the 1990s, before Super-K discovered neutrino oscillation, the AMANDA physicists had been imagining ways to study oscillation themselves. As with Super-K, they would employ the atmospheric neutrino beam, and the idea would be to watch neutrinos change from one flavor to another as they traversed the Earth. Since IceCube views the entire northern sky, the neutrinos it detects travel different distances through the planet in order to reach it. Those arriving from the direction of the north pole traverse the longest distance, and those from nearest the horizon the least. Since the chance that a neutrino will oscillate increases the farther it travels in dense matter, different flavor mixtures will arrive at IceCube from the different directions.

This is pure particle physics, as opposed to astrophysics. The Super-K discovery has spawned a global, multi-billion-dollar effort to examine every detail of oscillation, mainly because, as I've said, it's the only physics so far discovered that violates the standard model. And the IceCube physicists have carved out an exclusive niche for themselves, since they can study neutrinos at higher energies than any other experiment—orders of magnitude higher than any neutrino produced by an accelerator, or detected by Super-K, since that instrument's Greisen-type design limits it to lower energies. And higher energies hold particular interest, because the study of oscillation is basically the study of the weak nuclear force, and anomalous behavior in weak interactions, if there is any, is more likely to occur at higher energies.

I've also pointed out that Per Olof Hulth had a playful approach to physics. He was also in a good position to take advantage of that inclination, since he wasn't as bogged down in the project management as, say, Francis Halzen was. When it began to look as though they'd have some room to play with IceCube, he came up with a proposal for making it better at measuring oscillation and several other things.

Enough was known about oscillation that they could run some well-informed Monte Carlos in order to design an optimum configuration, and what they came up with was a more densely packed sub-array, in which the strings would be closer together and the DOMs on each string closer to each other than in the standard IceCube array. This would open up an intermediate energy range, between those of Super-K and IceCube. It would also make the instrument more sensitive to low-mass WIMPs in the Sun and galactic center and to supernovae (areas of study that were actually more interesting to Per Olof—and Francis—than oscillation).

Buford Price, who has a similarly creative approach to science, had been leading an increasingly sophisticated study of the ice at the South Pole for more than a decade by then. One of the features his group had identified is a relatively dense layer of dust at depths of 2,000 to 2,100 meters, which was laid down during a particularly cold, dry spell at about the mid-point of the most recent ice age, roughly 65,000 years ago. To avoid the scattering and absorption of light in the dust layer, the collaboration decided to locate Per Olof's sub-array below it, at the deepest levels of IceCube, where the ice also happens to be the clearest. This sub-array was an in-fill, in the center of the IceCube array: it incorporated several standard IceCube strings as well. Per Olof obtained a grant of $9 million from the Knut and Alice Wallenberg Foundation in Sweden, a longtime supporter of

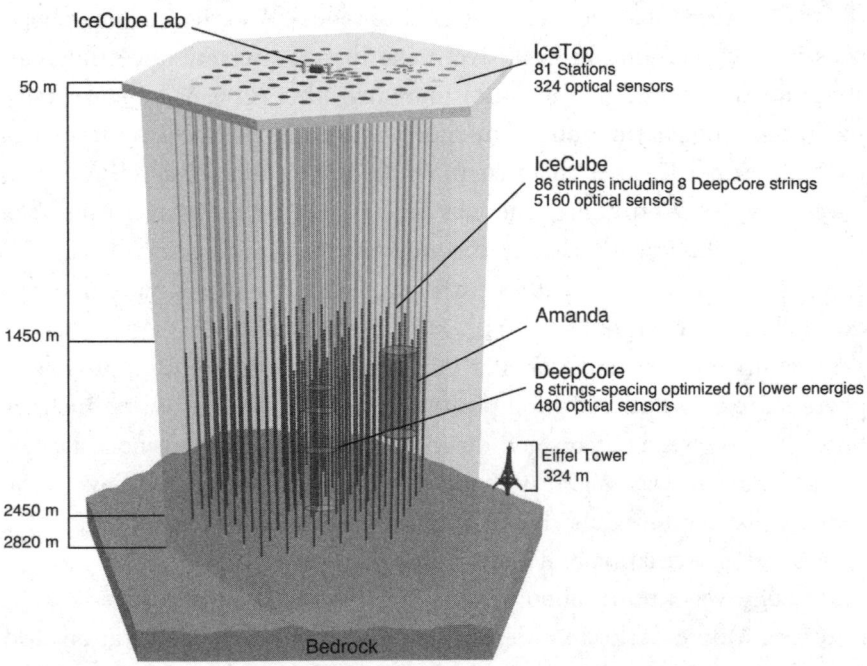

A schematic of IceCube at its completion in December 2010.
IceCube Collaboration

AMANDA and IceCube, in order to fabricate the six specialized strings for the sub-array. They decided to name it DeepCore.

The first DeepCore string was deployed in 2009–10, the year they brought the total to seventy-nine. The following year, 2010–11, their last year of deployment, they dropped the last five DeepCore strings into the Ice along with the last two standard IceCube strings and brought the total to eighty-six, eighty of which were pure IceCube, their original goal. Jim Yeck had gotten his scope back and more.

The final season was a victory lap—easy after the previous year's twenty strings. The seventh string of the season and the very last for IceCube was deployed on December 18, 2010, exactly a week before Christmas—quite a contrast from my millennial season, when we didn't deploy the *first* string until December 19.

Per Olof had come up with another playful idea, and that was to attach a video camera to the bottom of the string—they called it the Swedish Camera—in order to see what the ice looked like going down and watch the water refreeze after drilling. The second purpose was of special interest: they wanted to get an idea of how many bubbles formed as the ice grew back and find out what other strange things might be happening. (They discovered that a column of bubbles forms in the center of the hole and that the optical modules are pushed to the side of the hole as the ice refreezes. The modeling of the instrument has become so detailed that these features are now incorporated into their Monte Carlo and data analysis software.)

Kurt Woschnagg remembers "a Christmas-y feeling" during that last deployment. He just happened to be around when people started to assemble and decide who should take the various roles. Kurt usually did something in the background like monitoring the depth of the string, but for the first time ever, he was assigned to "position one." He'd be one of the people actually attaching the DOMs to the cable, and he was very happy to be where the action was on that special day. Gary Hill also made sure to be there the entire time. Since he led one of the deployment teams, he kept himself free of specific tasks in order to keep an eye on the big picture.

It took awhile to hook up the camera, and it wasn't that fine-tuned a deployment team to begin with, plus they wanted to savor the experience. They had set up a video link so that their colleagues in the real world could tune in, and Gary would wander into the room where that camera was every once in a while to commentate. People from all over the station were sticking their noses in just to watch. "Everyone's in a good mood and it's progressing nicely," remembers Kurt. "Every now and then we would stop, and people [would] sign DOMs and . . . take pictures." Robert Schwarz dropped by. So did an old friend named Paul Sullivan, who had wintered with Robert and Gary in 1997.

A Swedish crew was operating the Swedish Camera, so Kurt, who had grown up in Sweden, joined them after he finished his shift at position one. Ryan Bay, a protégé of Buford Price who had now converted more-or-less completely from particle astrophysics to climatology, was raptly monitoring the camera's display for the layers of dust and volcanic ash that were signatures of specific "time horizons," known geologic events of the distant past, as the camera dropped past them. Since the drillers had already finished their work, they were drinking champagne.

As the end approached, the TOS became very crowded and even more festive. A silence fell as they brought the last DOM into the room and attached it in a slow, ceremonial way. They all chanted, "DOMs away!" in unison and started the drop.

Gary remembers feeling that he'd blinked and IceCube was over. It seemed that he'd been in the same room deploying the first string just the other day. While AMANDA brought unexpected adventures every season, on IceCube, he says, "Every year down there was almost as if you'd turn up on the plane, you hop off the plane, you walk into the station, and then you'd see the same people, and by that time it felt like you'd never left. You'd get into your room; even if it was a different room to other seasons, you know, in a few minutes you felt like you'd never left the place. . . . [You'd] set up the drill camp, see all the drillers, a lot of them the same from year to year, go through a few disasters here and there with the drill where every year, you know, things would go wrong. And so my actual deployment experience at South Pole is sort of one continuous time down there, right? It all gets compressed into, 'Well, we did the first string, and, Oh!, now it's the last one.' . . .

"And then," he concluded in low storytelling voice, "as quickly as it all began, it was all over."

"There were speeches," remembers Kurt. "It was just a—that was a fantastic feeling actually, when that was going on. It was just fun. We didn't really want to—*hah!*—want to see the last one disappear. We really wanted to extend that moment." (See photograph 31.)

A big celebration had gotten under way outside, but Kurt declined the champagne and stayed inside to help monitor the drop: "That I really wanted to do. I wanted to be part of those who positioned the last string at its final depth."

When it was all done, they sent a message to Francis: "We made the impossible look easy." He posted it on the door to his office. "That," he says, "is my summary of the construction of IceCube."

The strange thing about the end of construction was that it made the instrument less of a tourist attraction at Pole. When you could show people the drill and the drilling and the hole and the cabling and the DOMs lined up on their shelves, all the distinguished visitors, DVs, wanted to stop by for a tour. When it was done, there was just this funny-looking building filled with computers, sitting on the Ice. Jim Haugen, who was usually the one who showed the DVs around, had seen this coming and secretly shipped eighty-six flags to the station, three feet wide by eight feet tall. Eighty were bright orange to mark the IceCube strings; six were red for DeepCore. There was already an array of small black flags out there marking the IceTop tanks.

The eighteenth was a Saturday. At about six p.m., when the last string was positioned and tied off for freeze-in, most people walked back to the station to continue celebrating and enjoy their Sunday off. Jim had been awake for about twenty-four hours at that point, and so had Gary, who had found himself a spot on an empty DOM shelf and taken a nap. Sven Lidström and a couple of other drillers also stayed out on the Ice in order to keep an eye on the quiescent drill, which hadn't been put to bed yet. There was still water in the hoses that had to keep circulating.

Jim pulled the flags from their hiding spot and revealed his plan. Then he, Gary, Sven, and one or two other drillers "went out under the dead of a blizzard," says Gary, "and went around—*hah!*—this whole bloody array for several hours, *zhzhzhzhzh*, with a drill and sticking these damn flags in." They couldn't see each other for the blowing snow, so they wandered around randomly in the huge hexagon, somehow managing to fill the whole thing in.

"So when people woke up for Sunday breakfast? They looked out and now what you saw was this beautiful, giant field of eighty-six flags," says Haugen. "Eighty orange flags and six red flags for the DeepCore flapping in the breeze where twelve hours ago there was nothing. So it was cool."

John Jacobsen's pattern in recent years had been to fly to the Ice sometime in January, after the hardware was hooked up, get the software up and running with the new year's crop of strings, and teach the winterovers how to run the instrument. Each year's instrument was named for its size, so the previous year's, which comprised seventy-nine strings, had been known as IC79. On the eighteenth of January 2011, John broadcast an e-mail under the title "IC79 est morte, vive IC86."

> Dear Collaborators,
>
> We are pleased to announce the first flight of IC86 running as part of the full Data Acquisition System.
>
> On-Ice lead Greg Sullivan launched the first run. . . .
>
> These first IC86 runs are very exciting to us 'Cubers at Pole, several of whom have been involved with the project since its earliest days or before (AMANDA).

Greg, a professor at the University of Maryland, had joined at the beginning of IceCube. John, of course, was one of the real old-timers.

There were still bugs to work out, equipment to install, and commissioning to be performed, so it wasn't until May that the winterovers switched IceCube into full data-taking mode. The dream of a kilometer-scale neutrino telescope, first held by the DUMAND pioneers some thirty-five years earlier, was now a reality.

Francis had been waiting for this moment for twenty-three of those years. He broke his usual silence on the collaboration-wide e-mail list and declared, "This is a great occasion. We will finally have a stable detector that we can fine-tune to death and get 0.1 degree angular resolution and 15% energy resolution!

"Thanks. You know who you are (even I know)."

The interesting thing is that this epochal moment occurred close enough to the one-hundred-year anniversaries of both the balloon flight on which Victor Hess discovered cosmic rays and Roald Amundsen's first footprints at the South Pole.

22. Crossing the Threshold

At its outer boundaries, experimental particle physics is largely a matter of groping in the dark. Since its practitioners are searching for a true signal of an unknown nature in the presence of a noisy background—or, as Fred Reines put it, "listening for a gnat's whisper in a hurricane"—there is always the danger of convincing themselves that they're seeing something that isn't really there. To deal with this problem (and recognizing what Nobel Laureate Eugene Wigner once called "the unreasonable effectiveness of mathematics" in describing the natural world), they have come up with a mathematical definition of discovery, based on probability and statistics. Assuming that you've designed your experiment correctly and your analysis makes sense, a true discovery would be an event that is highly unlikely to have occurred as a result of a random fluctuation in the background. And the bar is set rather high. The agreed-upon threshold for a claim of discovery is that the odds that it was produced by random chance are less than one in about three million.

Because of the way they do the math, however, physicists talk about it in another way. They speak in terms of the characteristic range of the noise or fluctuations in an experiment, which statisticians call their standard deviation. If you were trying to measure the height of a boat on a windblown sea, for example, the standard deviation might be proportional to the average height of the waves. If the boat was a small dingy in a heavy storm, it would be hard to measure its height—it might even be hard to see the thing. If it was a tall ship on a calm day, it would be much easier. The Greek letter sigma, σ, is the mathematical symbol for standard deviation. In particle physics parlance, saying that your result stands "five-sigma" proud of the noise in your experiment (or, loosely speaking, that the ship is five times taller than the average wave) is equivalent to saying that the odds that it was a chance fluctuation are less than one in three million—or that you have made a discovery.

In April 2011, the University of Wisconsin held a splendid gala for the commissioning of IceCube. Eminent cosmic ray physicists and cosmologists, pioneers in neutrino astronomy, representatives of the funding agencies, and other friends were flown in from around the world. They gathered in a Frank Lloyd Wright–inspired conference center at the edge of Madison's Lake Monona and were treated to a day of commemorative events, culminating in an elegant banquet in the city's classy new arts center. Two symposia were held in conjunction with the gala, one on particle astrophysics and the other on science in Antarctica. John Learned was there. Grigorii Domogatsky was invited but had to graciously decline: he couldn't get a visa in time, because he had just returned from his annual spring field trip to Lake Baikal. Congressman David Obey, he of the kosher pork, gave a speech; and one of the more memorable lines of the day was delivered by Anders Karlqvist, former director of the Swedish Polar Research Secretariat, which has been funding IceCube and AMANDA since the early nineties. "Now it is time for the scientists to deliver," he said. "Let me remind you that my organization is located in the same building as the Royal Swedish Academy of Sciences. I look forward to welcoming three of you (for it cannot be more than three) as you walk in the door. Welcome to Stockholm!"

For the scientists in the audience, it is probably safe to say that the highlight of the day was when Francis Halzen went out on a limb and predicted that IceCube would have a five-sigma result within five years. He has nothing if not nerve.

While most of the sound and fury had swirled around the building of the instrument for the previous decade, the physicists had been making significant progress on the scientific front as well. Plenty of graduate students and post-docs had helped put AMANDA's data to good use, and they had all hopped on IceCube in 2005, when it began producing data. Thanks to the digital optical module, they could detect up-going muons with just their first hard-won string, and they built upon that small beginning as the instrument grew.

To give an idea of how much more sensitive IceCube is than AMANDA at detecting astrophysical or extraterrestrial neutrinos, the ones they're really looking for, five years of data from the larger instrument is equivalent to about a million years from the smaller one. And IceCube detects about 100,000 atmospheric neutrinos a year, whereas AMANDA-B10, the ten-stringed instrument they used to prove feasibility, saw only about 700. As IceCube grew, therefore, the collaboration could study the atmospheric neutrino beam with greater and greater precision. With the early, smaller incarnations of the instrument, they also searched (a loaded word) for point sources and for the cascade and double-bang signatures of electron and tau neutrinos.

Their instrument is also a cosmic ray detector. Between IceCube and IceTop, they observe four or five billion cosmic ray muons raining down from the southern sky every year—and it so happened that this brought their first small surprise. It turns out that cosmic rays don't stream in uniformly across the sky; there's a waviness in their arrival directions.

The Swedes systematically expanded and enhanced their search for dark matter WIMPs or neutralinos. In 2006, employing AMANDA, the collaboration published a search for neutralino annihilations in the Sun, which happens to be the best place to look, for various reasons. In 2009, with the 22-string edition of IceCube, they produced a second study of the Sun that set stringent enough limits on the existence of WIMPs to begin threatening the viability of a certain class of supersymmetry theories. Not only did this add to the creeping unease about supersymmetry in general, it narrowed the possibility that neutralinos might comprise any significant fraction of the cold dark matter. Even before its completion, IceCube was making waves.

As always, the holy grail was a point source or neutrino star, and since the mid-1990s, the most promising candidate for that has been the gamma ray burst. These enormously bright and distant explosions occur about once a day somewhere in the visible universe, and the explosion in infrastructure that has taken place in particle astrophysics over the past several decades has produced two marvelous telescopes for looking at them. Both are satellite-based and operated by NASA. The Fermi Large Area Telescope produces a map of the entire gamma ray sky every three hours, and Swift is specifically designed to detect gamma ray bursts. It has a special "burst alert telescope," which detects and determines the position of about one hundred bursts a year. After a detection it "swiftly" reorients itself to observe the afterglow in the X-ray, ultraviolet, and visible bands. (Unfortunately, the satellite isn't that swift, as it takes almost two minutes to spin into position. There's an entire class of GRB that lasts less than two seconds, while the longest last several hours.) Both of these telescopes produce catalogs of the whens and wheres of the bursts they've detected, which the IceCube scientists can use to search for neutrinos. The great advantage here is that since gamma ray bursts are localized both in space and in time, there is essentially no background. Even one high-energy neutrino coming from the direction and within some short time window of a GRB would be statistically significant all by itself, and two might very well constitute a five-sigma discovery.

The big question about these violent creatures is what they actually are. The short bursts are thought to result from the merger of a binary system, a neutron star or black hole absorbing some companion revolving around it, and the longer ones from an unusually powerful supernova or "hypernova," the collapse of an extremely massive star. The end result of either of these events would be a rapidly spinning black hole, and the bursts of gamma rays are probably emitted during the brief time of the collapse. What separates GRBs from the usual supernova is that they emit their light in laser-like beams or "jets," shooting out in opposite directions along the axis of rotation of the collapsing system—from opposite poles, you might say. These flashes of light can be extremely distant. Some come from so-called cosmological distances, which means that they occurred billions of years ago, when the universe was comparatively young. The very fact that we can see such distant events means that they must be incredibly bright, probably even brighter than supernovae.

Back in the mid-nineties, John Bahcall and his brilliant young Israeli protégé, Eli Waxman, proposed a "fireball model" for GRBs that gave estimates for the number of neutrinos that would be emitted along with the gammas. The notion was that protons produced during the gravitational collapse would be accelerated to very high energies by the enormous magnetic fields of the spinning system and would then interact with the gamma ray field, producing pions that would decay quickly into other particles, including neutrinos. Hence, the neutrinos would emerge in the jets along with the gammas—and at extremely high energies, even higher than that of the famous Oh-My-God particle, 10^{20} electron-volts. The two theorists even proposed at that early date that some future kilometer-scale neutrino telescope should be able to observe GRB neutrinos. Other fireball models, involving neutrons rather than protons, were also proposed.

The models suggested that particles besides neutrinos might also be emitted at ultrahigh energies: protons, neutrons, and helium and other nuclei. And this held out the possibility of answering the oldest and most intriguing question in cosmic ray physics: where do ultrahigh-energy cosmic rays come from? Eli Waxman had proposed GRBs as the answer back in 1995, two years before he and his mentor filled in the details of their fireball model. But as the one-hundred-year anniversary of Victor Hess's seminal balloon flight approached, this overarching question remained unanswered.

John Bahcall was not only one of the foremost theoretical astrophysicists of his time, he was also admired for his big ideas, his integrity, his sense of fun, and his personal warmth. Neutrino astrophysics wasn't the only field that lost a visionary when he died of a rare blood disorder in 2005, at the young age of seventy. Eli took up the mantle as touchstone theorist in neutrino astrophysics and as IceCube neared completion, began boldly predicting that his faith in GRBs would soon be confirmed.

Francis had been on the trail for more than a decade at that point. Rellen Hardtke, the student whom he suspected of pulling his leg when she showed him the Peacock events that August morning in 1998, wrote her doctoral thesis on a GRB search with AMANDA. They had hoped to see GRB neutrinos even then, of course, though they knew it was a long shot because AMANDA wasn't sensitive enough to detect the fluxes predicted by the models. The first incarnation of IceCube that was, forty-string IC40, was completed in early 2008, by which time the IceCube group at the University of Maryland had joined in. A grad student there by the name of Kevin Meagher, working under professor Kara Hoffman, conducted one study, and Francis's student Nathan Whitehorn conducted another. In January 2011, about three weeks after IceCube's last string was deployed, they uploaded a report on these studies to the arXiv preprint server at Cornell University, the preferred way for high-energy physicists to communicate their latest results to their colleagues. Their search was still a search: one year's data from IC40 had yielded no evidence of neutrino emission by gamma ray bursts.

At the same time, Francis and two fellow theorists, Maria Concepcion Gonzalez-Garcia and her post-doc Markus Ahlers, had been coming at the problem from a different angle. (Gonzalez-Garcia held appointments at the University of Barcelona and the State University of New York at Stony Brook. Ahlers worked with her at the latter.) In March, the three published a second paper on the arXiv, under the title "GRBs on Probation," demonstrating that the simple fact that IC40 had detected no extraterrestrial neutrinos indicated that gamma ray bursts probably weren't the source of ultrahigh-energy cosmic rays. If they were, the three surmised, there would be a diffuse flux of equally high-energy neutrinos, raining in from every direction, as the cosmic rays do, and the neutrino flux could be calculated from the cosmic ray flux using the fireball model. Carrying out the calculation, they demonstrated that IC40 was sensitive enough to detect the predicted neutrino flux. Since it hadn't, they deduced, GRBs probably don't produce ultrahigh-energy cosmic rays either.

This is an excellent example of the way a null result, or new limit, can propel science forward. Everyone would have *preferred* to see GRB neutrinos, and it would have been spectacular if they had, but nature is indifferent to the preferences of humans. As the author of a perspective in *Nature* wrote, "Carl Sagan's antimetabole 'absence of evidence is not evidence of absence' immediately comes to mind: not detecting an expected signal can be as powerful a diagnostic as the arguably more desirable discovery, and may send us back to re-evaluate how we arrived at our expectations in the first place."

GRBs were only on probation, however; they were not yet expelled. There was enough wiggle room in both the models and the results, factors of two here and there, to keep the fireball alive.

Eli Waxman was invited to speak at the particle astrophysics symposium that accompanied IceCube's inaugural gala, and since all this news was hot off the press at the time, his talk was much anticipated. Eli's a bit of a prizefighter up there. His simplest statements come across as challenges somehow. While he was willing to admit that IceCube was "getting dangerous," he stuck to his guns. The instrument wasn't sensitive enough yet, he told us. The Cubists had missed a factor of two in their analysis. The fireball would survive.

Such cocksureness on the part of theorists gets under the skin of the working experimentalist. There was noticeable grumbling and rolling of the eyes as Eli spoke. Some said he wasn't particularly concerned with experimental evidence. John Learned, who has his own rather well-informed ideas about how GRBs might work, became especially argumentative.

The other shoe dropped about a year later, when the collaboration added another year's data to their GRB search, from the next iteration of their instrument, IC59. Nathan Whitehorn again took the lead in Madison, and grad student Peter Redl, advised by professor Greg Sullivan, took the lead in Maryland. Again they failed to find neutrinos, now associated with two years of GRBs from the catalogs. Their revised upper limit on the flux was at least a factor of four below the fireball predictions and possibly a factor of ten or even thirty. (Nathan explains it more simply: they would have expected to see about twenty neutrinos and they saw none.) "This implies either that GRBs are not the only sources of cosmic rays with energies exceeding 10^{18} electron-volts or that the efficiency of neutrino production is much lower than has been predicted," the collaboration wrote in *Nature*.

In the thick of it all, in November 2011, as he and Peter Redl were putting the finishing touches on the *Nature* manuscript, Nathan admitted that this was a bit disappointing. "It would have been really nice to see something; I'm not going to say that's not true. And IceCube will eventually see some astrophysical neutrinos, and that will be very exciting. . . . But one of the nice things about this particular thing is that you had actually a null result that was really meaningful in this case, right? You have a very distinct prediction and you can say something about it, you know, just as powerfully as if you had seen something. . . . And IceCube is pushing into this region in general on a lot of fronts, not just gamma ray bursts, where if we continue to see nothing then that has some really interesting meaning."

Four years later, with that much more sensitivity and a larger and more capable detector to boot, IceCube still has not seen GRB neutrinos. The biggest mystery in cosmic ray physics endures, and the theorists have gone back to the drawing board. Eli's latest guess is a type of galaxy called a starburst, which has an unusually high rate of star formation.

Life became easier when construction ended. As Francis had said in his e-mail, they now had a stable instrument that they could fine-tune to death. But they did much more than simply polish the lens of their new telescope. When they were released from the enormous burden of adding to it every year, they channeled at least that much energy—and probably more—into scientific creativity. It was like a bunch of kids being let out onto the playground. There was a buzz in the air at the fall collaboration meeting in

Uppsala, less than six months after the inaugural. Francis told me, "Ice-Cube is so incredibly multidimensional in ways that we discover every week." They were pushing out the boundaries in both astrophysics and pure particle physics.

They had just detected their first cascades, which meant that after a decade of relying solely on muon neutrinos they were now opening up the possibility of detecting the electron and tau flavors. Per Olof's vision, DeepCore, was looking as though it would be capable of studying the details of neutrino oscillation as well as the accelerators could, and be the only game in town at high energies. (They would open many eyes—and roil the accelerator community—at a neutrino conference in Kyoto the following spring by presenting five-sigma evidence for the highest-energy oscillations ever detected.) And in connection with oscillation, the prospects looked good for solving the longstanding mystery of the sterile neutrino, a potential fourth flavor of the particle, whose existence had been hinted at by several experiments over the previous two decades. Francis had been focusing on that prospect for about a year, and was now realizing that "if you have to design an experiment to see sterile neutrinos, you design IceCube."

"And it would be ten years away," I responded.

"Or more. . . . By the way, that's an interesting point. I always raised this when we were, you know, begging for money. If instead of a neutrino telescope you called this now a sterile neutrino experiment, a hundred million? That's nothing in the particle physics context. The problem was, we were a cosmic ray experiment. You know, the smallest Fermilab experiment costs many, many times that."

Francis, too, had been let out onto the playground. This was just one of his recent excursions into theory and phenomenology—and not just about IceCube. He still visited CERN for about one month a year, where he worked with data from the Large Hadron Collider between bicycle rides in the foothills of the Alps.

It seems as though the Madisonian powers-that-were had been waiting until IceCube succeeded before establishing it as an official research center on the campus. When they did, however, they did it in style, establishing a fellowship in honor of John Bahcall. One of the first John Bahcall Fellows was Markus Ahlers, who'd been lead author on the "GRBs on Probation" paper. And Ahlers, Francis, Maria Gonzalez-Garcia, and two other theorists were *also* collaborating on another game in the IceCube playground: cosmogenic or GZK neutrinos.

The *G* is none other than Kenneth Greisen, father of the Kamiokande strain of neutrino detector and so much else in cosmic ray physics. The *Z* is Georgii Zatsepin, who conceived of underground Cherenkov detection as early as Greisen and Moiseĭ Markov, but never bothered to write it down. The *K* is Vadim Kuzmin, an esteemed Russian cosmic ray theorist and cosmologist. In the 1960s, these three, along with Ven Berezinsky, the erudite Russian who later surprised the Americans in DUMAND with his knowledge of George Orwell, had realized that the interaction of ultrahigh-energy cosmic rays with the soup of photons that makes up the cosmic microwave background should produce neutrinos by the usual mechanism: a proton interacts with a photon to produce a charged pion, which decays into either a muon or an electron and one or two, in this case, ultrahigh-energy neutrinos.

The flux and energy spectrum of GZK neutrinos should depend upon both the cosmic ray spectrum and the density of the cosmic microwave soup. Ahlers, Francis, and their theorist colleagues employed the latest data from various cosmic ray instruments to estimate that these neutrinos should have energies between about 10^{15} electron-volts, aka a peta-electron-volt or PeV, and about a million PeV. That's 10^{21} electron-volts: *ten times* the energy of the Oh-My-God particle. And the peak in GZK neutrino production should occur at between a hundred or a thousand PeV. The advantage of probing such stupendous energies is that the atmospheric neutrino spectrum cuts off at a few tenths of a PeV, so you're way past that and there should be very little background, as in the case of GRBs.

Aya Ishihara has been interested in the extreme universe since her graduate school days at the University of Texas at Austin, where she worked on the STAR experiment at the Relativistic Heavy Ion Collider at Brookhaven (which Jim Yeck built). The idea behind STAR (Solenoidal Tracker At RHIC) was to create a kind of mini Big Bang. In 2005, she shifted to a post-doc with Albrecht Karle in Madison, where she began searching for neutrinos at the very highest energies and the GZK variety in particular. After three years in Madison, she moved back home to Japan, where she took up a second post-doc in the IceCube group at Chiba University. Like so many of the people on this project, Aya perseveres. By 2012, she had been searching for ultrahigh-energy neutrinos unsuccessfully for about seven years.

Over time and working on successive incarnations of the instrument, she developed simulations and fashioned a set of "cuts" aimed at picking out the kind of neutrino event she was looking for. By now, the collaboration was much more sophisticated and, for better and worse, bureaucratic than it had been in the AMANDA days. She was a member of the "diffuse-atmospheric" working group, which batted around emerging ideas and preliminary studies of neutrino populations that bombard us from all angles in the sky. Once an idea has ripened in a working group it is elevated to the "analysis call," a phone/video conference that takes place once a week. Here, the presentations are a bit more formal and representatives from all over the collaboration have a chance to weigh in. One of the purposes is to hone the question being asked in an analysis in order to give it the best possible chance of yielding a meaningful answer, and one of the best ways of doing that from a statistical standpoint is to ask the question only once. The petitioner might do a preliminary study on a so-called burn sample, a small portion of the dataset, and if the results are promising, get permission to "un-blind" the full thing and see what she's got. Then a paper might be written, at which point the publications committee kicks in.

In those days, the analysis call took place at an awkward time for those who lived in the Far East, "half-past midnight in Japan or something," says Aya with a laugh, "so it's a kind of uniting us, how the things develop in IceCube" (another indication of her stick-to-itiveness—and her generosity: she spreads the credit around).

Since the Earth happens to be relatively opaque to ultrahigh-energy neutrinos, Aya optimized her cuts to pick out muon neutrinos coming from directions just below the horizon, that is, just barely up-going. This way she would still have shielding from the down-going cosmic ray background, but the neutrinos would traverse a relatively short distance inside the planet and thus have a better chance of reaching the detector. She also cut on the total brightness of the event: how many photomultiplier tubes were lit up and just how lit up they were. This is a measure of the total energy of the corresponding neutrino. She placed this cut to select events above about a tenth of a PeV, the top end of the atmospheric background.

Aya wasn't expecting much in May 2012 when she received permission to un-blind the previous two years' data. She'd already done this three times; it was becoming routine. However, the last set of data she had un-blinded had come from IC40, the forty-stringed instrument, and now she was dealing with IC79 and IC86, true kilometer-scale detectors. As far back as the writing of the IceCube proposal a decade earlier, Francis and several others had suspected that they would need the full detector in order to see cascades.

Aya was searching only for muon tracks, and again she saw none. But two absolutely beautiful cascades snuck by her cuts. They snuck by in a larger sense as well, since they were *down-going*: they came from the southern sky. The simple algorithm she had used to determine the direction of what she assumed would be muon tracks had been confused by these two huge, globular flashes of light. It reconstructed them as traveling upward when they were actually traveling down. As is so often the case in science, this mistake was a stroke of luck. Had they been reconstructed as down-going she never would have seen them.

The two events had such personality and were examined in such detail that rather than refer to them by their official designations ("18545/63733662" doesn't roll off the tongue), a grad student in Madison by the name of Jakob Van Santen decided to name them after the *Sesame Street* duo Bert and Ernie. Bert had exploded in the detector in early August 2011, just three months after the full instrument had taken flight, and Ernie had rung in the New Year a little late, on January 3, 2012. Both were enormous. They filled significant portions of the array. They were larger than the football stadium on the campus at Madison, where football is a big deal (see photograph 32).

So, why, after focusing on up-going events for almost three decades, were they now so excited about two down-going ones? Because they knew they were neutrinos and they *suspected* that they were extraterrestrial or astrophysical, the first neutrinos originating outside our solar system that anyone had detected since the miraculous supernova of 1987—and the first *high-energy* astrophysical neutrinos ever detected. They knew they were neutrinos for two reasons. First, IceTop, the surface array for detecting and vetoing down-going cosmic ray air showers, was silent during both events. Second, both were fully contained in the detector. This meant that the interactions that caused them had to have occurred inside the array and therefore had to have been initiated by neutrinos. It isn't clear, incidentally, which flavor of neutrino causes a cascade. It could arise from the straightforward inverse beta decay of an electron or tau neutrino with a nucleon, or it could arise from a so-called neutral current interaction in which any of the three flavors interacts without losing its life. (Neutral currents are what gave Dave Cline the nickname "A.C.D.C." back in the early 1980s, when he first thought he found them, then that he hadn't, and finally that he actually had.)

Both Bert and Ernie packed enough punch to make it unlikely that they were atmospheric. They were, in fact, the highest-energy neutrinos that had ever been detected. They came in at about one PeV each, roughly ten times the atmospheric cut-off. They were not energetic enough to be GZK neutrinos, so in a sense Aya had struck out again. However, she remembers stumbling across those two events as a once-in-a-lifetime experience, the kind physicists live for.

Isn't it wonderful that IceCube's first real surprise came from cascades and from "behind the telescope"?

It is hard in this internet age to keep an intriguing result quiet. Francis began getting e-mails right away. Being a fan of the pithy correspondence, his favorite was the four-word inquiry he got from Luis Anchordoqui, a colorful Argentinian who had collaborated with him and Markus Ahlers on several of the recent theory papers. "Is it fucking true?" asked Anchordoqui. "Yes," Francis responded.

The crucial question was whether Bert and Ernie were extraterrestrial. When Aya asked that question of her data, employing statistical techniques hammered out through lengthy negotiation in the analysis call, she found that there were about three chances in a thousand that the two events might simply have been unusually energetic atmospheric neutrinos. That may sound good, but it's only 2.8 sigma, not significant enough to claim discovery—and not even significant enough to claim "evidence." That word is used when you reach three sigma—and even then, there is an adage that half of all three sigma results are wrong!

Francis holds the opinion, incidentally, that Aya had at least found "evidence" for and possibly even "observed" (equivalent to "discovered") the very first high-energy extraterrestrial neutrinos. Aside from seeing Bert and Ernie as being far too energetic to be atmospheric, he realized when he inspected them in the event viewer that they were in a "different league" altogether from the most energetic atmospheric neutrinos they had seen before. Pointing out that the labored statistical reasoning that had emerged from the analysis call ignored the fact that the veto from IceTop had been silent during both events, he adds wryly that "statistics is a lot easier than physics." If neutrinos of such high energy had been created by cosmic rays hitting the atmosphere, other particles, including muons, would almost certainly have been created along with them.

So Francis landed in the optimistic camp, as usual. Naturally, of course, there were pessimists as well.

Nathan Whitehorn was the kind of graduate student that passes through a professor's life once or twice in a career. Francis says that "if somebody is going to make it big in this experiment, I predict it's him."

One of the reasons Nathan enjoyed working on the project was that "especially by comparison with these large accelerator experiments, Ice-Cube is actually a very simple detector." (One has the impression that it might seem simpler to him than to most people.) He liked the fact that as a student he could not only grasp all the steps in the analysis chain, from the moment light was detected to the point where he had a dataset he could do physics with, he could actually play a role in ensuring that they worked and improving them. He contributed to what you might call the "guts" of the experiment in several ways, not necessarily related to his thesis work. He wrote a software program called millipede, for example, for estimating the energy of muons passing through the detector, that has become a standard tool. (When he was working on millipede, he came across a series of papers that the DUMAND collaboration had produced when they were trying to do the same thing two decades earlier. They had never had a detector to work with, of course, but now, with IceCube built, Nathan told me, he could "do the thing that they meant to do. . . . You see this ghost of the other experiment.")

Nathan was in an excellent position to capitalize upon Aya's discovery, since, even before she made it, he had been working on ways of using the veto power of IceTop to open the view toward the southern sky. He was also free, since he'd defended his thesis by then. He was staying on as a temporary post-doc in Madison, while his wife, a biologist, completed her thesis.

He and Claudio Kopper, a John Bahcall Fellow in Madison who specialized in Monte Carlos, began thinking out loud together, sketching ideas on whiteboards and the like. "It was a pretty great opportunity," remembers Claudio. "After we saw these two high-energy cascades, both Nathan and I thought, 'Well, this is exciting and there have to be more and we have to find those.'"

They settled upon a disarmingly simple idea.

On a Saturday, about a month after Aya's discovery, the two of them, independently and at home, according to Claudio, "just wrote some sample code that would filter out, like, starting tracks and cascades." There were subtleties to it, but the basic idea was simply to pick out high-energy events that were born, as Bert and Ernie were, inside the detector. Over the next several months, the two post-docs went about presenting their "starting-track analysis" in a working group and then at the analysis call, and as it dawned on the rest of the collaboration that they had come up with a brilliant idea, the old fear of discovery kicked in and they were subjected to ever more intense scrutiny. Two of the more critical members of the collaboration were assigned to review their proposals, Spencer Klein from Lawrence Berkeley Laboratory and Christopher Wiebusch, now a professor at Aachen University, who had played a similar role fourteen years earlier when the Peacock events were discovered. Claudio and Nathan were put through their paces.

In October 2012, when they finally got permission to un-blind, their analysis revealed twenty-six more events at energies near and above the atmospheric cut-off. Seven were muon tracks, the other nineteen were cascades, and most came from the southern sky. There were now twenty-eight in all, including Bert and Ernie. This was "evidence," at a hair less than five sigma, for extraterrestrial neutrinos—not quite "observation," but that would be splitting hairs.

Since point source fever reigns in the collaboration pretty much all the time, Nathan and Claudio had brought in Madison's point source expert early on: Naoko Kurahashi Neilson, a post-doc working with Albrecht Karle. Searching for a point source is in some sense the next step after finding the events, so the collaboration had wisely kept this threesome blind to the events' directions by keeping them blind to the times when they occurred. (Since the Earth spins, you need the time in order to know which way IceCube was pointing.) Nevertheless, this "evidence" ramped the fear of discovery up almost to the level of the AMANDA days. Ty DeYoung, who is now a professor at Michigan State University, observes that it also sparked a comparable burst in creativity. And he should know, since he was responsible for much of the creativity the first time around.

Claudio says that the real work started after the un-blinding, when they tried to determine the energies and directions of the events (inside the detector, that is; they were still blind to the times).

Their most impressive accomplishment was to figure out a way to reconstruct the directions of cascades. Claudio and Jakob Van Santen, the grad student who had named Bert and Ernie, took the lead. (Albrecht was adviser to both.) What makes this task especially difficult is that the cigar-shaped shower of electrons and positrons that creates the light in a cascade is only ten or twenty meters long, nothing like the track of a muon, which can run along for more than a kilometer. And since the light detectors in IceCube are so far apart—the strings are separated by sixty meters and the detectors on a string by seventeen—you don't have enough pixels, so to speak, to get any sort of image of the original shower; you have to wait until its light reaches your detectors. What you end up with, as in Bert and Ernie, is a nearly spherical blob of light the size of a football stadium, in which the tiny shape of the shower is completely lost.

Claudio and Jakob discovered, amazingly, that the individual bunches of photons emitted by a shower, wave-packets they're called, do carry information about its direction. Without going into the details, if the packets are going in the same direction as the shower, they will be deformed in one way as they travel through the ice, and if they're "swimming upstream" against the direction of the shower, they'll be deformed in another way. Thus directional information is carried by the *shapes* of the individual light pulses as they hit the detectors. This is the sort of insight that can make science seem like magic sometimes.

It's also where Dave Nygren's digital optical module comes in. For it is capable of recording the shapes of the pulses and sending them to the surface of the Ice. And it turns out that Francis made an important contribution here as well. Recording and saving the shapes, or waveforms, added complexity to the DOM. It would have been much easier just to take a couple of measures of the height and shape of the pulses and send them up. So at first Dave questioned whether the waveforms ought to be saved. He remembers organizing a software workshop in Berkeley way back in 1998, when they were designing the first fifty DOMs for AMANDA's string 18. "We tried to ask the question, are these waveforms really useful? Didn't really know," he says. "But Francis said, 'Yes, the waveforms are useful,' so the decision point was made."

Although Francis doesn't remember that exchange, he points out that phenomenologists often know what is needed from a measurement better than experimentalists do: "I mean, I knew what data I needed. . . . Nygren knew how to get it."

Jakob and Claudio went to an astonishing level of detail: They used the properties of the ice that had been elucidated by Buford Price's group—and even the properties of the refrozen ice in the holes around the DOMs!—to simulate how light would have traveled to each of them. They used the waveforms at every single DOM that was involved in an event—354 in Ernie's case—and rotated hypothetical showers around in every possible direction until they found the one that gave the best match to the multitude of pulse shapes. In the end, they could nail down the direction of a cascade to within about ten degrees—nowhere near as good as a muon track, which they could resolve to less than half a degree, but still very helpful.

This effort also gobbled up an impressive amount of computer time. Technology had long since left the Cray supercomputers in Berkeley behind. The current state of the art was the so-called Linux cluster, a bank of personal computers, aka central processing units or CPUs, running on the Linux operating system. After they had determined the directions of the twenty-one cascades, Claudio looked into the "cluster statistics" and found that they had used about *sixty-two CPU-years'* worth of computing time—three years for each event! In other words, it would have taken that long if the calculations had been carried out on one computer. Since the computers in the cluster worked in parallel, it only took only a couple of weeks to a month.

In mid-February 2013, they un-blinded the times and hence directions to produce their first-ever map of the extraterrestrial neutrino sky. It looked as though there were "hot-spots"—hints of point sources.

Francis, Albrecht, and the then spokesperson of the collaboration, Maryland's Greg Sullivan, had scheduled a phone conference with their NSF managers to discuss an important but mundane funding issue. When they shared the news, they were summoned to Washington. (The NSF folks were greatly relieved: they had two multi-hundred-million-dollar projects going at the time, the other being LIGO, and neither had much to show for it.) It was over a wine-infused dinner on the evening of the meeting, evidently, that the bureaucrats and scientists together decided to fast-track a letter to *Nature*. (It might have been Francis who mooted the idea: he had asked Nathan to begin drafting such a letter about a month earlier.)

It wasn't a *bad* idea. The results were important, perhaps explosively so, and the news had already snuck out. Outsiders and popular science magazines had begun inquiring by e-mail, and theorists, a febrile lot, tend to jump on every new experimental result. Several had revealed Nathan's "non-discovery" of GRB neutrinos on the arXiv before the Cubists had announced it themselves. They didn't want to be scooped again.

The next morning (too early perhaps, considering the excitement and wine of the previous evening) Greg informed the rest of the collaboration of a plan to submit a letter within one or two weeks and get it published, hopefully, within a month. Unfortunately, he neglected to include a draft.

Panic ensued. E-mails flew. Even the proposed title, "Evidence for Extraterrestrial Sources of High-Energy Neutrinos," provoked heated arguments—"sources" being the main sticking point. Not far below the surface, there was not only fear but considerable resentment. There were now about 220 people in the collaboration, many of whom had no idea what had been going on in Madison. Some suspected they were being sold a bill of goods by the "Evil Empire." Some had petty personal concerns like getting tenure or their next round of funding. Some (the usual suspects) were so fundamentally pessimistic that they would never be convinced. And some, to be honest, couldn't understand it all, even when it was explained to them. But most simply wanted the time to review the paper thoroughly and carefully, which was undoubtedly a good idea.

The dark cloud hanging over this drama was the knowledge that several years earlier, the Pierre Auger collaboration, which ran an enormous cosmic ray detector in western Argentina, had announced to great fanfare that they were 99 percent sure they had found the source of ultrahigh-energy cosmic rays: active galactic nuclei—and had then had to retract their claim when more data, that is, better statistics, washed their preliminary finding away.

The veneer of bureaucracy notwithstanding, when it got right down to it, IceCube was as anarchistic as AMANDA was. They still worked by consensus. Everyone had to agree—or at least not disagree too strenuously—before the new results could go public. And that meant there was plenty of emotion involved. People needed to be heard.

Francis did not engage in argument. He worked behind the scenes, pulling from the front end, while numerous others dug in their heels and dragged along behind. "This is the real thing, except to some of us of course who fear that a discovery may affect their lifestyle," he wrote to me on the day of Greg's announcement. "I showed our results privately to a friend at Goddard [Space Flight Center in Maryland] today. She had no doubt about what she was seeing. This is a vicious loop when you are conservative. The next step is that you can never see anything. No matter how you turn this, it is a discovery."

He may have been overcorrecting a bit on the positive side, but that was his role. The skeptical role was also important, and beneath his frustration he knew that. IceCube was also in the luxurious position of having no competition, so there was no need to hurry.

Luckily, they had a steady hand in the chair of their publications committee, Olga Botner, who had been Kurt Woschnagg's doctoral adviser in Uppsala before both had migrated from the accelerators to what was then AMANDA. Olga had been a force of quiet reason in the collaboration ever since she'd joined in 1998. Not only does she possess a formidable intelligence, she strikes a good balance between optimism and conservatism, she's soft-spoken, she exudes trustworthiness, and she's had plenty of experience. She sits on the Nobel physics committee. Also, and appropriately at this important moment, this capable and self-effacing woman was elected spokesperson of the collaboration less than a week after Greg's announcement, although he remained in the post for the time being.

She and Francis worked together to tamp down the noise. They took the review of the paper private, and the review panel began to pound it into shape, sharing drafts with the rest of the collaboration as appropriate, a good way to bring them along.

The collaboration had inadvertently put themselves in the spotlight by inviting the world to an IceCube Particle Astrophysics Symposium to take place on the heels of their next collaboration meeting, scheduled for Madison in early May. As Greg told them all by e-mail, "At some point our very credibility is at stake, since . . . the community knows that we are doing (have done) [another] search, and obviously wonder[s] if we are competent enough to perform and report on a follow-up analysis with funding for over 200 people and a \$270M detector." (Francis put it more succinctly: "The whole world knows. That's why we have to release this.")

The pressure increased when Aya submitted her paper on Bert and Ernie to *Physical Review Letters* and posted it simultaneously on the arXiv. (*PRL* publishes breaking physics news. They don't insist on embargos until press time, as *Science* and *Nature* do.) This attracted a flurry of international press coverage and won Aya the annual Young Scientist Prize from the cosmic ray physics division of the International Union of Pure and Applied Physics, the preeminent global umbrella organization. The first neutrino astrophysicist ever so honored, she would accept the prize at the International Cosmic Ray Conference in Rio de Janeiro the following summer.

Part of the problem was that the story of the twenty-eight events was too complicated to tell in the brief format of a *Nature* letter. It was like trying to fit a large woman into a tiny corset. Even some members of the collaboration had a hard time following the argument, as it relied extensively on so-called supplemental material, which would need to be published on the journal's Web site when the letter appeared. But Nathan was fixated on getting another paper into this respected journal, and he and Claudio, the other primary author, were not converging with the rest of the collaboration. It also turned out that the collaboration's internal review panel was a bit too large and included a couple of intransigent skeptics. About a week before the collaboration meeting, Francis, the political master, broke the logjam by working diplomatically with Nathan and Claudio to rewrite the paper in a normal physics format. He shared the resulting draft privately with Olga, and after some back and forth the two agreed that "this was the right paper." The spokesperson and principal investigator make a powerful pair. (Interestingly, neither has been to Pole. "IceCube is like a space experiment," Olga tells me. "You don't have to go to lead the project.")

On the first day of the collaboration meeting, the seventh of May, Olga went out on a limb and announced that the paper was *almost* ready for collaboration-wide review, the last step before journal submission, and proposed that they present the results publicly at the following week's symposium.

Part of the reason she was willing to take this step was that they had caught a break a day or two earlier, when the traditionally skeptical Spencer Klein had presented a spectacular new cascade in a "pre-meeting" before the full meeting began. "Big Bird" was born right in the center of the detector and fully contained, and came in at about two PeV, twice the energy of Ernie, the previous record holder. Francis called this "our life insurance." "If you didn't believe anything, you cannot explain three PeV events as something coming from the atmosphere." He also pointed out that this was yet another accident in what was becoming "an accumulation of accidents." Bert and Ernie had snuck into Aya's hands by chance, of course, and one of Spencer's students had stumbled upon Big Bird in the burn sample from 2012, 10 percent of the data, which she had been allowed to un-blind for a preliminary study. Since the rest of that year's data was still under wraps, there was a ten-to-one chance that she would have missed it.

Although Big Bird turned Spencer and several others into believers, many others still held out—and they had a point. When a harried Olga graciously gave me a moment of her time during the collaboration meeting, she showed me the crucial graph from the paper and pointed out that if you ignored Bert and Ernie (Big Bird could not be included in a study from the previous year), the other twenty-six events weren't that far from the atmospheric background. "We have to make sure we're not just seeing what we want to see," she said.

She was holding endless side meetings to fine-tune the message and perfect the graphs that Claudio, Nathan, and Naoko were hoping to present at the symposium.

In addition to the feeling of a threshold being crossed, there was an air of poignancy at this gathering. The old guard was leaving. Buford Price was not there, having long since shifted his focus from the astrophysics aspect to the glaciology and climatology. Per Olof was absent for medical reasons. Christian Spiering was there, voicing his usual skepticism, but he had retired the previous week, having reached the mandatory age in Germany. Over dinner at a sushi place near the capitol building, he told me he'd probably attend the fall collaboration meeting in Munich, but not many more in the United States.

At the traditional banquet, I sat at one end of a square table, two people on a side, next to Tom Gaisser, the main cosmic ray theorist in the collaboration and the motivating force behind IceTop. Greg Sullivan and the newly elected deputy spokesperson, Ty DeYoung, sat perpendicular to us on our right. Albrecht sat to our left next to Allan Hallgren, a stalwart of the Swedish contingent since 1992. At Allan's shoulder, directly across the table from me, sat his wife, Olga, and Francis sat next to her, across from Tom. Tom and I were both moved at the sight of Olga and Francis speaking softly and amiably together. They looked older than they had when we'd first met them, much longer ago in Tom's case than in mine. He had first met Francis in the mid-seventies. Olga wore an understated black dress lined with thin horizontal ruffles. Francis had removed his suit jacket to reveal a black vest over a light blue collarless shirt with thin vertical ribbing. He wore round horn-rimmed glasses. The two were oddly reminiscent of the farm couple in the painting *American Gothic*. Tom borrowed Albrecht's camera and took a shot (see photograph 33).

They are civilized enough not to give long speeches at these banquets. During the gala in 2010, Francis had set the tone by opening with, "I'm from Belgium, where we take food seriously" and speaking for less than a minute. As this banquet came to a close, Greg stood up to pass the baton to Olga. Making light of his failed attempt to fast-track the paper, he said that the experience of being spokesperson had taken him back to his early interest in Buddhism and Taoism. He'd learned that "the harder I push, the slower things go."

He handed the microphone to Olga, who thanked them all for the honor of serving as spokesperson and expressed her confidence in their present findings and future promise.

Francis spoke last, the longest-running of the old-timers. It had been twenty-five years since he had presented the paper in Lodz, Poland, that had formed the basis for AMANDA's letter of intent. He wasn't going anywhere. "IceCube is my joy, every night," he said.

These meetings traditionally end with a collaboration report delivered by the spokesperson. Olga had to leave early for some business in Europe, so Greg took her place. This was fitting. After delivering his "State of the Collaboration Address," he presented a plan to reveal their new results at the symposium in several days and follow it up with a journal article—*not*, most likely, in *Nature*. They would claim "evidence" for the first extraterrestrial neutrinos ever detected, what astronomers call "first light" in the nascent field of neutrino astronomy. They would present their map of the high-energy neutrino sky, but they would make no claims as to hotspots, point sources, "neutrino stars" or anything of the like—a sensible and conservative tack. Press releases would accompany both the talks, which would be given by Claudio, Nathan, and Naoko, and the journal article. Since their main goal was to get the greatest possible impact out of the latter, Greg expressed misgivings about this double-press-release strategy. Then he did a wise thing. He invited discussion.

It seemed that everyone had been waiting for this moment, whether they realized it or not. Several people griped about the lack of transparency in producing the paper. Some doubted the results. Hardly anyone was happy with the publicity plan. But no one objected strenuously, even though quite a few shuffled out of the room looking down at their feet or shaking their heads, once those who wanted to speak had spoken and the conversation had come to a natural end.

They seemed thrilled at their discovery and scared at the same time. They were in uncharted territory.

Epilogue: The Dawn of Multi-Messenger Astronomy

People think science converges. It doesn't converge. It answers questions that raise more questions that are usually more interesting but also more difficult.

—FRANCIS

THE PAPER APPEARED IN SCIENCE THE FOLLOWING NOVEMBER, ABOUT NINE months behind Greg's original plan. A dramatic muon event made the cover, superimposed upon an elegant white background. ("Mr. Snuffleupagus," though the journal disregarded this whimsy.)

Francis, Olga, and the three young authors were inundated with phone calls. The *New York Times* featured the story at the top of its Web page for about a day, under a photograph of the IceCube Laboratory standing in blue- and violet-hued isolation on the wind-carved and manifestly frigid ice sheet. (Sven Lidström had taken the photograph, probably in the "time of the long shadows," sunrise or sunset, during one of his winterovers.) Stories were carried by the BBC, *Time*, NBC, National Public Radio, and other prominent outlets worldwide.

In mid-December, *Physics World*, the members' magazine of the British Institute of Physics, named the "first observations of high-energy cosmic neutrinos" its breakthrough of the year. This made even Francis nervous. When he accepted the award he kept thinking, "What are we going to do if this is wrong?" There was an extremely remote possibility that their twenty-eight high-energy neutrinos actually were atmospheric, born from the decay of charm particles produced by cosmic rays in the atmosphere. It wasn't a particularly realistic possibility, but Francis found himself obsessing about it anyway. For the next several years, he spent much of his "physics time" working on charm, and finally, in 2016, demonstrated conclusively that it could not explain their results.

The accolades continued. In February, Nathan Whitehorn was named a "Young Star" by the Astrophysics Division of the American Physical Society. The following year, 2015, Francis received the Giuseppe and Vanna Cocconi Prize from the European Physical Society, "for his visionary and leading role in the detection of very high-energy extraterrestrial neutrinos, opening a new observational window on the Universe," and the prestigious Balzan Prize.

The International Balzan Foundation awards four prizes each year, "two in literature, the moral sciences and the arts, and two in the physical, mathematical and natural sciences and medicine," but the specific subject areas change year to year. In science disciplines that don't receive a Nobel, the Balzan is considered the equivalent. In its citation, the Foundation wrote, "Francis Halzen is an inspiring example of a scientist who, though coming from a different research discipline, had the foresight to see where the next breakthrough in our understanding of the universe is likely be found, and who has the energy and leadership to realize this vision." While this assessment was probably true, Francis noted at the beginning of his acceptance speech that he was "humbly aware that I receive this Prize thanks to the talent and dedication of the technicians, engineers, scientists and administrators who made the IceCube project a reality."

The prize carries 750,000 Swiss francs, equivalent to about $770,000, and half the money is to be spent on research, "carried out preferably by young humanists and scientists." He used it to set up a Balzan Fellowship in Madison.

At the end of his brief speech, Francis mentioned the late John Bahcall's two secrets about physics: that it does not proceed logically and that "we would do it even if we were not paid for it."

"The same is true even without awards," he said in closing, "but I must admit that this one is very much appreciated."

Indeed, in an essential sense, the accolades *were* beside the point. The collaboration was too caught up in the thrill of the chase to pay them much mind. Even before their epochal paper was accepted by *Science*, Nathan and Claudio had run their starting-track analysis on a third year of data and found nine more high-energy events. This pushed them over the five-sigma threshold, so that they could now claim "observation" of astrophysical neutrinos (and Francis got his "five sigma in five years"). About a year later, a student of Albrecht's named Chris Weaver found "evidence," at 3.7 sigma, for extraterrestrial muon neutrinos pouring in from the northern sky. This was not a starting-track analysis, it was a vanilla-flavored search for up-going muon tracks. Since it meshed in nearly all details with the starting-track analyses, it served as separate confirmation—a difficult thing to do in expensive, one-of-a-kind experiments like this one.

When the excitement finally wore off, it dawned on Francis that their discovery was more momentous than they had realized. They had been so fixated on finding a point source—and disappointed that they hadn't—that they had not stopped to consider the implications of what they *had* found. Their measured flux of high-energy neutrinos was both "diffuse," meaning there were no real hotspots in the sky, and "isotropic," meaning the neutrinos seemed to arrive more or less evenly from all directions. Thus the majority were not only extraterrestrial, they were extragalactic: they originated outside the Milky Way. Had they been born inside our disk-shaped galaxy there would have been a hotspot in the direction of the galactic center, perhaps, or along the line of the galactic plane. On further consideration, Francis realized that the flux was "exceptionally high by astronomical standards" (even though twenty-eight in two years doesn't seem like a lot) and that it dovetailed well with the diffuse flux of high-energy gamma rays measured by the Fermi satellite telescope—and to a lesser extent that of ultrahigh-energy cosmic rays. Standard particle physics led him to surmise that all three of these cosmic messengers may originate from the same array of sources, be they starburst galaxies, galactic clusters, active galactic nuclei, gamma ray bursters, a combination of them all, or whatever else. In other words, the extragalactic universe appears to be awash in high-energy photons, neutrinos, protons, and other nuclei, generated by untold numbers of what Francis calls "mega-LHCs," or Large Hadron Colliders.

It should come as no surprise that what he was actually doing when he came to these realizations was thinking ahead. He was planning for the next generation of the instrument, which they now call IceCube-GEN2. It wasn't that new an idea, actually; he had presented it to their NSF managers during the exciting collaboration meeting that had culminated in taking their breakthrough discovery public. But there are few things better than writing for getting the gears of the mind in motion, so he had taken deeper stock of their findings as he and Markus Ahlers had been working together on a GEN2 white paper.

This is the way it goes. GEN2 will be to IceCube as IceCube was to AMANDA—with one important difference: there won't be much increase in cost. Thanks to the extraordinary clarity of deep Antarctic ice, they can expand the size of the instrument to ten cubic kilometers (a factor of ten) by adding roughly the same number of strings, eighty or so, but placing them much farther apart, something like 250 meters. This would result in a disk-shaped instrument with about the same "thickness" as IceCube, about one kilometer, but a footprint on the ice sheet of ten square kilometers. (Jim Haugen will have a lot more running around to do if he wants to plant another set of flags after the last GEN2 string is dropped in the Ice. He'll also be a lot older. The present plan calls for GEN2 to be completed in 2031.)

Owing to its shape, GEN2 will be able to resolve smaller celestial objects in the horizontal plane than IceCube does. And owing to its size, it will not only detect ten times the number of astrophysical neutrinos per year, its reach will extend to higher energies. It is more likely that Aya Ishihara will find the quarry she has been seeking for more than ten years now, cosmogenic or GZK neutrinos. And just as it took the step in size from AMANDA to IceCube to detect their first cascades, so will the next step increase their chances of detecting the double-bang signatures of high-energy tau neutrinos, in which one cascade is generated when the neutrino dies to produce a tau particle and a second bursts forth some distance away when the tau decays.

The little particle continues to be a coy mistress, giving up its secrets slowly. As they contemplate the meaning of the extraterrestrial neutrinos that continue to trickle in, the collaboration has come to understand that the reticence of the particle, the very quality that allows it to bring news from distances and times that the photon cannot, compounds the difficulty of seeing a "neutrino star." Owing to the particle's extreme penetrating power, when you look at the neutrino sky, you see almost to the edge of the universe, which is equivalent to the beginning of time. Since all the sources from here to there are "visible," they act as a diffuse background, not unlike the *nearby* diffuse background from neutrinos created in the atmosphere. Reminding us that IceCube's resolving power is some fraction of a degree, Eli Waxman points out that "with this acceptance angle, the detector window is open to sources from more than 10,000 galaxies." Thus, for a high-energy neutrino star to be "visible," it will need to be relatively nearby and relatively bright. I don't know about you, but in me this line of reasoning evokes an immense sense of space.

Their latest search for a neutrino star employs seven years of IceCube data, reaching back to the days of their forty-stringed instrument, and they still haven't found one. GEN2 would make this holy grail about ten times more attainable.

Perhaps the most fundamental implication of the birth of this new astronomy is that it employs a previously unemployed messenger to bring us news from cosmological distances and times. And this seems to be where astronomy as a whole is headed. The biggest breakthrough in the field thus far this young century has been the detection of the first gravitational wave, a disturbance in space-time first predicted by Albert Einstein a century ago, by the Laser Interferometer Gravitational-Wave Observatory, in September 2015. LIGO is in some sense the sibling of IceCube, since it went through its birthing pains at the about same time and was similarly midwifed by the National Science Foundation. Unnoticed in the worldwide and well-deserved acclaim for the LIGO discovery was the fact that IceCube released a joint study with LIGO and several other instruments simultaneous to LIGO's announcement, revealing that the instrument at the South Pole had detected no neutrinos arriving in tandem with the gravitational wave.

A friend of mine recently pointed out that these two instruments stretch the idea of what a telescope is altogether. "Wait!" he asks. "Where is the imaging plane? What does the picture 'look' like? How exactly does 'looking' work in this new sense?" It is exceedingly odd that the IceCube scientists never have and never will see their instrument, that there is no imaging plane as there is in a camera or standard telescope, and yet, thanks to the wonders of modern computing, their instrument is quite visual and sometimes even intuitive to use. LIGO is even weirder. It measures the stretching and contracting of two perpendicular laser beams, four kilometers long.

LIGO's gravitational wave seems to have been caused by just the sort of cosmic earthquake that should produce high-energy neutrinos in abundance: the distant merging of two relatively large black holes. Several years ago, LIGO and IceCube signed an agreement that allowed them to share so-called alerts with each other. When one detects an event of interest, they will share it privately with the other, so that they can look over their data to find out if they saw something that might have been related. In fact, one of the central initiatives in IceCube since the completion of their instrument has been to make similar arrangements with a variety of instruments: gamma ray telescopes like Fermi, cosmic ray instruments like Auger, and so on. Astronomy is becoming a "multi-wavelength" or "multi-messenger" endeavor. Every wavelength of photon is being employed, from the radio band through the infrared and visible to the ultraviolet and on through X-rays to gamma rays. Now neutrinos and even gravitational waves have been enlisted. The latest thing is to broadcast alerts from these different messengers in real time, so that if IceCube detects what seems to be a gamma ray burst, for example, it can alert the Fermi satellite to point in that direction and see what it sees. What if LIGO were to detect a gravitational wave at the same time? The possibility of understanding the many violent creatures that roam the universe is greatly enhanced.

In April 2016, NSF director France Córdova held a retreat with her senior managers aimed at developing a list of grand ideas to inform the course of the agency over the next several decades. Multi-messenger astrophysics was one of the five overarching initiatives they identified. The head of the foundation's math and physical sciences directorate observed that they've been hoping to combine optical, particle, and gravitational observations for a long time, but that now they "have all the pieces."

Meanwhile, the Fellowship of the Cube is having a grand time. There are plenty of games in their playground as it is. If GEN2 or a big multi-messenger initiative comes along it will only add to the fun. In 2015 alone they did multi-messenger studies with four different instruments. They carried out a joint search with the ANTARES neutrino telescope in the Mediterranean, which is still at the AMANDA stage but gaining momentum. They searched for magnetic monopoles and dark matter. Let us not forget that they've had an increasingly sophisticated supernova watch in play since the early days of AMANDA, and they're still hooked up to SNEWS, the SuperNova Early Warning System.

Very sadly, Per Olof Hulth succumbed to cancer in February 2015. "The model of a kind but tough Swedish gentleman," in Buford Price's words, the outpouring of affection from the rest of the collaboration demonstrated that Peo was widely admired and loved. He had led the first European group to join AMANDA. He had husbanded the project through the bubbles. His student, Adam Bouchta, had detected the first up-going neutrino candidates with AMANDA-B4. He had helped launch IceCube as its first spokesperson, "with the same style and flair characteristic of his science," as Francis put it. And his vision for DeepCore continues to bear fruit. It has proven capable of world-class dark matter and oscillation studies, and the idea of extending it to even lower energies has now been proposed.

PINGU, the Precision IceCube Next Generation Upgrade, an even denser infill array within DeepCore, stands to be competitive with several multi-billion-dollar accelerator experiments in the study of neutrino oscillation, where the hottest topic today is the so-called mass hierarchy. The discovery of oscillation indicates that each neutrino flavor is actually a mixture of three neutrino "mass states," and the hierarchy question has to do with the relative masses of these states. Not only could this line of inquiry lead to a theory of neutrino flavor, it might help answer Ettore Majorana's age-old question about the neutrino and its antiparticle. It also has implications for theories of the early universe and how it evolved.

This is pure particle physics, not astronomy, and IceCube made a splash in that field as well in 2015, by coming very close to ruling out the existence of a fourth neutrino flavor, the sterile variety. (Here, one of the smallest things one can imagine has implications for the largest, since it turns out that the existence or non-existence of sterile neutrinos has implications for the large-scale structure of the universe.) There was more than the usual internal strife around this finding, so they didn't get the paper out until 2016. When they finally did, to Francis's tongue-in-cheek irritation, "not seeing sterile neutrinos attracted as much attention as cosmic neutrinos." There was even a prominent article about this non-discovery in the *Wall Street Journal*. Since they have another, more stringent sterile neutrino study coming fresh on the heels of this one, he can't wait for the theorists to figure out ways to get around their first result, so IceCube can give them a "left-right" with the second . . . sort of like they did with gamma ray bursts.

At the age of seventy-two, Francis is as enthusiastic as ever. I haven't seen him this excited since the days of the Peacock events (which he still considers the most exciting discovery they have made). He's working with new students, he's brimming with ideas, and he looks forward to leading the way on GEN2. Indeed, it's hard to imagine who could take his place.

In the traditional year-end e-mail he sent to the collaboration in December 2015, he made sure to reiterate that the prizes he had received actually recognized the collaboration as a whole and thanked them all for saving him from the daily life of a theoretician. After enumerating the many accomplishments of the year, he referred to IceCube as "the experiment that keeps on giving" and apologized: "I got carried away, but who wouldn't."

Francis has a strong hunch that they'll find a point source with Ice-Cube. There are two very plausible candidates in the running right now. I think I've been hearing him say this for about fifteen years now, and he's saying it with the same joy and lack of irony that he did the first time. He often points out that the people in his family tend to live into their nineties and that he doesn't plan on retiring until he reaches a hundred. That gives him another twenty-eight years.

✦ ✦ ✦

You can find out more about IceCube and this book at
telescopeintheice.com.

Acknowledgments

In the nineteen years that it has taken me to write this book, nearly everyone associated with AMANDA and IceCube, both at the South Pole and in the "real world," has supported me in large ways and small. This came in part from innate generosity and in part because they knew this story needed to be told. There have been scores of you. I cannot possibly name or even remember you all. But thanks.

Francis Halzen has been my most constant and generous ally. It began with his welcoming response to a phone call I placed to his office in March 1998 and continues even now that the book is written. No writer could hope for more helpful or entertaining a source of information and insight. I count eighteen lengthy, transcribed interviews with Francis, and part of the reason they took so long was that we laughed so much. That doesn't include meals together, phone calls, and innumerable e-mails. He has been remarkably open and large-minded. I know Francis well enough to know that he hasn't told me everything, but he has never turned down a request for information or refused to answer a question, and I can think of only one instance in which he restricted anything he told me in any way (it was inconsequential). Usually, he has given more than I have asked. As Serap Tilav once said of him, "The big man has no worries. A big man has no fear." The remarkable example he has set ethically and in rising above all pettiness is largely responsible for the success of IceCube and has been an inspiration to me, both personally and in the writing of this book. I shall miss having this excuse to talk to him. I hope we can keep it up in some way.

Bob Morse has also been a tremendous help, giving numerous interviews, sharing valuable documents and photographs, embodying "Wisconsin-style physics" and the "glory days" at Pole, and telling fabulous stories. Not least, he arranged for me to work at Pole in 1999. Many thanks, Bob.

The late Bruce Koci introduced me to this project and, I am proud to say, became a friend. This book benefits greatly from his generosity, his remarkable memory, his quiet insight, and his "disheveled wisdom." I would also like to thank his widow, Ann Guhman, for her gracious support.

Gary Hill has been my go-to guy for all kinds of facts and information—about Pole, about the science, about the history. He was especially helpful in uncovering the story of the "first nus." John Jacobsen provided brilliant insights from a unique vantage point—and marvelous photographs. Christian Spiering was a generous host in Zeuthen and Berlin many years ago and shared his first-hand knowledge of Baikal and the Russian side of things, as well as his comprehensive knowledge of the history of neutrino astronomy in general.

What can I say about John Learned? Aside from the fact that there would not have been a story here without him, he always seemed to drop in—in Madison, Zeuthen, or wherever—at propitious times, and was always willing share the overall perspective and long view of the field that only the "pied piper of neutrino astronomy" could. John has even more stories than Bob Morse. They could not all be told here. Perhaps they deserve another book.

Buford Price, Terry Millar, Jim Yeck, Albrecht Karle, and Jim Haugen have given extended interviews and gone the extra mile in other ways. "Queen" Kim Kreiger, Megan Madsen, and Catherine Vakhina have been the glue that has held things together for me on my many visits to Madison. Of special help in Antarctica were Stuart Klipper, Martin Lewis, Darryn Schneider, and John Wright.

Special thanks to Tan-Quy Tran of the MIT Physics Reading Room for providing me with dozens of scientific articles over a period of about six years.

The following people sat generously for interviews: Olga Botner, Adam Bouchta, Jeff Cherwinka, the late Dave Cline, Farshid Feyzi, Tom Gaisser, Azriel Goldschmidt, Kael Hanson, the late Per Olof Hulth, Aya Ishihara, Claudio Kopper, John Lynch, Dave Nygren, Bob Paulos, Jerry Przybylski, Greg Sullivan, Serap Tilav, Nathan Whitehorn, John Wiley, Ralf Wischnewski, and Kurt Woschnagg. Carlos le los Heros popped up several times in his amiable way with nice tidbits about the AMANDA days. When Ty DeYoung finally opened up (it took about thirteen years), he provided just the sort of wise and sober insight I expected from him. I also benefited from helpful exchanges with Steve Barwick, Kara Hoffman, Lutz Koepke, Jerry Marty, Naoko Kurahashi Neilson, Sandip Pakvasa, Elisa Resconi, Leo Resvanis,

George Smoot, Todor Stanev, Eli Waxman, Trevor Weekes, and Christopher Wiebusch.

I *think* I'd like to thank David Dobbs, a fine writer and fellow Vermonter, for getting me into this mess. About seven years ago, I mentioned a few book ideas to him and he pounced on this one, prompting me to focus on it for real. I know I'd like to thank Doug Linnel for his excellent line drawings, John Lehet for help with all things internet, Chin Woon Ping for getting me books from the Dartmouth libraries (and for general camaraderie), Evelyn Malkus and Bob Stokstad for help with photographs, and Troy Reeves of the Oral History Program at the University of Wisconsin–Madison for providing me with their long interview of Francis Halzen.

I am indebted to Jason Newman, Mike Tennican, and David Zahn for reading the first bumpy draft of the manuscript—and to Jason for reading the second draft, too! It was incredibly interesting to watch it bounce off three very different minds; they all had an effect on the final product. I hope I have implemented Mike's deeply considered ideas to his satisfaction. And Jason is probably the most generous reader I have ever encountered in the sense of appreciating everything and reading between the lines. His perception, transmitted during two delightful evenings at the Pine bar, has worked its way into the book's DNA. Here's to more such evenings, ranging to other subjects.

I am also indebted to the participating institutions in AMANDA and IceCube, particularly DESY Zeuthen; Uppsala University; the University of California, Berkeley; and especially the University of Wisconsin–Madison. They have defrayed the cost of conferences and collaboration meetings and occasionally even travel and lodging. Madison also hired me for a writing project that doubled as book research.

Many thanks to the people at St. Martin's Press for agreeing to publish this book, and especially to senior production editor Donna Cherry for her professionalism and grace under pressure and assistant editor Laura Apperson for her good-natured help with the photographs.

Finally, my deepest gratitude to Jane Gelfman, my agent, for believing in me and this book, for working far too hard on our behalves, and for becoming a real friend. Getting to know her may be the best thing about this project for me. I have met very few people in my life with her class and her wisdom.

Acronyms

A³RI	Antarctic Astronomy and Astrophysics Research Institute (in Madison)
AMANDA	Antarctic Muon And Neutrino Detector Array
AMANDA-A	The first, shallow AMANDA detector, deployed over the 1993–94 Antarctic season
AMANDA-B	The second, deeper AMANDA detector
AMANDA-B4	The first, four-string iteration of AMANDA-B, deployed over the 1995–96 Antarctic season
AMANDA-B10	The second, ten-string iteration of AMANDA-B, deployed over the 1996–97 Antarctic season
ATHENE	ATmospheric High Energy Neutrino Experiment
CCFMR	The Cline, Camerini, Fry, March, Reeder research group at the University of Wisconsin
CERN	Conseil Européen pour la Recherche Nucléaire (European Organization for Nuclear Research), the European high-energy physics laboratory on the outskirts of Geneva
DeepCore	A small infill array in IceCube for detecting lower-energy neutrinos
DOE	United States Department of Energy
DUMAND	Deep Underwater Muon And Neutrino Detector
GASP	Gamma-ray Astronomy at the South Pole
GISP2	Second Greenland Ice Sheet Project
GRB	Gamma ray burst
HEP	High-energy physics
HPW	Harvard-Purdue-Wisconsin experiment
IceTop	A two-dimensional surface array on the surface of the Ice above IceCube
ICRC	International Cosmic Ray Conference
IGY	International Geophysical Year
IMB	Irvine-Michigan-Brookhaven experiment
JULIA	Joint Underwater Laboratory and Institute for Astroparticlephysics
Kamiokande	Kamioka Nucleon Decay Experiment
LBL	Lawrence Berkeley National Laboratory
LIGO	Laser Interferometer Gravitational-Wave Observatory
LHC	Large Hadron Collider at CERN

NESTOR	NEutrinos from Supernovae and TeV sources, Ocean Range
NSF	National Science Foundation
PI	Principal investigator
PICO	Polar Ice Coring Office
PSL	Physical Sciences Laboratory at the University of Wisconsin
RAMAND	Radio Antarctic Muon And Neutrino Detector
RHIC	Relativistic Heavy Ion Collider
RICE	Radio Ice Cherenkov Experiment
SAGENAP	Scientific Assessment Group for Experiments in Non-Accelerator Physics
SNO	Sudbury Neutrino Observatory
SPASE	South Pole Air Shower Experiment
SSEC	Space Science and Engineering Center (in Madison)
SSC	Superconducting Super Collider
Super-K	Super-Kamiokande
TOS	Tower Operations Structure (for IceCube)
UNCLE	Undersea Cosmic Lepton Experiment
UNDINE	UNderwater Detection of Interstellar Neutrino Emission
UNICORN	UNderwater Interstellar Cosmic-Ray Neutrinos
UW	University of Wisconsin
VULCAN	Very Un-Likely Cherenkov Array Name

Timeline

1911	Roald Amundsen and his men leave the first footprints at the South Pole.
1912	In a hydrogen balloon above Germany, Victor Hess detects the first cosmic rays.
1927	Charles Drummond Ellis and Wilhelm Orthmann demonstrate that the energy spectrum in radioactive beta decay is continuous. This seems to indicate that energy is lost in the decay process.
1929	Several experiments reveal a conundrum related to the spin of the nitrogen nucleus.
1930	Wolfgang Pauli proposes a possible solution to both of these conundrums by hypothesizing the existence of a new particle, the neutrino.
1932	James Chadwick discovers the neutron, which explains the nitrogen anomaly. Carl Anderson discovers the first antiparticle, the positron, in a shower of cosmic rays.
1933	Enrico Fermi develops a successful quantum theory for beta decay, which suggests that the neutrino is responsible for the "missing" energy.
1935	James Chadwick wins Nobel Prize in Physics for discovery of the neutron. Irène and Frédório Joliot Curie win chemistry Nobel for "their synthesis of new radioactive elements."
1936	Victor Hess and Carl Anderson share the Nobel Prize in Physics.
1937	Anderson discovers the muon, again in a shower of cosmic rays. Ettore Majorana suggests that the neutrino might be its own antiparticle.
1938	Christmas Eve: based on experimental results she has seen in a letter from her colleague, Otto Hahn, Lise Meitner realizes that Hahn has "split the atom." Meitner's nephew, Otto Frisch, later names the process "fission."
1939	Hans Bethe produces a general theory of energy production in stars, which predicts that the Sun should emit neutrinos.
1942	Enrico Fermi engineers the first manmade nuclear chain reaction.
1944	Otto Hahn receives the Nobel Prize in Chemistry for the discovery of fission.
1945	May: Bruno Pontecorvo proposes a chemical method for detecting neutrinos. Over the next five years, he makes many contributions to the understanding of the neutrino, including the idea that it may come in different flavors. July 16: Trinity, the first manmade nuclear explosion.
1947	Pion discovered in photographic emulsions high in the Pyrenees and Andes.
1950	Bruno Pontecorvo defects to the Soviet Union.
1951	Ray Davis begins building the first of many neutrino detectors based on Pontecorvo's chemical detection method.

mid-50s	The first high-energy particle accelerators come on line.
1956	Fred Reines and Clyde Cowan detect the neutrino at the Savannah River nuclear plant in South Carolina.
	Tsung-Dao Lee and Chen-Ning Yang Lee suggest that the weak nuclear force, intimately related to the neutrino, may violate parity, the law of mirror symmetry.
1957	Chien-Shiung Wu proves experimentally that the weak force does violate parity. Two other experiments quickly follow suit.
	Lee and Yang win physics Nobel.
	Bruno Pontecorvo proposes neutrino oscillation.
1958	Moiseĭ Markov and his student Igor Zheleznykh propose a concept for a neutrino telescope to be placed in "an underground lake or deep in the ocean."
	Wolfgang Pauli dies.
1960	Kenneth Greisen proposes an idea for a detector similar to Markov's to be "located in a mine far underground."
1962	Leon Lederman, Mel Schwartz, and Jack Steinberger detect the first muon neutrinos.
1964	Theorist John Bahcall joins Ray Davis in proposing a large detector based on Pontecorvo's method in order to study neutrinos emitted by the Sun.
1965	Davis starts building a solar neutrino detector in the Homestake Gold Mine in South Dakota.
	Separate groups led by Goku Menon and Fred Reines detect the first naturally occurring (atmospheric) neutrinos in gold mines in India and South Africa.
1967	Ray Davis observes solar neutrinos at Homestake, but only one-third as many as Bahcall has predicted. This discrepancy becomes known as "the solar neutrino problem."
1968	John Learned defends his doctoral thesis, describing some of the earliest "fishing for muons" in a natural body of water: preliminary research for a neutrino telescope in the manner of Markov.
	Learned joins the Cline, Camerini, Fry, March, Reeder group at the University of Wisconsin–Madison.
1971	At a conference in the French Alps, Vernon Barger offers Francis Halzen a six-month position at UW-Madison. Halzen moves to Wisconsin in the fall.
1973	DUMAND committee, comprising U.S., Soviet, Japanese, and European scientists, founded at an International Cosmic Ray Conference (ICRC) in Denver.
1975	Geodesic dome erected at South Pole station.
	Martin Perl and colleagues discover the tau lepton.
mid-70s	With experimental confirmation of the existence of quarks, the standard model of particle physics is put in place.
1976	Seminal DUMAND workshop in Honolulu provides the basis for virtually every Greisen- or Markov-type instrument that will be built for the next four decades, including AMANDA and IceCube.
1977	Francis Halzen promoted to full professor at Madison.
1977–78 Antarctic Season	Bruce Koci makes his first trip to Antarctica to drill on the Ross Ice Shelf.

1978 or 1979	Bob Morse joins the HPW experiment in Utah, a neutrino detector in the manner of Greisen.
1979	Construction starts on the IMB detector in Ohio, another Greisen-type detector. Ross Ice Shelf Project morphs into the Polar Ice Coring Office (PICO). Soviets invade Afghanistan.
1980	DUMAND funded. Hawaii DUMAND center established. John Learned becomes technical director. This is the first attempt to build a true neutrino telescope in the manner of Markov.
1981	President Ronald Reagan ends Soviet participation in DUMAND. Soviet scientists begin working on a neutrino telescope in Lake Baikal, Siberia.
1982	Construction begins on Kamiokande, a Greisen-style instrument in the Japanese Alps.
1983	A group led by Carlo Rubbia detects the W and Z intermediate vector bosons, the carriers of the weak force, at CERN. Accelerator physics enters "the desert." Groups in England and Germany announce the detection of very high-energy gamma rays from Cygnus X-3. Although this result turns out to be false, it prompts a small migration of high-energy physicists from the accelerators back to cosmic rays.
1987	February: A thirteen-second flash of neutrinos from Supernova 1987a passes through IMB, Kamiokande, and a Russian detector in the Caucasus. SPASE (South Pole Air Shower Experiment) begins. On a visit to the University of Kansas, Francis Halzen learns about a Russian neutrino experiment in Antarctica and conceives of "DUMAND on Ice." He gets in touch with John Learned.
1988	Learned sends a letter of intent for DUMAND on Ice to John Lynch at NSF. Lynch responds in the negative. First International Workshop on Neutrino Telescopes in Venice. Lederman, Schwartz, and Steinberger share Nobel Prize in Physics for their discovery of the muon neutrino.
1989	John Wiley becomes dean of UW-Madison grad school. Conference on Astrophysics in Antarctica at Bartol Institute in Delaware. Many future AMANDA and IceCube scientists participate. November: Berlin Wall falls.
1989–90 Antarctic Season	Bob Morse travels to South Pole to install GASP: Gamma Astronomy at South Pole. Doug Lowder and Andrew Westphal caught in Antarctica attempting an unauthorized feasibility study for DUMAND on Ice.
1990	January: At the 21st ICRC in Adelaide, Francis Halzen and Bob Morse of UW-Madison, Buford Price of Berkeley, and Steve Barwick of UC Irvine agree to form the AMANDA collaboration. August: First "ice fishing for muons." The AMANDA collaboration deploys a test string on the summit of the Greenland Ice Sheet with the help of driller Bruce Koci. October: German unification.
1991–92 Antarctic Season	First AMANDA drilling campaign is "a learning experience": they lose the drill in the ice.

1992	AMANDA receives its first round of funding.
	Swedish groups from Stockholm and Uppsala universities join AMANDA.
1993	Baikal collaboration deploys its third string in the lake and thus wins the "3-string race" in neutrino astronomy.
1993–94 Antarctic Season	AMANDA collaboration deploys AMANDA-A, four strings in shallow, bubbly ice.
1994	David Nygren of Lawrence Berkeley Laboratory conceives of a digital optical module (DOM) for use in Markov-type neutrino telescopes.
1995	A group from Zeuthen, in the former East Germany that is already collaborating on Baikal, also joins AMANDA.
	Fred Reines and Martin Perl share Nobel Prize in Physics for discovering the neutrino and tau lepton, respectively.
1995–96 Antarctic Season	AMANDA-B4, comprising four strings, deployed in deep, clear ice.
1996	Baikal collaboration announces the detection of three "gold-plated" neutrinos, the first ever by a neutrino telescope. This is a great step forward, but it is not astronomy, since the neutrinos probably came from Earth's atmosphere.
	After sixteen years of mishaps, DUMAND is terminated.
	Dave Nygren joins AMANDA.
1996–97 Antarctic Season	AMANDA-B10, ten strings, deployed.
	Jerry Przybylski deploys two prototype DOMs on one of these strings.
1997	At a collaboration meeting in Berkeley, Adam Bouchta presents AMANDA's first neutrino candidates, detected by AMANDA-B4.
	Congress and President Bill Clinton approve funding for a new station at the South Pole.
1998	March: IceCube Neutrino Facility Workshop in Irvine (the first IceCube meeting).
	June: Super-Kamiokande collaboration, the successor to Kamiokande, announces evidence that neutrinos oscillate. This implies that they have mass, in violation of the standard model of particle physics.
	August 25: Gary Hill and Phil Romenesko find two gold-plated neutrinos in data from AMANDA-B10.
	August 26: Fred Reines dies.
	November: The Madison group presents an automated analysis yielding nine neutrinos at a collaboration meeting in Madison. This the first indication that AMANDA is a working neutrino telescope.
1999	May: In a meeting at NSF headquarters, the officers of the foundation formally embrace IceCube.
	November: IceCube proposal submitted to NSF.
1999–00 Antarctic Season	Final AMANDA drilling campaign results in AMANDA-II, nineteen strings. String 18 is comprised entirely of DOMs.
2000	The DONUT collaboration at Fermilab announces the first direct detection of tau neutrinos.
	October: IceCube approved under the Clinton administration.

2001	January: John Wiley becomes Chancellor of UW-Madison.
	February: IceCube loses funding when the new president, George W. Bush, issues a "no-new-starts" diktat to NSF. With the encouragement of John Wiley, the Wisconsin Alumni Research Fund loans $4.8 million to IceCube.
	An external advisory committee chaired by physicist Barry Barish recommends the DOM as the basic detector technology for IceCube.
	March: AMANDA collaboration publishes a paper in *Nature* establishing the feasibility of ice-based neutrino astronomy.
	June: Scientists at the Sudbury Neutrino Observatory (SNO) announce the solution to the solar neutrino problem and erase the last lingering doubts about neutrino oscillation.
2002	National Science Board formally funds IceCube.
	Masatoshi Koshiba, leader of the Kamiokande collaboration, and 88-year-old Ray Davis share half the Nobel Prize in Physics "for pioneering contributions to astrophysics, in particular for the detection of cosmic neutrinos."
2003	New South Pole station first occupied.
	Jim Yeck becomes Project Director of IceCube.
2004–05 Antarctic Season	First IceCube deployment season results in disaster when Swedish driller Sven Lidström is severely injured in a drilling mishap.
	First IceCube string successfully deployed.
2005–06 Antarctic Season	Back on track, the collaboration deploys eight strings.
2006	Bruce Koci dies of non-Hodgkin's lymphoma.
2010–11 Antarctic Season	December: IceCube collaboration deploys its eighty-sixth and final string.
	January: E-mail from John Jacobsen, a computer whiz at Pole, announces "first flight" of full IceCube detector.
2011	April-May: At the IceCube commissioning gala in Madison, Francis Halzen predicts a "five-sigma" discovery within five years.
	May: Full IceCube instrument begins taking data.
	August: "Bert," the highest-energy neutrino ever detected, explodes in the middle of the IceCube array.
2012	January: "Ernie," more energetic than Bert, explodes in the detector.
	April: IceCube collaboration announces "an absence of neutrinos associated with" gamma ray bursts, effectively ruling out all current GRB models.
	May: IceCube collaborator Aya Ishihara discovers Bert and Ernie.
	July: Scientists at CERN announce discovery of the Higgs boson, the last undetected particle in the standard model of particle physics.
	October: IceCube collaborators Nathan Whitehorn and Claudio Kopper uncover twenty-six more events at energies near Bert and Ernie's.
2013	May: At a symposium in Madison, the collaboration reveals their twenty-eight events to the world, characterizing them as "evidence" of the first extraterrestrial high-energy neutrinos ever detected.
	November: Paper on the twenty-eight events appears in *Science*, announcing the birth of neutrino astronomy.
	December: *Physics World* (UK) names IceCube's discovery of cosmic neutrinos its breakthrough of the year.
2015	February: Per Olof Hulth of Stockholm University dies of pancreatic cancer.
	September: Francis Halzen wins the Balzan Prize.
	November: Takaaki Kajita of Super-K and Arthur McDonald of SNO share the Nobel Prize in Physics "for the discovery of neutrino oscillations, which shows that neutrinos have mass."

Notes

Introduction: Making Mistakes

1   **The universe can't exist . . . something like that**   Herzog 2007; philosophicalsociety.com 2012.

1   **detected high-energy neutrinos coming from outer space**   Aartsen et al. 2013b.

2   **"weirdest" of the seven wonders of modern astronomy**   Musser 1999.

3   **so successful that it's beginning to feel like a straightjacket**   Cho 2013.

4   **goods on ships . . . the English Channel**   Francis Halzen got this story from a science historian who walked up to him after he, Francis, had given a talk before a large audience at an American Physical Society meeting in Washington.

5   **our celestial science . . . remains to be discovered**   This quote comes from an entertaining review of such surprises entitled "Serendipitous Astronomy" (Lang 2010).

7   **Soon the sun . . . fox in his tail**   Bowen 2005, 17–18.

9   **initiate . . . Antarctica and extend . . . dimensions**   AMANDA Collaboration 1998.

11   **one up-going neutrino**   Sutton 1992, 209; Babson et al. 1990.

11   **Sailors . . . unforgiving medium**   Roberts 1992.

13   *Nobel Dreams*   Taubes 1986.

This Crazy Child

17   **most exciting eight-year period**   There are innumerable stories and anecdotes about the birth of quantum mechanics and nuclear physics and the remarkable characters who participated in these dramas. Most have been told many times. I have elected to repeat only the most relevant here. My favorite book about those times is *Faust in Copenhagen* by Gino Segrè (2007). Abraham Pais's exquisite and unique *Inward Bound* (1986) is basically the bible for the history of atomic, nuclear, and particle physics up until the time it was written, and chronicles the development of the ideas in a way that only a theorist of Pais's insight could. On the subject of Pauli himself, I recommend his *Writings on Physics and Philosophy* (1994), especially the biographical introduction by Enz; the pithy profile in *Physics Today* by von Meyenn and Schucking (2001), and Peierls's biographical memoir for the Royal Society (1960).

18   **most ambitious . . . Bohr-Sommerfeld theory**   Peierls 1960, 176.

18   **No one . . . appraisal**   Einstein 1922; English translation of this quote: von Meyenn and Schucking 2001, 44.

18   **institute had been founded . . . outside Germany**   Pais 1982, 450.

18   **I will never . . . successor**   Enz 2000, 22; von Meyenn and Schucking 2001, 43.

19   **greater scientist, though not as great a man**   von Meyenn and Schucking 2001, 43.

19   **letters were copied**   Ibid., 45–6.

19   **very conscience . . . turbulent sea**   Bohr 1960, 4.

19   **supreme judge**   Enz 1994.

19–20 **I know . . . merciless criticism**   Segrè 2007, 149.

20   **inspired by Pauli's insights and suggestions**   von Meyenn and Schucking 2001, 47.

20   **We always benefitted . . . met with his approval**   Bohr 1960, 3.

20   **Scourge of God**   Segrè 2007, 79.

20   **After the second . . . feminine**   Ibid., 81.

20   **Dr. Jekyll and Mr. Hyde**   This is the name of a chapter in Miller (2010).

21   **changed his name**   Miller 2010, 20. Pauli's father's given surname was Pascheles.

21   **unclear when he learned of his Jewish heritage**   Ibid., 22.

23   **festering for more than twenty years**   Sutton 1992, 10–12. The second chapter of Sutton's excellent book gives a more detailed account of the experimental developments than I deemed appropriate here.

23   **proving the existence of a spectrum**   Chadwick 1914.

23   **ruled out secondary processes . . . going missing**   Ellis and Wooster 1927.

23   **confirmed and extended Ellis and Wooster's result**   Meitner and Oorthman 1930.

24   **He didn't see . . . relativity, would not**   I'm barely revealing the tip of the iceberg of this fascinating, quasi-philosophical debate. Bohr and Pauli had been disagreeing about energy conservation for about six years at that point. Enz (1981) reviews their debate with great insight, and so (it goes without saying) does Pais (1986, 309–13).

24   **I must say . . . let the stars shine in peace!**   Bohr 1929; Pauli 1929; Peierls 1983, 222; Sutton 1992, 20–1; Pais 1986, 312.

24   **classically non-describable two-valuedness**   Pauli 1925; Pauli 1946; see also Pauli 1945.

24   **identifying this property as spin**   Uhlenbeck and Goudsmit 1925 and 1926.

25   **several experiments**   Kronig 1928; Rasetti 1929; Heitler and Herzberg 1929.

25   **tentative solution to both**   There is evidence that Pauli had been discussing his idea with others prior to the writing of this letter. Abraham Pais (1986, 315) has uncovered a letter that Heisenberg wrote to Pauli on December 1, which refers to "your neutrons." Pauli composed his letter to the radioactive ladies and gentlemen three days later.

25   **I have . . . electron is constant**   "Wolfgang Pauli wrote a highly sophisticated and beautiful German," writes Charles Enz in the preface to a collection of Pauli's writings on physics and philosophy (Pauli 1994) and he expressed himself with uncommon precision. When it came to translating his works into English, Pauli encouraged literal translation over smooth or vernacular phrasing. I am using Robert Schlapp's translation of the letter, which is found in Pauli (1958a).

26   **those who say he invented both**   Brown 1978; Pontecorvo 1982.

26   **It is difficult . . . neutrino invention by Pauli**   Pontecorvo 1982.

27   **later generations . . . elusive particle**   Wu 1960.

27   **this crazy child . . . behaved crazily**   Enz 1994, 19; Pais 1986, 314.

27   **I've done a terrible . . . detected**   Hoyle 1967. The story of the bet is almost certainly true, but the quote from Pauli is probably impossible to verify. It is almost always reported as "I have done a terrible thing. I have postulated a particle that cannot be detected." But the only source I can find for that is Fred Reines's Nobel lecture (1995), in which he asserts that Pauli said it at Caltech, probably at the conference he attended in Pasadena in the summer of 1931. This makes a certain amount of sense, since Walter Baade was working at the Mt. Wilson Observatory, near Pasadena, at that time, but the reason I prefer Hoyle's version is that it's only second-hand, whereas Reines's is third-hand at best and several more decades removed.

27   **public for the first time**   AAAS 1931, 111; APS 1931.

27–28   **Physicists will accept . . . mathematical creation**   Kaempffert 1931.

28   **he entered . . . laugh from audience**   Brown 1978.

28   **never inconvenienced Pauli himself in the least**   Peierls 1960, 185.

28   **even quite practical . . . liberated and lightened**   Enz 1994, 17.

29   **brief stop in Göttingen just at the time of the accident**   Gamow 1966, 64.

29   **"accident" never happened**   Peierls 1960, 185.

29   **was still cautious . . . Mussolini**   Pais 1986, 317 and 323n104.

29   **at once showed a lively interest . . . conserved only statistically**   Pauli 1958a, 199.

29   **threatened with dismissal from the ETH**   Miller 2010, 120.

29   **And what shall we say . . . lunatic asylum**   Jung and de Laszlo 1958, xii, xiv; Miller 2010, 127.

29–30   **chock-full . . . myself**   Atmanspacher and Primas 1997, 371.

30   **rolled around on the floor as he poured out his stories**   Miller 2010, 129.

30   **Chadwick discovered the neutron**   Chadwick 1932.

30   **'Neutrino,' . . . through Fermi**   Bonolis 2005, 497n5. Bonolis is in turn quoting Ugo Amaldi in his preface to Battimelli and Paoloni (1998). She notes that "the word arose in a humorous conversation at the Istituto di via Panisperna. Fermi, [Edoardo] Amaldi, and a few others were present and Fermi was explaining Pauli's hypothesis about his 'light neutron.' To distinguish this particle from the Chadwick neutron, Amaldi jokingly used this funny name."

30   **Anderson at Caltech detected the positron**   Anderson 1932; Anderson 1933.

30 **Victor Hess made the first measurements** Hess's "step grandson" Bill Breisky (2012), wrote a wonderful article about his grandfather and the centenary celebration of his discovery in the *New York Times.*

31 **mountaineers, mine workers, divers and air riders** Auger 1985. See also Korff 1985.

31 **The different . . . 'in iron'** Brown and Rechenberg 1996, 69.

31 **as much matter as matter is matter** Pais 1986, 15.

31 **Dirac equation** Dirac 1928a, 1928b.

31 **most beautiful equation in all of physics** Pais 1986, 290; AIP 1962; Dirac 1977; Overbye 2002.

31 **postulated the existence of a positively charged "anti-electron"** Dirac 1931.

31 **first theory of so-called quantum electrodynamics** Dirac 1927.

32 **some scientists began to suspect** Iwanenko 1932.

32 **a general clarification** Pauli 1958a, 200.

32 **players in the beta ray saga were there** There is a photo of the attendees of this historic conference at http://en.wikipedia.org/wiki/File:Solvay1933Large.jpg.

32 **produced a positron instead of an electron** Curie and Joliot, 1934; Pais 1986, 400; Pauli 1958a, 200.

33 **Ellis and his student W. J. Henderson presented results** Later published by Henderson alone (1934).

33 **didn't give in for another three years** Bohr 1936. Bohr tended to keep conversations open, as he relished contradiction and the unanswerable question. Abraham Pais (1982, 24) points out that "no one resisted the photon [the quantum of light, proposed by Einstein] longer than Bohr"; and the textbook author David Griffiths (2008, 24n) adds that "he mercilessly denounced Schrödinger's equation, discouraged Dirac's work on the relativistic electron theory . . . , ridiculed Yukawa's theory of the meson, and disparaged Feynman's approach to quantum electrodynamics. Great scientists do not always have good judgment—especially when it concerns other people's work—but Bohr must hold the all-time record."

33 **Pauli wrote many years later** Pauli 1958a, 200.

33 **I gave my ideas . . . in the discussion** It is typical of Pauli's modesty and disregard for recognition that the only record of his neutrino postulate that he bothered to publish appears as a brief discussion point in an exchange that followed a talk on nuclear structure that Heisenberg delivered at this 1933 Solvay conference. It was written in Pauli's exquisite French (1934). An English translation by C. S. Wu is found in Brown (1978).

34 **and then for three months . . . gifted in this respect** Atmanspacher and Primas 1997, 371.

34 **They contain . . . archetypal images** Jung 1967, 5–182 ("Tavistock Lectures," delivered in London in 1935); Miller 2010, 129–30.

34 **identity of the dreamer . . . kept scrupulously anonymous** In his writings, Jung referred to Pauli at various times as "a scientifically educated young man," "a great scientist," and "a very famous man, who lives today."

34 **searching for ways to extend it** Pauli's efforts in this arena took place independently of Jung as well. Atmanspacher and Primas (2006) have written a masterful review.

34 **the no-man's-land . . . our times** Miller 2010, 147, 291n.

35 **that the observer . . . completely detached** Pauli 1957b, 131–2.

35 **complementary aspects . . . ordering principles** Atmanspacher and Primas 1997, 381; 2006, 17.

35 *The Interpretation of Nature and the Psyche* Jung and Pauli 1952.

35 **synchronistic [manifestation] . . . non-rational** Atmanspacher and Primas 1997, 378.

Infancy and Youth

36 **electron and neutrino might also be related** The French physicist Francis Perrin had guessed that there might be a symmetry between the electron and the neutrino shortly after Pauli sent his letter to the Radioactive Persons in Tübingen (Perrin 1930).

37 **first example of a modern field theory** Weisskopf 1972, 17; Brown 1978.

37 **greatest editorial blunder they had ever made** Close 2010, 24.

37 **in three more specialized—and less visible—physics journals instead** Fermi's first short paper on the theory (1933), written in his native Italian, was a brief announcement similar to the one he had intended to publish in *Nature* in English. He then published a more complete treatment in

Italian (1934a) with a translation in German (1934b). See also Bonolis (2005, 497n7). Wilson (1968) eventually translated this classic paper into English.

37   **very small with respect to the mass of the electron**   Fermi 1934a, 1934b.

37   **Peierls and . . . Bethe, showed**   Bethe and Peierls 1934.

38   **ingenuity of experimentalists**   Peierls 1983.

38   **One can only imagine . . . Fascist Italy in 1934**   Bernstein 2001a, 7.

39   **Bohr managed to find her a position in Stockholm**   Rhodes 1986, 236.

39   **that if you really do . . . great energy**   Ibid., 258.

39   **billion trillion nuclei in a single gram of uranium**   Ibid., 259.

39   **by analogy with the binary fission of bacteria**   Ibid., 263.

40   **Ettore Majorana**   I construed this brief synopsis of Majorana's life from a slightly less brief synopsis by Bruno Pontecorvo (1982, 226–7).

41   **paper he published in 1937**   Majorana 1937.

41   **he and his student Seth Neddermeyer detected the particle**   Neddermeyer and Anderson 1937.

41   **three Italians . . . demonstrated**   Conversi et al. 1945; 1946; 1947.

41   **a tall . . . tennis champion from Pisa**   Rhodes 1986, 217.

41   **escaping Paris . . . Toulouse**   Bonolis 2005, 489; Schecter and Schecter, 55.

42   **that the appearance . . . decent occupation**   Pontecorvo 1982, 228, 230.

42   **technical report**   Pontecorvo wrote a second report on neutrino detection in November 1946, which was declassified at the same time as the first. The second is the one that has gone down in history (Pontecorvo 1945; 1946).

42   **remained classified for four years**   The paper's declassification was prompted by an independent proposal of essentially the same method, which Luis Alvarez of the University of California at Berkeley published in the open literature in 1949 (Alvarez 1949; Lande 2009, 23).

42   **by far the best**   Pontecorvo 1982, 230.

42   **a powerful reactor . . . the Sun**   Ibid., 231.

43   **theory of energy production in stars**   Bethe 1939. Hans Bethe produced an immense body of seminal work in the course of his long life. (He died in 2005 at the age of 99 and worked nearly to the end.) The citation for his Nobel Prize in Physics, which he received in 1967, recognized his entire opus up to that point while singling out this high point in particular.

43   **nuclear reactions . . . cause the sun to shine**   Bahcall 1996a.

44   **That was indeed . . . lots of questions**   Pontecorvo 1982, 232.

44   **series of discoveries**   Hincks and Pontecorvo, 1948a, 1948b, 1949, 1950a, 1950b. Some of these results were confirmed by others (Pontecorvo 1982, 236n36–7).

45   **For people working . . . always been present**   Pontecorvo 1982, 234.

47   **It made front-page news in Britain**   Turchetti 2003, 410.

47–48   **foremost expert**   Turchetti 2003; Turchetti 2012.

48   **the pressure put on him by security forces during vetting**   Turchetti, 2003, 414, 414n107.

48   **he had defected . . . peaceful uses of atomic energy**   Richards 1992.

48   **Eleven . . . Nobel Prizes based on Pontecorvo's theoretical work**   Frank Close (2010, 32) counts nine, and two more have been awarded since he wrote his book.

48   **had he lived long enough**   Longevity turns out to be the basic game with the Nobel Prize, since you have to be alive to get one.

49   **muon neutrino was discovered in 1962**   Danby et al. 1962.

49   **intuited . . . ghostlike property**   Pontecorvo 1958.

49   **tau particle . . . discovered**   Perl et al. 1975.

49   **its neutrino was finally detected**   Kodama et al. 2001.

49   **Pontecorvo's 1958 conjecture was proven true**   Fukuda et al. 1998.

From Poltergeist to Particle

(This was the subtitle of Fred Reines's 1995 Nobel lecture. The text of the talk appeared in *Reviews of Modern Physics* the following year [Reines 1996].)

50   **Ray Davis**   Biographical sources for Ray Davis: Davis 2002, Lande 2009, Bahcall 1996a.

50   **indirect method for detecting the particle**   Davis 1952.

50   **Davis's stated goals**   Davis 1955; Lande 2009.

51 **published his results in 1955** Davis 1955.
51 **the story of how one of the paper's reviewers was unimpressed** Davis 2002, 61.
52 **Fred Reines** This biographical background, such as it is, comes from Kropp et al. 2009.
53 **got off to working on neutrinos** The anecdotes about Reines's months staring at a blank pad and his decision to work with Cowan come from his Nobel lecture (1995, 1996).
53 **Certainly . . . should not work** Both letters will be found in Reines's Nobel lecture (1995, 1996).
54 **first set of experiments** Cowan 1964.
54 **"probable" detection of the free neutrino** Reines and Cowan 1953.
54 **evidence would not . . . appear "genuine"** Cowan 1964.
54 **On the way down . . . lose him now** Barker 1979; Enz 1994, 19.
54 **received a tip** Reines 1995; 1996.
55 **detected only about three inverse beta decays per hour** Cowan et al. 1956.
55 **We were done . . . made them so** Cowan 1964.
55 **letter never arrived** Reines told the story of the night letter in his Nobel address (1995; 1996).
55 **Hoyle remarked** Hoyle 1967.
56 **several experiments . . . birth defect** Enz 1994.
56 **grossly overestimated the detection efficiency** Reines 1979, 16n13. (Note that this admission is conveniently obscured in a footnote.)
56 **snapping into line with the new theory** Reines and Cowan 1959; also Pauli 1958a.
57 **did nothing to defuse the controversy** I learned about this controversy from Sandip Pakvasa and John Learned. It is briefly mentioned in LANL 1997.
57 **a man of imposing physical stature** Kropp et al. 2009.
57 **talented in . . . oversized shoes** Ibid.; Wheeler and Ford 1998, 281.
57 **Bernstein refers** Bernstein 2001b, 6.
58 **Physicists . . . tampered with lightly** The quotes come from an excellent profile of Lee and Yang that Bernstein wrote for *The New Yorker* in 1962. Bernstein believes this might have been "the first really serious scientific article" that the magazine ever published (2001b, 6–7).
58 **Lee and Yang were energetic collaborators** The details in this paragraph come from Bernstein (1962).
58 **classic paper** Lee and Yang 1956.
59 **produced a definitive result, confirming Wu's** There are a number of written accounts of these events, most of which attribute the idea of the experiment to Lederman, who appears to be a gifted self-promoter (Bernstein 1962; Close 2010; Lederman 1993, 256–73). It seems that Garwin played an equal role in conceiving the experiment and the larger role in designing the apparatus (Jerome Friedman, private communication; AIP 2001).
59 **Parity wasn't just dead; it was annihilated** The papers reporting these three results all appeared in *Physical Review:* Wu et al. 1957; Garwin et al. 1957; Friedman and Telegdi 1957.
59 **front page of the *New York Times*** Schmeck, Jr. 1957.
60 **elegant experiment at Brookhaven** Goldhaber et al. 1957.
60 **Physicists now realize** Wilczek 2009.
60 **admitted to having a "mirror complex"** Meier 2001, 163 (from a long and detailed letter to Jung that Pauli wrote on August 5, 1957).
61 **there was not . . . particular subject** Ibid., 162 (letter from Pauli to Jung, August 5, 1957).
61 **finally proved that they did** Pauli 1955; Pauli 1958b.
61 **on everybody's lips** Meier 2001, 162 (letter from Pauli to Jung, August 5, 1957).
61 **fixed point around which all else turned** Miller 2010, 243.
61 **Now, 'mirroring' is an archetype . . . engaged in something** Atmanspacher and Primas 2006, 8, 8n20.
61 **very impressive dream** The indented quote and Pauli's remarks about the dream come from his August 5, 1957, letter to Jung (Meier 2001, 160–6).
62 **very impressed . . . Chinese young lady** Meier 2001, 161 (letter from Pauli to Jung, August 5, 1957).
62 **I do *not* believe . . . symmetric results** Bernstein 1962, 84.
62 **two theoretical papers by Lee and Yang** Lee et al. 1957; Lee and Yang 1957.

62   **Witnesses say** Enz 1973.
63   **I congratulate you . . . still persecutes me** Wu 1960; Enz 1973; Miller 2010, 240.
63   **very upset . . . quite a while** Miller 2010, 238; Meier 2001.
63   **Now after the first shock . . . right-and-left symmetric?** Pauli 1957a.
63   **Pauli's last piece of scientific writing** Pauli 1958a.
63   **gift for her eightieth birthday** Segrè 2007, 263.
63   **Another of his obsessions . . . fine-structure constant** Indeed, Arthur I. Miller (2010) has written an entire book on Pauli's obsession with this number.
63–64 **theoretical interpretation . . . atomic physics** Pauli 1948.
64   **the number . . . meaning for Pauli** Enz 1994, 23.

Wisconsin-Style Physics

67   **Reines once told Francis** Halzen 2015.
67   **Very few . . . idea in writing** Reviewing the dim prehistory of deep underwater neutrino detection in his invaluable "personal history of the DUMAND project," physicist Arthur Roberts (1992, 269–70) writes, "It seems clear that the idea of an ocean neutrino detector had occurred to many physicists independently, though none (other than Markov) seem to have committed it to print." Roberts does not mention Greisen, but then Greisen's concept did not involve the deep ocean.
67   **diploma thesis of one Igor Zheleznykh** Zheleznykh 1958. The European diploma thesis is more-or-less equivalent to a master's thesis in the United States.
67   **Zheleznykh readily admits** Zheleznykh 2006.
67   **Markov first presented it** Markov 1960.
67   **Greisen . . . proposed a similar idea** Greisen 1961.
67   **setting up Cherenkov radiation** Markov 1960.
67–68 **journal article published in January of the following year** Markov and Zheleznykh 1961.
68   **high-energy neutrino astronomy . . . radio waves** Greisen 1961.
68   **review article on cosmic ray showers** Greisen 1960.
68   **muon neutrinos . . . would be detected two years later** Danby et al. 1962.
68   **conceived by Bruno Pontecorvo in Russia before it occurred to Lederman's team** Pontecorvo 1958; Pontecorvo 1959; Schwartz 1960.
69   **"plum pudding"** I got this phrase from John Learned.
70   **cascade** The standard physics term is *shower*.
73   **Oh-My-God particle** Wolchover 2015.
73   **even a neutrino** Halzen, Vazquez, et al. 1994.
73   **two hundred billion secondary particles and decay products** Pierre Auger Collaboration 2011.
75   **1960 review article** Reines 1960. This article appeared in the same issue of the *Annual Review of Nuclear Science* as Greisen's classic paper.
75   **he began visiting . . . early as 1966** Learned 1991.
75   **"goggafangers" . . . "big boss bug catcher"** Kropp et al. 2009, 17.
76   **published their results about two weeks before Reines's did** Achar et al. 1965; Reines et al. 1965; Reines 1967; Menon 1967; Kropp et al. 2009.
76   **everyone else chalked it up as a tie** In a comprehensive review of the history of neutrino astronomy, Christian Spiering (2012, 520) suggests that this was an "historic race which had no losers but two winners."
76   **only 167 atmospheric neutrinos over the six years that they took data** Reines 1995, 217; Reines 1996, 324.
77   **small research vessel the *Undula*** Rossi 1985; Regener 1931.
78   **first baby step** Learned's thesis work eventually appeared in a journal article (Davis and Learned 1973).
78   **Compton had proven** Compton 1933.
78   **first accurate measurements of the muon's lifetime** Rossi 1939; Rossi and Hall 1941. Rossi's muon experiments are said to have "inaugurated the current era of particle physics" (Bonolis 2011).
79   **years before the latter began to produce high enough energies to be of significant use** The accelerators began to provide surprises only after 1953, when the Cosmotron, the first true

high-energy accelerator, went on line at Brookhaven. This was quickly followed by the Berkeley Bevatron, where the antiproton, the antineutron, and a number of strange particles were discovered over the course of the fifties. CERN, which was conceived largely in response to cosmic ray physics, wasn't founded until 1954.

80 **discovery of the pion was confirmed** Lattes, Muirhead et al. 1947; Lattes, Occhialini et al. 1947.

80 **One of his causes . . . faculty positions** AIP 1995.

82 **neutrino physics . . . Accelerator Laboratory** Galison 1983, 490.

82 **add more weight to his proposal** Ibid., 491.

82 **story of the E1A experiment is told elsewhere** Taubes 1986, chapters 1 and 2; Galison 1983.

83 **radical, crazy people** Taubes 1986, 13.

83 **once visited Rubbia . . . shoulder-length hair** Ibid., 18.

84 **on the basis of a paper** Halzen and Michael 1971.

86 **two most influential papers** Cline, Halzen, and Luthe 1973; Cline, Halzen, and Waldrop 1973.

86 **considered one of the most influential of Cline's illustrious career** Cousins and Rosenzweig 2016.

Peaceful Exploration by Interested Scientists Throughout the World

88 **"Serpents in an Astrophysical Eden"** Roberts 1992, 283; Kampa 1979. It also turns out that underwater neutrino detectors can be used to find whales. In 2005, some Italian physicists mounting an ultimately unsuccessful attempt to build a neutrino telescope named NEMO (NEutrino Mediterranean Observatory) off the coast of Sicily ran a feasibility study for an acoustic method of neutrino detection using four sensors more than two miles deep. The idea was to find out if it was quiet enough down there to "hear" neutrino interactions. As it happens, the frequency range of the expected neutrino signal overlaps those of sperm whale and dolphin calls, and, in a delightful surprise to whale experts, they discovered that hundreds of sperm whales were passing within about sixty miles of their listening post every year (AAAS 2007).

88 **quirky "personal history" of DUMAND** Roberts 1992. I am borrowing heavily from this paper.

88 **satirical songs . . . about physics and the academic life** In "Take Away Your Billion Dollars," which he composed shortly after the war, Roberts bemoaned the effect of big money and military interest on the practice and teaching of physics (Haverford College 2002):

Take away your army generals; their kiss is death, I'm sure.
Everything I build is mine, and every volt I make is pure . . .

89 **he responded with an exposition of "jugular science"** Overbye 1998.

89 **Kotzer . . . edited its proceedings** Kotzer 1976.

89 **UNCLE** New Scientist 1978.

89 **study for communicating with the subs with neutrinos** Perry 1978.

90 **telecommunication with neutrino beams** Sáenz et al. 1977. Clyde Cowan worked on Cherenkov detection himself at The Catholic University of America (Überall and Cowan 1965; Cowan et al. 1966). He and Überall refer to a similar design that had been previously suggested by Keuffel (1963 and 1965).

90 **DUMAND . . . a concept, encompassing underwater and underground neutrino astronomy in all its forms** John Learned still speaks of it that way in certain contexts, and so does Igor Zheleznykh (2006).

91 **Roberts presented their first thoughts** Roberts et al. 1977.

91 **UNDINE** Undines are elemental or mythic beings, usually female, associated with water. A mermaid would be one example. They first appear in the writings of Paracelsus, the medieval alchemist.

92 **Adherents began to gather** Roberts 1992, 308.

92 **UNDINE was returned . . . other suitors** Ibid., 274.

93 **offered "several thousand phototubes for DUMAND"** Ibid., 308.

94 **Russians already had . . . Pontecorvo and Davis** INR 1997.

97 Rubbia went out of his way . . . Nobel lecture Rubbia 1984.
98 more interested in supersymmetry than in the particles they ended up finding Taubes 1986.
99 contributing a theory paper Halzen 1981.
99 DUMAND symposium Stenger 1981.
99 the severing of the Russian link . . . available for us Roberts 1992, 275n2.

Science at Its Best
100 glass sphere capable of withstanding at least 500 times atmospheric pressure Ibid., 261.
100 oceanographers were amazed . . . order of 100 Ibid., 278.
101 snap loading Ibid., 280.
101 the explosive bolts . . . failed to release when fired Ibid., 272.
101 They took a total . . . muons' angular distribution Babson et al. 1990.
103 The Russians proceeded methodically The information on Baikal's progress comes from Spiering (2012) and the Baikal Collaboration (1992).
104 monopole, first postulated by Dirac Dirac 1931.
105 proposed one of the first credible theories of grand unification Georgi and Glashow 1974.
105 Pauli employed group theory Pauli 1955.
105 diamonds might not be forever Robinson 1979.
105 theory had been used to estimate the proton's lifetime Goldman and Ross 1979.
106 listening for a gnat's whisper in a hurricane LANL 1997.
107 Glashow pronounced the experiment "an unmitigated failure" Taubes 1986, 160.
107 IMB "had some candidates, but they weren't elected" Bernstein 2001b, 79.
108 there's a Hyper-K in the offing Normile 2015.
108 Supernova 1987a The *a* designates this as the first supernova to be observed that year. Twelve others were observed in 1987, but they were all much dimmer and seen only through telescopes.
109 The following night . . . no star before Simonsen 2009.
110 Duhalde had taken . . . knew it well Schorn 1987b, 470.
110 When McNaught developed the negative, the supernova was there I pieced together the story of the optical detection of Supernova 1987a with the help of Shorn (1987a, b, and c), Woosley and Phillips (1988), Sutton (1992, 184–6), Bernstein (2001, 64–66), Simonsen (2009), and Kirshner (2002).
110 the swift rise to brilliance Schorn 1987b, 471.
110 constrained the current theoretical models Ibid., 755.
111 Especially during . . . last even longer Woosley and Phillips 1988, 751.
111 A six-second flash . . . Neutrino Observatory in the Caucasus Baksan detection: Alekseev et al. 1987 and 1988; Chudakov et al. 1987. Although I have been focusing on Cherenkov detectors, the 1980s saw a rush to build other types of neutrino detector as well. The Liquid Scintillation Detector (LSD), built in a tunnel under Mont Blanc by a Swiss-Italian team, detected a flash of what seemed to be neutrinos about five hours before IMB and Kamiokande detected theirs. IMB and Kamiokande did not see the LSD flash (double entendre intended), and LSD did not see theirs. The LSD signal seemed real (don't they all?); and, in fact, the LSD collaboration was the first to claim they had detected neutrinos from the supernova; but their claim is now seen as a false alarm. It created much confusion early on and people still argue about it, but it is now generally believed that LSD experienced a very unlucky fluctuation.
112 Michigan folks . . . at a ski resort in France Kirshner 2002, 43.
112 IMB announced . . . after Kamiokande did Kirshner 2004.
112 two collaborations published back-to-back papers IMB: Bionta et al. 1987; Kamiokande: Hirata et al. 1987 and 1988. (Both instruments had been updated by this time. They were now known as IMB-II and Kamiokande-II.)
113 When all is said and done . . . most violent events in the universe Woosley and Phillips 1988, 754.
113 Kamiokande almost missed the flash Sutton 1992, 192.
113–114 Core collapse . . . transformed into an astronomy Burrows 1989.
114 Landau first postulated Landau 1932.

114 **Baade . . . joined fellow astronomer Fritz Zwicky in proposing** Baade and Zwicky 1934.
114 **For many decades . . . stalwart theorists** Burrows 1987, 36.
116 **recent study** Pilachowski and Pace 2015; Croswell 2015.

Solid-State DUMAND

121 **Francis was born** The basis for this account of Francis's early life is a series of interviews he did for the Oral History Project at the University of Wisconsin (University of Wisconsin-Madison Archives 2007). I followed up with questions of my own.
127 **influential paper on cosmic rays** The paper was published the following year (Gaisser and Halzen 1985).
127 **groups from Kiel, Germany, and the University of Leeds presented evidence** Kiel: Samorski and Stamm 1983; Leeds: Lloyd-Evans et al. 1983.
127 **detected the first true high-energy gammas from the Crab Nebula** Weekes et al. 1989.
128 **radio method had been conceived by the Russian physicist** Askaryan 1961 and 1965; Markov and Zheleznykh 1986.
128 **My interest peaked . . . Russian colleagues** The quote is from an informal paper Francis once posted on the internet (Halzen 1995a). The solution was published in two papers (Halzen, Zas, and Stanev 1991; Zas, Halzen, and Stanev 1992).
128 **Francis, . . . Learned, and . . . Stanev later estimated** Halzen, Learned, and Stanev 1990.
129 **another line in the water in Arkansas** GRANDE (Gamma-Ray And Neutrino DEtector) was a vision for a Cherenkov detector inside an opaque plastic bag filled with purified water, to be located in an abandoned quarry in Arkansas. It never happened.
130 **letter of intent, based on a paper** Halzen and Learned 1988.
131 **halls once graced by Galileo himself** Reines 1992.
132 **Neutrino Detection in Clear Polar Ice** March, Halzen, and Learned 1988.
133 **first true high-energy gammas were detected from the Crab Nebula** Weekes et al. 1989.
133 **instrument finally detected pulsed gamma rays . . . from the Crab** This was the MAGIC (Major Atmospheric Gamma Imaging Cherenkov) telescope in the Canary Islands (Aliu et al. 2008).
135 **Pomerantz chaired a conference at Bartol** Mullan, Pomerantz, and Stanev 1990.
135 **Morse presented** Morse and Gaidos 1990.
135 **that field investigations begin . . . neutrino bandpass filter** Halzen, Learned, and Stanev 1990.
135 **balloon experiment** Westphal et al. 1990.
136 **LC-130 Hercules transport plane, aka "Skibird"** People who work in the polar regions tend to become aircraft aficionados, not only because it's the main way they travel to their work sites but also because the history of polar exploration in many ways tracks the history of airplane flight. When I was at Pole in late 1999, a so-called Basler 67, a reworked, ski-equipped, tail-dragging, DC-3 Dakota, came in, loaded with tourists who had coughed up $75,000 each to take this special side-trip on their regular Antarctic tour, and almost everyone at the station went out to the airstrip to admire and photograph the plane.
137 **received a call . . . never heard of them** Halzen 1995a.

Enter Bruce

140 **I never was in the drilling business to be a driller** I have cribbed some of these interviews of and stories about Bruce from my book *Thin Ice* (2005).
141–42 **It was our first day . . . gentle champion** Frost 2006.
142 **his loyalty . . . soft-spoken nature** Madison.com 2006.
146 **letter to *Nature*** Lowder et al. 1991.
147 **it's pretty clear . . . turned out not to be true** Francis's remarks here are an amalgam of two different statements, one he made to an historian in Madison (University of Wisconsin-Madison Archives 2007) and the other he made to me, over dinner at an Italian restaurant in Madison in November 1998.
149 **His work suggested . . . 1,100 meters** Gow and Williamson 1975.
151 **Great God! . . . reward of priority** The quote from Scott's diary comes from Huntford (1999, 497), one of the best books about adventure and exploration that has ever been written. The

book demonstrates that Scott, far from being the romantic hero he has been made out to be, was actually a pompous bungler, who should have been court-martialed for reckless endangerment of his men, had he and they survived.

153   **many people go back year after year**   New Zealander Anthony Powell (2014) gives a wonderful feel for life at McMurdo and for wintering over there in his film *Antarctica: A Year on Ice.*

156   **increase in carbon dioxide . . . first confirmed**   Bowen 2005, 118–119.

The Crossover

165   **Torneträsk**   The Torneträsk project was named PAN: Particle Astrophysics in Noorland.

168   **Wisconsin Idea**   Stark 1995.

168   **support it across the political spectrum**   In February 2015, Wisconsin governor Scott Walker created an uproar when he rewrote the university's mission statement, essentially erasing the Wisconsin Idea, in a budget that also called for sharp cuts in state aid to the university system. It was widely believed that he was burnishing his conservative credentials for what turned out to be a futile run for the U.S. presidency. Evidently, he didn't realize how much the people of his state cherished the Idea. The backlash was general, it came from all over the state, and Walker quickly backtracked (Evans 2015; Bosman 2015).

170   **Koci went off on other adventures**   Bowen 2005, chapters 13–15.

170   **altered the course of thinking**   Ibid., 262; Broecker 1995.

A Supernova of Science

175   **NEutrinos from Supernovae and TeV sources, Ocean Range**   The name of the instrument was subsequently changed to "Neutrino Extended Submarine Telescope with Oceanographic Research."

176   **attempt to deploy the first string for DUMAND-II**   Grieder et al. 1995.

177   **NT-200**   Sokalski and Spiering 1992.

177   **Christian writes**   Spiering 2012, 531–2.

179   **told him about a paper**   Barkov and Lipenkov 1984.

181   **interesting enough to find publication in *Science***   Askebjer et al. 1995.

182   **paper Domogatsky had pointed out to him**   Barkov and Lipenkov 1984.

182   **paper presenting the theory**   Price 1995.

182   **developed one of his own**   AMANDA Collaboration 1995.

184   **more than three hundred meters**   Askebjer et al. 1997.

184   **took her colleagues almost a year to accept the idea**   Tilav et al. 1997.

185   **Francis . . . Jacobsen, and . . . Zas had shown**   Halzen, Jacobsen, and Zas 1994. This paper expanded upon some earlier work by the DUMAND collaboration (Pryor et al. 1988).

188   **brief spike . . . might signal the formation of an exotic quark star**   Abbasi et al. 2011c.

188   **would have detected about *twenty thousand* neutrinos from the same event**   Halzen, Jacobsen, and Zas 1996.

189   **with minimal maintenance or investment in manpower and operation**   Ibid.

189   **first-ever up-going muons**   These were not "gold-plated" events that could be ascribed definitively to neutrino interactions, so the Baikal collaboration stopped short of claiming that they had detected the first atmospheric neutrinos (Belolaptikov et al. 1995 and 1997).

189   **hardware trigger**   Wischnewski et al. 1997.

189   **updated paper**   Halzen, Jacobsen, and Zas 1996.

190   **99.8 percent**   Actually, as it turns out, the uptime of the supernova data acquisition system is different from that of the detector as a whole. In a talk at an IceCube collaboration meeting in the spring of 2015, John Kelley, who was in charge of detector operations, reported that the supernova uptime is about 98.7 percent.

Doubling Down

196   **Comanche war parties**   For an excellent and unsentimental portrait of Comanche war parties, see Gwynne (2010).

198   **Cherenkov telescope high on a mountainside in the Canary Islands**   The instrument was named HEGRA (High-Energy Gamma Ray Astronomy).

200   **scattering dropped dramatically between one and two thousand meters**   Hulth et al. 1997. The collaboration never published a journal article about this specific optical finding, because it

wasn't all that interesting from a physics standpoint and the problem could only be solved by Monte Carlo. A short article eventually appeared in the proceedings of a workshop on Simulation and Analysis Methods for Large Neutrino Telescopes that was held in conjunction with a collaboration meeting at Zeuthen more than two years later (Lowder and Woschnagg 1999).

201 **Buford Price announced triumphantly** Lowder et al. 1996.
202 **the SAGENAP committee wrote** Williams et al. 1996.
203 **detected the first-ever "gold-plated" . . . up-going neutrinos** Balkanov et al. 1998 and 1999.

Glory Days
207 **master's thesis by . . . Stuart Kleinfelder** Kleinfelder 1992 and 2003.
215 **iceman** Schwarz 2015.

Night on the Ice
219 **a worker was thrown . . . from the rest of the population** Cockrell 2009.
225 **cold, dry air . . . TB clinic** Hutchison 2001.
225 **characterized by . . . 'Antarctic stare'** Palinkas 2000.
225–26 **Problems are not swept . . . eight-month-long night** Cockrell 2009.
226 **first observed . . . by . . . Frederick Cook** Palinkas 2000.
226 **similarity between wintering over and spending extended periods of time in space** Palinkas 1989.
226 **template for the laws that may govern the colonization of the Moon and distant planets** NSF administrator John Lynch (1990) gave a talk on this subject at the Bartol Conference in 1989, the one at which Francis Halzen's talk launched Doug Lowder and Andrew Westphal on their unauthorized and ill-fated AMANDA feasibility study in West Antarctica.
226 **Polar T3 syndrome, apparently related to winterover syndrome** Palinkas et al. 1997.
226 **the more severe the environment, it seems, the lower the incidence of depression** Palinkas 2000.
227 **tempering effect** Ibid.

The First Nus
231 **pick out five "interesting events" by eye** Jacobsen 1996a.
231 **Christopher got his feet wet in this business** Wiebusch 1994.
232 **a bit 'too hot off the press'** Rankin et al. 1998.
232 **first credible up-going muon tracks ever detected by AMANDA** Bouchta 1998.
233 **presented the three gold-plated events** Balkanov et al. 1998 and 1999.
235–36 **building the case for a larger instrument** Halzen 1995b, 1995c, and 1996.
236 **says Francis, inscrutably** University of Wisconsin-Madison Archives, Oral History Project 2007.
238 **another workshop** Spiering 1999.
239 **most anticipated talk** Totsuka 1999; Fukuda et al. 1998.
239 **new theoretical estimates for neutrino emission by the Sun** Holmgren and Johnston 1958; Cameron 1958; Fowler 1958.
240 **back-to-back papers . . . in *Physical Review Letters*** Bahcall 1964; Davis 1964. Davis told the story of his experiment's beginnings in his Nobel lecture (2002). The story is also told by Lande (2009) and Bahcall and Davis (1982).
240 **Sun would emit about 7.5 SNU** This is actually the prediction that obtained when Davis published his first results (Bahcall et al. 1968).
240 **7.5 neutrinos only every half-million seconds** Close 2010, 79.
240 **bought only about ten minutes of advertising on commercial television** Ibid., 76–7.
240 **revealed his first results** Davis 2002, 71; Davis et al. 1968.
241 **Bahcall would point out** Bahcall 1996b, 480. Bahcall also referred to the "problem" as a cause for rejoicing at the watershed meeting for IceCube that took place near NSF headquarters in Arlington, Virginia, on May 10, 1999.
241 **Bahcall's prediction** Bahcall et al. 2001.
241 **Davis's measurement** Cleveland et al. 1998.
241 **he was the only . . . successful experiment** Bahcall and Davis 1982.
241 **the flux of observable . . . neutrino flux** Pontecorvo 1967.

242 **oscillate into tau neutrinos, which were essentially invisible to the detector** Sobel and Suzuki 2008.

243 **SNO collaboration announced the solution** Ahmad et al. 2001. The finding was further confirmed the following year (Ahmad et al. 2002).

243 **prisoners . . . I feel like dancing!** Chang 2001.

244 **for the discovery of neutrino oscillations, which shows that neutrinos have mass** Royal Swedish Academy of Sciences 2015.

The Peacock and Eva Events

247 **included in *Best American Science Writing 2000*** Francis's essay (1999) first appeared in *The Sciences*, a publication of the New York Academy of Sciences.

251 **the bit about Dr. Johnson's dog** Learned is actually botching a sexist remark from Samuel Johnson here. "Sir, a woman's preaching is like a dog's walking on his hinder legs," Boswell quotes Johnson as saying. "It's not done well; but you are surprised to find it done at all." (Insight courtesy of Jason Newman.)

254 **found seventeen neutrino candidates** DeYoung et al. 1999.

254 **solid competing analysis** Biron et al. 1999.

255 **confirmed the nine-month-old Madison analysis** The situation was a bit more complicated than I'm putting it here: the Zeutheners had eliminated certain days of the year when they deemed that the detector had been acting oddly, so only fourteen of the original seventeen events were in their raw dataset. Of those fourteen, thirteen passed all the cuts in the Zeuthen analysis, and when they deigned to use their eyes to examine the fourteenth event, they realized it was "looking good" (Biron et al. 1999, 55) and that it had been eliminated because one of their cuts was too stringent.

256 **talk of lawsuits** This account of the COBE controversy comes from the book by Mather and Boslough (2008, xix, 232–45).

257 **report they produced in April** Augustine et al. 1997.

257–8 **Argentina once sent a woman . . . both conceived *and* born in Antarctica** Romero 2016.

260 **article presenting the up-going events they had discovered with AMANDA-B4** Andres et al. 2000.

263 **professors had proven incapable of managing the project** A sanitized version of the LIGO debacle is told by Cho (2016).

Y2K at Pole

276 **No More Frontier Attitude** Wright 2012, 28.

Sometimes You Get What You Ask For

283 **a rumored $2 million** This account of the COBE story comes from the book by Mather and Boslough (2008, 232–45).

287 **Observation of high-energy neutrinos using Čerenkov detectors embedded deep in Antarctic ice** Andres et al. 2001. A subsequent, improved analysis was published the following year (Ahrens et al. 2002b).

No New Starts

288 **Within days of George W. Bush's inauguration** Bowen 2007, 159.

288 **nicknamed "no-new-starts"** Mervis 2001a.

291 **observed that a Ph.D. physicist generates about $1 million** This observation was made by Nima Arkani-Hamed, a theorist at the Institute for Advanced Study, in an interview included among the "extras" in Levinson (2015).

292 **There shouldn't be . . . neutrino sources** This general approach to detecting dark matter particles was first proposed by Katherine Freese (1986).

293 **paper . . . eventually published in 2002** Ahrens et al. 2002c.

293 **searched for supernovae** Ahrens et al. 2002a.

293 **searched for a point source** Ahrens et al. 2003a.

293 **searched for a diffuse flux** Ahrens et al. 2003b.

294 **searched for neutralinos coming from the Sun** Ackermann et al. 2006.

300 **propose that . . . turn the mine into a major underground research laboratory** Malakoff 2001.

300 **elbowing their way to the head of the line** Mervis 2001a.

301   **five major research facilities were funded**   Mervis 2001b.
301   **article in the** *Chronicle of Higher Education*   Borrego 2003.

The Coming of Yeck
313   **told** *Science* **magazine that "he had lost the confidence"**   AAAS 2002.
314   **memorandum formalizing their recommendation**   Bement, Jr. 2004.

Failure and Success
327   **Nygren's team soon demonstrated**   Stokstad 2005.

As Quickly as It All Began . . .
336   **two-part thesis**   Ganugapati 2008.
342   **increasingly sophisticated study of the ice**   Buford and his fellow ice property enthusiasts have not only provided information of an applied physics nature that has been crucial to the operation of IceCube, they have also contributed to the fields of glaciology and climatology (Aartsen et al. 2013c).

Crossing the Threshold
347   **listening for a gnat's whisper in a hurricane**   LANL 1997.
347   **unreasonable effectiveness of mathematics**   Wigner 1960.
349   **detect up-going muons with just their first hard-won string**   Achterberg et al. 2006.
349   **equivalent to about a million years from the smaller one**   Gary Hill pointed this out during a talk at an IceCube collaboration meeting in Berkeley in 2012.
349   **waviness in their arrival directions**   Abbasi et al. 2010.
349   **search for neutralino annihilations in the Sun**   Ackermann et al. 2006.
349   **second study of the Sun**   Abbasi et al. 2009.
350   **fireball model**   Waxman and Bahcall 1997.
351   **fireball models, involving neutrons rather than protons**   Rachen and Mészáros 1998.
351   **Waxman had proposed GRBs as the answer**   Waxman 1995.
351   **arXiv**   The arXiv is a valuable repository of scientific papers, freely available to the public, that can be used to publish results quickly, before going through the lengthy peer-review process required of a journal article. This report was officially published in April in *Physical Review Letters*, which has a relatively speedy review process (Abbasi et al. 2011a).
352   **GRBs on Probation**   Ahlers et al. 2011.
352   **Carl Sagan's antimetabole . . . first place**   Hartmann 2011.
353   **collaboration wrote in** *Nature*   Abbasi et al. 2012.
353   **Four years later . . . IceCube still has not seen GRB neutrinos**   Aarsten et al. 2016a
353   **Eli's latest guess**   Waxman 2014.
354   **focusing on that prospect for about a year**   Halzen 2011.
354   **cosmogenic or GZK neutrinos**   Ahlers et al. 2010.
363   **announced to great fanfare**   Abraham et al. 2007.
364   **paper on Bert and Ernie**   When *Physical Review Letters* published Aya's paper in their July 9 issue (Aartsen et al. 2013a), the editors chose "Bert" to be the "cover boy."

Epilogue: The Dawn of Multi-Messenger Astronomy
368   **paper appeared in** *Science*   Aartsen et al. 2013b.
369   **demonstrated conclusively**   Halzen and Wille 2016.
369   **his brief speech**   Halzen 2015.
370   **"observation" of astrophysical neutrinos**   Aartsen et al. 2014a. Eli Waxman (2014) wrote an accompanying perspective.
370   **extraterrestrial muon neutrinos**   Aartsen et al. 2015.
370   **exceptionally high by astronomical standards**   Aartsen et al. 2014b.
370   **universe appears to be awash . . . Large Hadron Colliders**   Adrian Cho (2015) has written a feature article in *Science* about this "cosmic convergence," in which he relies heavily on Eli Waxman and his seemingly ever-changing "Waxman-Bahcall limit" for a theoretical interpretation. But according to Francis, you don't need such complicated thinking to explain what IceCube, the cosmic ray instruments, and the Fermi space telescope are measuring. All

you need is a cosmic beam dump that produces equal parts gamma rays, neutrinos, and protons.

371　**GEN2 white paper**　Aartsen et al. 2014b.

371　**she has been seeking for more than ten years now**　As I write, Aya has just completed yet another "search" for ultrahigh-energy neutrinos, this one spanning seven years (Aartsen et al. 2016c).

372　**Waxman points out**　Waxman 2014.

372　**latest search for a neutrino star**　Aartsen et al. 2016d.

372　**detection of the first gravitational wave**　Abbot et al. 2016.

372　**joint study with LIGO and several other instruments**　Adrián-Martínez et al. 2016.

372　**a friend**　David Zahn.

373　**have all the pieces**　Mervis 2016.

374　**PINGU**　IceCube-Gen2 Collaboration 2016. (Pingu is the name of a penguin living in Antarctica who stars in an animated children's television series. Francis and some others lobbied against this unfortunate name and lost.)

374　**didn't get the paper out until 2016**　Aartsen et al. 2016b. See also Schmitz 2016.

374　**article about this non-discovery in the *Wall Street Journal***　Hernandez 2016.

References and Bibliography

AAAS (American Association for the Advancement of Science). 1931. General reports of the first Pasadena meeting of the American Association for the Advancement of Science and associated societies. *Science* **74**:103–23.

———. 2002. Eisenstein Leaves NSF. *Science* **296**:1219.

———. 2007. Whale sensing. *Science* **315**:1199. [doi: 10.1126/science.315.5816.1199d]

Aartsen, M. G., et al. (IceCube Collaboration). 2013a. First observation of PeV-energy neutrinos with IceCube. *Phys. Rev. Lett.* **111**:021103. [doi: 10.1103/PhysRevLett.111.021103] [arXiv:1304.5356]

———. 2013b. Evidence for high-energy extraterrestrial neutrinos at the IceCube detector. *Science* **342**:1242856. [doi: 10.1126/science.1242856] [arXiv:1311.5238]

———. 2013c. South Pole glacial climate reconstruction from multi-borehole laser particulate stratigraphy. *J. Glaciol.* **59**:1117–28.

———. 2014a. Observation of high-energy astrophysical neutrinos in three years of IceCube data. *Phys. Rev. Lett.* **113**:101101. [doi:10.1103/PhysRevLett.113.101101] [arXiv:1405.5303v2]

———. 2014b. IceCube-Gen2: A vision for the future of neutrino astronomy in Antarctica. [arXiv:1412.5106v2]

———. 2015. Evidence for astrophysical muon neutrinos from the northern sky with IceCube. *Phys. Rev. Lett.* **115**:081102. [doi:10.1103/PhysRevLett.115.081102] [arXiv:1507.04005 [astro-ph.HE]]

———. 2016a. An all-sky search for three flavors of neutrinos from gamma-ray bursts with the IceCube Neutrino Observatory. *Astrophys. J.* **824**:115. [doi: 10.3847/0004-637X/824/2/115] [arXiv:1601.06484]

———. 2016b. Searches for sterile neutrinos with the IceCube Detector. *Phys. Rev. Lett.* **117**:071801 [doi:10.1103/PhysRevLett.117.071801] [arXiv:1605.01990 [hep-ex]]

———. 2016c. Constraints on ultra-high-energy cosmic ray sources from a search for neutrinos above 10 PeV with IceCube. *Phys. Rev. Lett.* **117**:241101. [doi: 10.1103/PhysRevLett.117.241101] [arxiv.org/abs/1607.05886]

———. 2016d. All-sky search for time-integrated neutrino emission from astrophysical sources with 7 years of IceCube data. *Astrophys. J.* (forthcoming) [arXiv:1609.04981 [astro-ph.HE]]

Abbasi, R., et al. (IceCube Collaboration). 2009. Limits on a muon flux from neutralino annihilations in the Sun with the IceCube 22-string detector. *Phys. Rev. Lett.* **102**:201302. [doi:10.1103/PhysRevLett.102.201302] [arXiv:0902.2460 [astro-ph.CO]]

———. 2010. Measurement of the anisotropy of cosmic ray arrival directions with IceCube. *Astrophys. J. Lett.* **718**:L194–98. [doi:10.1088/2041-8205/718/2/L194] [arXiv:1005.2960 [astro-ph.HE]]

———. 2011a. Limits on neutrino emission from gamma-ray bursts with the 40 string IceCube detector. *Phys. Rev. Lett.* **106**:141101. [doi:10.1103/PhysRevLett.106.141101] [arXiv:1101.1448]

———. 2011b. First search for atmospheric and extraterrestrial neutrino-induced cascades with the IceCube detector. *Phys. Rev. D* **84**:072001. [doi:10.1103/PhysRevD.84.072001] [arXiv:1101.1692]

———. 2011c. IceCube sensitivity for low-energy neutrinos from nearby supernovae. *Astron. Astrophys.* **535**:A109. [doi: 10.1051/0004-6361/201117810] [astro-ph.HE]]

———. 2012. An absence of neutrinos associated with cosmic-ray acceleration in γ-ray bursts. *Nature* **484**:351–4. [doi: 10.1038/nature11068] [arXiv:1204.4219 [astro-ph.HE]]

Abbott, B.P., et al. (LIGO Scientific Collaboration and Virgo Collaboration). 2016. Observation of gravitational waves from a binary black hole merger. *Phys. Rev. Lett.* **116**:061102. [doi:10.1103/PhysRevLett.116.061102] [arXiv:1602.03837]

Abraham, J., et al. (Pierre Auger Collaboration). 2007. Correlation of the highest-energy cosmic rays with nearby extragalactic objects. *Science* **318**:938–43 [doi: 10.1126/science.1151124]

Achar, C. V., M. G. K. Menon, et al. 1965. Detection of muons produced by cosmic ray neutrinos deep underground. *Phys. Lett.* **18**:196–9.

Achterberg, A., et al. (IceCube Collaboration). 2006. First year performance of the IceCube neutrino telescope. *Astropart. Phys.* B:155–73.

Ackermann, M., et al. (AMANDA Collaboration). 2006. Limits to the muon flux from neutralino annihilations in the Sun with the AMANDA detector. *Astropart. Phys.* **24**:459–66. [arXiv:astro-ph/0508518]

Adrián-Martínez, S., et al. (Antares Collaboration, IceCube Collaboration, LIGO Scientific Collaboration, and Virgo Collaboration). 2016. High-energy neutrino follow-up search of gravitational wave event GW150914 with ANTARES and IceCube. *Phys. Rev. D* **93**:122010. [doi: 10.1103/PhysRevD .93.122010] [arXiv:1602.05411[astro-ph.HE]]

Ahlers M., L. A. Anchordoqui, M. C. Gonzalez-Garcia, F. Halzen, and S. Sarkar. 2010. GZK Neutrinos after the Fermi-LAT diffuse photon flux measurement. *Astropart. Phys.* **34**:106–15. [doi: 10.1016/j.astropartphys.2010.06.003] [arXiv:1005.2620]

Ahlers, M., M. C. Gonzalez-Garcia, and F. Halzen. 2011. GRBs on probation: Testing the UHE CR paradigm with IceCube. *Astropart. Phys.* **35**:87–94. [doi: 10.1016/j.astropartphys.2011.05.008] [arXiv:1103.3421 [astro-ph.HE]]

Ahmad, Q. R., et al. (SNO Collaboration). 2001. Measurement of the rate of $\nu_e + d \rightarrow p + p + e^-$ interactions produced by ^{8}B solar neutrinos at the Sudbury Neutrino Observatory. *Phys. Rev. Lett.* **87**:071301.

Ahmad, Q. R., et al. (SNO Collaboration). 2002. Direct evidence for neutrino flavor transformation from neutral-current interactions in the Sudbury Neutrino Observatory. *Phys. Rev. Lett.* **89**:011301.

Ahrens, J., et al. (AMANDA Collaboration). 2002a. Search for supernova neutrino bursts with the AMANDA detector. *Astropart. Phys.* **16**:345–59. [doi: 10.1016/S0927-6505(01)00154-2] [arXiv:astro-ph/0105460v1]

———. 2002b. Observation of high-energy atmospheric neutrinos with the Antarctic Muon and Neutrino Detector Array. *Phys. Rev. D* **66**:012005. [doi: 10.1103/PhysRevD.66.012005] [arXiv:astro-ph/0205109]

———. 2002c. Limits to the muon flux from WIMP annihilation in the center of the Earth with the AMANDA detector. *Phys. Rev. D* **66**:032006 [arXiv:astro-ph/0202370]

———. 2003a. Search for point sources of high-energy neutrinos with AMANDA. *Astrophys. J.* **583**:1040–57. [arXiv:astro-ph/0208006]

———. 2003b. Limits on diffuse fluxes of high energy extraterrestrial neutrinos with the AMANDA-B10 detector. *Phys. Rev. Lett.* **90**:251101. [arXiv:astro-ph/0303218]

AIP (American Institute of Physics), Niels Bohr Library & Archives. 1962. Interview of P. A. M. Dirac by Thomas Kuhn.

———. 1963. Interview of Sir Rudolph Peierls by John L. Heilbron. http://www.aip.org/history/ohilist /4815_1.html. (accessed July 25, 2014).

———. 1995. Interview of Robert March by Patrick Catt. http://www.aip.org/history/ohilist/5910 .html. (accessed October 31, 2011).

———. 2001. Interview of Richard Garwin by W. Patrick McCray. http://www.aip.org/history/ohilist /24292.html. (accessed September 29, 2011).

Alekseev, E. N., L. N. Alekseeva, I. V Krivosheina, and V. I. Volchenko. 1988. Detection of the neutrino signal from SN1987a in the LMC using the INR Baksan Underground Scintillation Telescope. *Phys. Lett. B* **205**:209–14.

Alekseev, E. N., L. N. Alekseeva, V. I. Volchenko, and I. V. Krivosheina. 1987. Possible detection of a neutrino signal on 23 February 1987 at the Baksan Underground Scintillation Telescope of the Institute of Nuclear Research. *JETP Lett.* **45**:589–92.

Aliu, E., et al. (MAGIC Collaboration). 2008. Observation of pulsed γ-rays above 25 GeV from the Crab pulsar with MAGIC. *Science* **322**:1221–4. [doi: 10.1126/science.1164718] [arXiv:0809.2998 [astro-ph]]

Alvarez, Luis. 1949. A proposed experimental test of the neutrino theory. Report UCRL-328, University of California Berkeley, 18 April. (unpublished). http://escholarship.org/uc/item/1sh4k6s2 (accessed September 23, 2014).

Amaldi, Edoardo. 1982. Beta decay opens the way to weak interactions. In Colloque International sur l'Histoire de la Physique des Particules, *J. de Phys.* **43**, suppl. C8:261–300.

AMANDA Collaboration. 1995. On the age vs. depth and optical clarity of deep ice at the South Pole. *J. Glaciol.* **41**:445–54. [arXiv:astro-ph/9501072]

———. 1998. Web site for IceCube neutrino detector workshop, University of California, Irvine, March 27–28. http://www.ps.uci.edu/~icecube/workshop.html (accessed February 9, 2011).

Anderson, Carl D. 1932. The apparent existence of easily deflectable positives. *Science* **76**:2389. [doi: 10.1126/science.76.1967.238]

———. 1933. The positive electron. *Phys. Rev.* **43**:491–44.

Andres, E., et al. (AMANDA Collaboration). 2000. The AMANDA neutrino telescope: Principle of operation and first results. *Astropart. Phys.* **13**:1–20. [doi: 10.1016/S0927-6505(99)00092-4] [arXiv:astro-ph/9906203v1]

———. 2001. Observation of high-energy neutrinos using Čerenkov detectors embedded deep in Antarctic ice. *Nature* **410**:441–3. [doi: 10.1038/35068509]

APS (American Physical Society). 1931. Minutes of the Pasadena meeting, June 15 to 20, 1931. *Phys. Rev.* **38**:579–92.

Askaryan, G. A. 1961. *Zh. Eksp. Theor. Fiz.* **25**:276.

———. 1965. *Soy. Phys. JETP* **48**:988.

Askebjer P., et al. (AMANDA Collaboration). 1995. Optical properties of the South Pole ice at depths between 0.8-km and 1-km. *Science* **267**:1147–50.

———. 1997. UV and optical light transmission properties in deep ice at the South Pole. *Geophys. Res. Lett.* **24**:1355–8. [doi: 10.1029/97GL01246]

Atmanspacher, Harald, and Hans Primas. 1997. The hidden side of Wolfgang Pauli: An eminent physicist's extraordinary encounter with depth psychology. *J. Sci. Explor.* **1**:369–86. (Originally published in 1996: *J. Consciousness Stud.* **3**:112–26.)

———. 2006. Pauli's ideas on mind and matter in the context of contemporary science. *J. Consciousness Stud.* **13(3)**:5–50.

Auger, Pierre. 1985. Experimental work on cosmic rays: Proof of the very high energies carried by some of the primary particles. In Yataro Sekido and Harry Elliot, eds. *Early history of cosmic ray studies: Personal reminiscences with old photographs.* Dordrecht, Boston: D. Reidel, 213–8.

Augustine, Norman, et al. 1997. *The United States in Antarctica: Report of the U.S. Antarctic program external panel.* Washington D.C.: National Science Foundation. http://www.nsf.gov/pubs/1997/antpanel/antpanel.htm (accessed October 22, 2015).

Baade, Walter, and Fritz Zwicky. 1934. *Phys. Rev.* **45**:138.

Babson, J., et al. (DUMAND Collaboration). 1990. Cosmic-ray muons in the deep ocean. *Phys. Rev. D* **42**:3613–3620.

Bahcall, J. N. 1964. Solar neutrinos. I. Theoretical. *Phys. Rev. Lett.* **12**:300–02.

———. 1996a. Ray Davis: The scientist and the man. *Nucl. Phys. B (Proc. Suppl.)* **48**:281–3.

———. 1996b. Solar neutrinos: Where we are, where we are going. *Astrophys. J.* **467**:475–84.

Bahcall, J. N., and R. Davis, Jr. 1982. An account of the development of the solar neutrino problem. In Charles A. Barnes, Donald D. Clayton, and David Schramm, eds. *Essays In Nuclear Astrophysics.* Cambridge: Cambridge University Press, 243–85.

Bahcall, J. N., M. H. Pinsonneault, and S. Basu. 2001. Solar models: Current epoch and time dependences, neutrinos, and helioseismological properties. *Astrophys. J.* **555**:990–1012.

Baikal Collaboration. 1992. *The Baikal neutrino telescope NT-200: Project description.* Sokalski, I., and Spiering, C. eds. Tech. Rep. Baikal-92-03. Zeuthen: Deutsches Elektronen Synchrotron (DESY); Dubna: Institute for Nuclear Research.

Baldin, A. M., and A. A. Komar. 1978. Moiseï Aleksandrovich Markov (on his seventieth birthday). *Usp. Fiz. Nauk* **125**:363–8 (*Sov. Phys. Usp.* **21**:544–8).

Balkanov V.A., et al. (Baikal Collaboration). 1998. Reconstruction of atmospheric neutrinos with the Baikal Neutrino Telescope NT-96. *Proceedings, 25th international cosmic ray conference, Durban, South Africa, July 30–August 6, 1997.* Singapore: World Scientific.

———. 1999. Registration of atmospheric neutrinos with the BAIKAL neutrino telescope NT-96. *Astropart. Phys.* **12**:75–86. [doi: 10.1016/S0927-6505(99)00078-X]

Barker, W. 1979. Pauli remembered (a letter). *Phys. Today* **32**:11.

Barkov, N. I., and V. Ya. Lipenkov. 1984. Kolichestvennaya kharakteristika struktury l'da do glubiny 1400m v rayone stansii Vostok v Antarktide [Numerical characteristics of ice structure down to a depth of 1400 m in the region of Vostok station, Antarctica]. *Mater. Glyatsiolog. Issled.* **51**:178–86.

Battimelli, G., and G. Paoloni, eds. 1998. *20th Century physics: Essays and recollections. A selection of historical writings by Edoardo Amaldi.* Singapore: World Scientific.

Becker-Szendy, Ralph, et al. 1990. Search for proton decay into e++p0 in the IMB-3 detector. *Phys. Rev. D* **42**:2974–6.

———. 1993. IMB-3: A large water Cherenkov detector for nucleon decay and neutrino interactions. *Nucl. Instrum. Meth. A* **324**:363–82.

Belolaptikov, I.A., et al. (Baikal Collaboration). 1995. Separation of upward moving muons in the Baikal underwater telescope. In A. Taroni, ed. 1997. *XXIV ICRC Rome 1995.* (24th International Cosmic Ray Conference: Proceedings) Vol. 1. Bologna: Ed. Compositori, 789–92. (*Il Nuovo Cim.* **19C** (1997) 623–804).

———. 1997. The Baikal underwater neutrino telescope: Design, performance and first results. *Astropart. Phys.* **7**:263–82.

Bement, Arden L., Jr. 2004. NSB-04-56 (Memorandum to the members of the National Science Board: IceCube Neutrino Observatory, 5 April). http://www.nsf.gov/od/opp/opp_advisory /briefings/may2004/icecube.doc (accessed March 10, 2016).

Bernstein, Jeremy. 1962. A question of parity. *The New Yorker,* 12 May.

———. 2001a. *Hitler's Uranium Club: The secret recordings at Farm Hall,* 2nd ed. rev. New York: Copernicus.

———. 2001b. *The merely personal: Observations on science and scientists.* Lanham, Maryland: Ivan R. Dee.

Bethe, Hans A. 1939. Energy production in stars. *Phys. Rev.* **55**:434–56.

———. 1990. *Supernova mechanisms. Rev. Mod. Phys.* **62**:801–66.

Bethe, Hans A., and Rudolph E. Peierls. 1934. The "neutrino." *Nature* **133**:532–3. [doi:10.1038/133532a0]

Bilenky, S. M. 2006a. Bruno Pontecorvo: Mister neutrino. In Milla Baldo Ceolin, ed. 2006. *Third NOVE International Workshop on: Neutrino Oscillations in Venice (7–10 February 2006, Venice, Italy).* Padova: Edizioni Papergraf. [arXiv:physics/0603039 [physics.hist-ph]]

———. 2006b. Neutrino Majorana. *Annales de la Fond. L. de Broglie* **31**:139–56.

Bilenky, S. M., T. D. Blokhintseva, I. G. Pokrovskaya, and M. G. Sapozhnikov, eds. 1997. *B. Pontecorvo, selected scientific works: Recollection on B. Pontecorvo.* Bologna: Società Italiana di Fisica, Editrice Compositori.

Bionta, R. M., et al. (IMB II Collaboration). 1987. Observation of a neutrino burst in coincidence with supernova 1987A in the Large Magellanic Cloud. *Phys. Rev. Lett.* **58**:1494–6.

Bionta, R. M., et al. 1983. Search for proton decay into e+ p0. *Phys. Rev. Lett.* **51**:27–30.

Biron, Alexander, et al. 1999. Separation of v_μ event candidates in AMANDA-B. AMANDA Internal Report 19990902, 27 September.

Bogatyrev, V. K. 1971. On the possibility of constructing large detectors for neutrino astronomy. *Sov. J. Nucl. Phys.* **13**:187.

Bohr, Niels. 1929. Letter to Pauli, July 1, 1929, with draft about β-decay. Niels Bohr Archives, Copenhagen.

———. 1936. Conservation laws in quantum theory. *Nature* **138**:25.

———. 1960. Forward to Fierz and Weisskopf 1960, 1–4.

Bonolis, Luisa. 2005. Bruno Pontecorvo: From slow neutrons to oscillating neutrinos. *Am. J. Phys.* **73**:487.

———. 2011. Walther Bothe and Bruno Rossi: The birth and development of coincidence methods in cosmic-ray physics. *Am. J. Phys.* **79**:1133. [arXiv:1106.1365 [physics.hist-ph]]

Borrego, Anne Marie. 2003. Waiting game at the NSF. *Chron. Higher Ed.* **49**:A20.

Bosman, Julie. 2015. 2016 Ambitions seen in Walker's push for university cuts in Wisconsin. *New York Times,* 16 February.

Bouchta, Adam. 1998. Muon analysis with the AMANDA-B four-string detector. Ph.D. thesis, Stockholm University.

Bowen, Mark. 2005. *Thin ice: Unlocking the secrets of climate in the world's highest mountains.* New York: Henry Holt and Company.

———. 2007. *Censoring science: Inside the political attack on Dr. James Hansen and the truth of global warming.* New York: Dutton.

Breisky, Bill. 2012. On its centenary, celebrating a ride that advanced physics. *New York Times,* 7 August.

Broecker, Wallace S. 1995. Cooling the tropics. *Nature* **376**:212–13.

Brown, Laurie M. 1978. The idea of the neutrino. *Phys. Today* **31**:23–8.

Brown, Laurie M., and Helmut Rechenberg. 1996. *The origin of the concept of nuclear forces.* Philadelphia: Institute of Physics Pub.

Browne, Malcolm W. 1998a. Mass found in elusive particle; universe may never be the same. *New York Times,* 5 June.

———. 1998b. Room in the universe for ancient beliefs and modern physics. *New York Times,* 9 June.

Burrows, Adam. 1987. The birth of neutron stars and black holes. *Phys. Today* **40**:28–37.

———. 1989. In J. Schneps et al. eds. *Proceedings of the 13th International Conference on Neutrino Physics and Astrophysics, Boston (Medford) June 5–11, 1988.* Singapore: World Scientific, 142.

Cameron, A. G. W. 1958. Modification of the proton-proton chain. *B. Am. Phys. Soc.* **3**:227.

CERN Communications Group. 2009. *LHC: The guide* (CERN-Brochure-2009-003-Eng). http://cdsweb.cern.ch/record/1165534/files/CERN-Brochure-2009-003-Eng.pdf (accessed May 24, 2011).

Chadwick, James. 1914. Intensitätsverteilung im magnetischen Spektrum der ß-Strahlen von Radium B + C. *Verhandlungen der Deutschen Physikalischen Gesellschaft* **16**:383.

———. 1932. Possible existence of a neutron. *Nature* **129**:312. [doi:10.1038/129312a0]

Chang, Kenneth. 2001. Sun's missing neutrinos: Hidden in plain sight. *New York Times,* 19 June.

Cho, Adrian. 2013. Boxed in. *Science* **341**:1056–9. [doi: 10.1126/science.341.6150.1056]

———. 2015. Cosmic convergence. *Science* **349**:465–7. [doi: 10.1126/science.349.6247.465]

———. 2016. Will Nobel Prize overlook LIGO's master builder? *Science* **353**:1478–9. [doi: 10.1126/science.353.6307.1478]

Chudakov, A. E., Ya. S. Elensky, and S. P. Mikheev. 1987. Characteristics of the neutrino emission from supernova SN1987A. *JETP Lett.* **46**:373–7.

Cleveland, B. T., et al. 1998. Measurement of the solar electron neutrino flux with the Homestake chlorine detector. *Astrophys. J.* **496**:505–26.

Cline, D., F. Halzen, and J. Luthe. 1973. High-transverse-momentum secondaries and rising total cross sections in cosmic-ray interactions. *Phys. Rev. Lett.* **31**:491–4. [doi: 10.1103/PhysRevLett.31.491]

Cline, D., F. Halzen, and M. Waldrop. 1973. Hadron collisions at high transverse momentum. *Nucl. Phys. B* **55**:157–66. [doi: 10.1016/0550-3213(73)90415-X]

Close, Frank. 2010. *Neutrino.* Oxford: Oxford University Press.

Cockrell, William. 2009. A mysterious death at the South Pole. *Men's Journal.* Dec 2009, Jan 2010 issue. http://www.mensjournal.com/magazine/a-mysterious-death-at-the-south-pole-20131125 (accessed October 21, 2016).

Compton, Arthur H. 1933. A geographic study of cosmic rays. *Phys. Rev.* **43**:387–403.

Conversi, M., E. Pancini, and O. Piccioni. 1945. On the decay process of positive and negative mesons. *Phys. Rev.* **68**:232.

———. 1946. Sull'assorbimento e sulla disintegrazione dei mesoni alla fine del loro percorso. *Il Nuovo Cimento* **3**:372.

———. 1947. On the disintegration of negative mesons. *Phys. Rev.* **71**:209–10. [doi: 10.1063/PT.3.3243]

Cousins, Robert D., and James B. Rosenzweig. 2016. David Bruce Cline (Obituary). *Phys. Today* **69**:69–70.

Cowan, C. L., Jr. 1964. Anatomy of an experiment: An account of the discovery of the neutrino. *Smithsonian Institution Annual Report,* 409–30.

Cowan, C. L., Jr., F. Reines, F. B. Harrison, H. W. Kruse, and A. D. McGuire. 1956. Detection of the free neutrino: A confirmation. *Science* **124**:103.

Cowan, C. L., H. Überall, and C. P Wang. 1966. Search for the intermediate boson in weak interactions. *Il Nuovo Cimento A* **44**:526–9.

Croswell, Ken. 2015. How neutrinos saved your teeth from cavities. *Science* Web site, 11 September. http://news.sciencemag.org/chemistry/2015/09/how-neutrinos-saved-your-teeth-cavities (accessed September 20, 2015).

Curie, Irène, and Frédéric Joliot. 1934. *Septieme conseil de physique Solvay 1933: Noyauz atomiques,* 154. Paris: Gautier-Villars.

Danby, G., et al. 1962. Observation of high-energy neutrino reactions and the existence of two kinds of neutrinos. *Phys. Rev. Lett.* **9**:36–44.

Davis, Howard F., and John G. Learned. 1973. Cosmic-ray muon integral intensity measurement under water. *Phys. Rev. D* **8**:8–12.

Davis, Raymond, Jr. 1952. Nuclear recoil following neutrino emission from beryllium 7. *Phys. Rev.* **86**:976–85.

———. 1955. Attempt to detect the antineutrinos from a nuclear reactor by the $Cl^{37}(v, e^-)A^{37}$ reaction. *Phys. Rev.* **97**:766–9.

———. 1958. An attempt to observe the capture of reactor neutrinos in chlorine-37. *Proc. First UNESCO Conf., Paris* **1**:728.

———. 1964. Solar neutrinos. II. Experimental. *Phys. Rev. Lett.* **12**:303–5.

———. 2002. A half-century with solar neutrinos. Nobel lecture. http://nobelprize.org/nobel_prizes /physics/laureates/2002/davis-lecture.html (accessed June 2, 2011).

Davis, Raymond, Jr., D. S. Harmer, and K. C. Hoffman. 1968. A search for neutrinos from the Sun. *Phys. Rev. Lett.* **20**:1205–09.

DeYoung, Tyce, et al. 1999. Search for neutrinos with the AMANDA-B10 string array—a first analysis of the 1997 data. AMANDA Internal Report 19990101, 8 January.

Dirac, P. A. M. 1927. The quantum theory of the emission and absorption of radiation. *Proc. Roy. Soc. A* **114**:243.

———. 1928a. The quantum theory of the electron. *Proc. Roy. Soc. A* **117**:610–24.

———. 1928b. The quantum theory of the electron. Part II. *Proc. Roy. Soc. A* **118**:351–61.

———. 1931. Quantized singularities in the electromagnetic field. *Proc. Roy. Soc. A* **133**:60. [doi: 10.1098/rspa.1931.0130]

———. 1977. In Charles Weiner, ed. *History of Twentieth Century Physics*. New York: Academic Press, 109.

Dumour, Earl. 1993. Interview with Arthur Roberts. *Computer Music Journal* **17**:17–22 (Also http:// www.haverford.edu/physics-astro/songs/roberts/roberts_interview.htm (accessed June 20, 2011).)

Einstein, Albert. 1922. Buchbesprechung: Pauli, W., jun. *Relativitatstheorie. Naturwissenschaften* **10**:184.

Ellis, Charles Drummond, and William A. Wooster. 1927. *Proc. Roy. Soc. A* **117**:109.

Enz, Charles P. 1973. W. Pauli's scientific work. In J. Mehra, ed. *Physicist's conception of nature*. Dordrecht-Holland: D. Reidel, 766–99.

———. 1981. 50 years ago Pauli invented the neutrino. *Helv. Phys. Acta* **54**:411–8.

———. 1994. Wolfgang Pauli (1900–1958): A biographical introduction. In Pauli 1994.

———. 2000. Wolfgang Pauli's 100th birthday. *Europhys. News* **31**:12.

Evans, Christine. 2015. Save the Wisconsin Idea. *New York Times*, 16 February.

Fermi, Enrico. 1933. Tentativo di una teoria dell'emissione dei raggi β. (Tentative theory of beta rays) *La Ricerca Scientifica* **4**:491–5.

———. 1934a. Tentativo di una teoria dei raggi β. *Nuovo Cimento* **11**:1–19.

———. 1934b. Versuch einer Theorie der β-Strahlen. I. *Z. Für Physik* **88**:161–71.

———. 1965. *The Collected Papers of Enrico Fermi, Volume 2*. Chicago: University of Chicago Press. Two volumes.

Fermi, Laura. 1954. *Atoms in the family: My life with Enrico Fermi*. Chicago: University of Chicago Press.

Fierz, Markus, and Victor Frederick Weisskopf, eds. 1960. *Theoretical Physics in the twentieth century: A memorial volume to Wolfgang Pauli*. New York: Interscience.

Fowler, W. A. 1958. Completion of the proton-proton reaction chain and the possibility of energetic neutrino emission by hot stars. *Astrophys. J.* **127**:551–6.

Freese, Katherine. 1986. Can scalar neutrinos or massive Dirac neutrinos be the missing mass? *Phys. Lett.* **167B**:295–300.

Friedman, J. I., and V. L. Telegdi. 1957. Nuclear emulsion evidence for parity nonconservation in the decay chain π^+-μ^+-e^+. *Phys. Rev.* **105**:1681.

Frost, Scott. 2006. Note on "Remembering Bruce Koci" Web page. http://icecube.wisc.edu/news/view /51 (accessed May 24, 2012).

Fukuda, Y., et al. (Super-Kamiokande Collaboration). 1998. Evidence for oscillation of atmospheric neutrinos. *Phys. Rev. Lett.* **81**:1562–7. [doi: 10.1103/PhysRevLett.81.1562]

Gaisser, T. K., and F. Halzen. 1985. "Soft" hard scattering in the teraelectronvolt range. *Phys. Rev. Lett.* **54**:1754. [doi: 10.1103/PhysRevLett.54.1754]

Gaisser, T. K., F. Halzen, and T. Stanev. 1995. Particle astrophysics with high-energy neutrinos. *Phys. Rep.* **258**:173.

Galison, Peter. 1983. How the first neutral current experiments ended. *Rev. Mod. Phys.* **55**:477–509.

Gamow, George. 1966. *Thirty years that shook physics: The story of quantum theory.* New York: Doubleday.

Ganugapati, Raghunath. 2008. Measurements of atmospheric muons using AMANDA with emphasis on the prompt component. Ph.D. thesis, University of Wisconsin–Madison.

Garwin, R. L., L. M. Lederman, and M. Weinreich. 1957. Observations of the failure of conservation of parity and charge conjugation in meson decays: The magnetic moment of the free muon. *Phys. Rev.* 105:1415.

Georgi, Howard, and Sheldon. L Glashow. 1974. Unity of all elementary-particle forces. *Phys. Rev. Lett.* 32:438–41.

Gershtein, S. S. 1997. Interesting recollections and reflections about Bruno Pontecorvo. In S. M. Bilenky, T. D. Blokhintseva, I. G. Pokrovskaya, and M. G. Sapozhnikov, eds. *B. Pontecorvo, selected scientific works: Recollection on B. Pontecorvo.* Bologna: Società Italiana di Fisica, Editrice Compositori, 440–54. (This seems to be identical to Gershtein, S. S. Recollections and reflections about Bruno Pontecorvo. http://pontecorvo.jinr.ru/gershtein.html (accessed March 10, 2011).)

Ginzburg, Vitalii L., et al. 1997. Georgii Timofeevich Zatsepin (on his eightieth birthday). *Phys.-Usp.* 40:547. [doi:0.1070/PU1997v040n05ABEH000241]

Goldhaber, M. 1974. *Neutrinos—1974* (AIP Conf. Proc. No. 22). New York: American Institute of Physics.

Goldhaber, M., L. Grodzins, and A. W. Sunyar. 1958. Helicity of neutrinos. *Phys. Rev.* 109:1015.

Goldman, T. J., and D. A. Ross. 1979. A new estimate of the proton lifetime. *Phys. Lett. B* 84:208–10. [doi: 10.1016/0370-2693(79)90286-7]

Goudsmit, Samuel. A. 1932. In *Convegno di fisica nucleare, October 1931,* 41. Rome: Reale Accademia d' Italia.

Gow, Anthony J., and T. Williamson. 1975. *Cold Regions Research and Engineering Laboratory research report 339.*

Greisen, K. 1960. Cosmic ray showers. *Ann. Rev. Nucl. Sci.* 10:63.

———. 1961. Possibility of high-energy neutrino measurements with cosmic rays. In *Proc. Int. Conf. on Instrumentation for High-Energy Physics (1960).* Berkeley: Interscience, 209–11.

Gribov, V., and B. Pontecorvo. 1969. Neutrino astronomy and lepton charge. *Phys. Lett.* 28B:493.

Grieder, P. K. F. (DUMAND Collaboration). 1995. DUMAND II: String 1 deployment, initial operation, results and system retrieval. *Nucl. Phys. B* (Proc. Suppl.) 43:145 8.

Griffiths, David. 2008. *Introduction to Elementary Particles.* 2nd ed. Weinheim: Wiley-VCH.

Gwynne, S. C. 2010. *Empire of the summer moon: Quanah Parker and the rise and fall of the Comanches, the most powerful Indian tribe in American History.* New York: Scribner.

Haidt, Dieter. 2006. The neutrino's 50th birthday. In *3rd international workshop on NOVE: Neutrino oscillations in Venice: 50 years after the neutrino experimental discovery.* Padova: Papergraf Spa.

Halzen, Francis. 1981. Prompt muons in very high-energy cosmic rays. In V. J. Stenger, ed. 1981. *DUMAND 80: Proceedings.* 2 vol. Honolulu: Hawaii DUMAND Center, 281–96.

———. 1995a. Ice fishing for neutrinos. http://icecube.berkeley.edu/amanda/ice-fishing.html (accessed April 3, 2012).

———. 1995b. The case for a kilometer-scale high-energy neutrino detector. In E. W. Kolb and R. Peccei, eds. 1995. *Proceedings of the Snowmass '94 Summer Study on Particle and Nuclear Astrophysics and Cosmology.* Singapore: World Scientific, 256.

———. 1995c. The case for a kilometer-scale high-energy neutrino detector. In A. Dar, G. Eilam, and M. Gronau, eds. 1995. *Neutrino 94. Proceedings, 16th international conference on neutrino physics and astrophysics, Eilat, Israel, May 29–June 3, 1994.* (*Nucl. Phys. Proc. Suppl.* 38:472.)

———. 1996. The case for a kilometer-scale high energy neutrino detector. In Milla Baldo-Ceolin, ed. 1996. *Proc. of the 7th international workshop on neutrino telescopes, Venice, Italy.* Padua: University of Padua.

———. 1999. Antarctic dreams. *The Sciences* 39:19–24. [doi:10.1002/j.2326-1951.1999.tb03422.x]

———. 2009. IceCube: The rationale for kilometer-scale neutrino detectors. 21st Rencontres de Blois, "Windows on the Universe." [arXiv:0910.0436 [astro-ph.HE]]

———. 2011. Sterile neutrinos and IceCube. From a talk given at NUFACT 11, the XIIIth Intl. workshop on neutrino factories, super beams and beta beams, 1–6 August 2011 at CERN and the University of Geneva. [arXiv:1111.0918v1 [hep-ph]]

———. 2015. Balzan Prize for astroparticle physics including neutrino and gamma-ray observation, acceptance speech, 13 November. http://www.balzan.org/en/prizewinners/francis-halzen/bern—13-11-2015-halzen (accessed August 12, 2016).

Halzen, Francis, et al. (AMANDA Collaboration). 1997. Status of the AMANDA South Pole neutrino detector. In H. V. Klapdor-Kleingrothaus, and Y. Ramachers, eds. *Proc. of the international workshop on aspects of dark matter in astrophysics and particle physics. Heidelberg, Germany (1996).* River Edge, NJ: World Scientific, 718–28. [arXiv:hep-ex/9611014v1]

Halzen, Francis, John Jacobsen, and Enrique Zas. 1994. Possibility that high-energy neutrino telescopes could detect supernovae. *Phys. Rev. D* **49**:1758–61. [doi: 10.1103/PhysRevD.49.1758]

———. 1996. Ultratransparent Antarctic ice as a supernova detector. *Phys. Rev. D* **53**:7359–61.

Halzen, Francis, and Spencer R. Klein. 2010. IceCube: An instrument for neutrino astronomy. *Rev. Sci. Instrum.* **81**:081101. [doi:10.1063/1.3480478]

Halzen, Francis, and John Learned. 1988. High energy neutrino detection in deep polar ice. In M. Giler, ed. 1989. *Proceedings of the 5th international symposium on very high energy cosmic ray interactions, Lodz, Poland (1988).* Lodz: University of Lodz, 90. (University of Wisconsin preprint MAD/PH/428.)

Halzen, Francis, John Learned, and Todor Stanev. 1990. Neutrino astronomy. In Mullan et al. 1990, 39–51.

Halzen, Francis, and Christopher Michael. 1971. Amplitude analysis of πN scattering at 6 GeV/c. *Phys. Lett. B* **36**:367. [doi:10.1016/0370-2693(71)90727-1]

Halzen, Francis, R. A. Vazquez, T. Stanev, and H. P. Vankov. 1994. The highest energy cosmic ray. Talk presented by F. Halzen at the *6th International Symposium on Neutrino Telescopes, Venice, Feb. 1994.* MAD-PH-832, C94-02-22. Also published in 1995: *Astropart. Phys.* **3**:151–6.

Halzen, Francis, and Logan Wille. 2016. On the charm contribution to the atmospheric neutrino flux. *Phys. Rev. D* **94**:014014. [doi:10.1103/PhysRevD.94.014014] [arXiv:1605.01409 [hep-ph]]

Halzen, Francis, E. Zas, and T. Stanev. 1991. Radiodetection of cosmic neutrinos: A numerical, real time analysis. *Phys. Lett. B* **257**:432–6. (University of Wisconsin preprint MAD/PH/606, Nov. 1990.)

Hartmann, Dieter H. 2011. A multi-messenger story. *Nature* **475**:303–4.

Haverford College. 2002. Songs from 1947 by Arthur Roberts. http://www.haverford.edu/physics-astro/songs/roberts/roberts1947.htm (accessed June 17, 2011).

Heitler, W., and G. Herzberg. 1929. *Die Naturwissenschaften* **17**:673.

Henderson, W. J. 1934. The upper limits of the continuous β-ray spectra of thorium C and C″. *Proc. Roy. Soc. A* **147**:572–82. [doi:10.1098/rspa.1934.0237]

Hernandez, Daniela. 2016. Phantom particles that would have redefined physics probably aren't real, scientists conclude. *Wall Street Journal,* 8 August.

Herzog, Werner, director. 2007. *Encounters at the end of the world,* DVD. Silver Spring: Discovery Films.

Hincks, E. P., and B. Pontecorvo. 1948a. Search for gamma-radiation in the 2.2-microsecond meson decay process. *Phys. Rev.* **73**:257–8.

———. 1948b. On the stability of the neutral meson. *Phys. Rev.* **73**:1122–3.

———. 1949. The penetration of μ-meson decay electrons and their bremsstrahlung radiation. *Phys. Rev.* **75**:698–9.

———. 1950a. *Can. J. Res.* **28A**:29.

———. 1950b. On the disintegration products of the 2.2-μsec. meson. *Phys. Rev.* **77**:102–20.

Hirata, K. S., et al. (Kamiokande II Collaboration). 1987. Observation of a neutrino burst from the supernova SN1987A. *Phys. Rev. Lett.* **58**:1490–93.

———. 1988. Observation in the Kamiokande-II detector of the neutrino burst from supernova SN 1987a. *Phys. Rev. D* **38**:448–58.

Holmgren, H. D., and R. L. Johnston. 1958. $H^3(\alpha, \gamma)Li^7$ and $He^3(\alpha, \gamma)Be^7$ reactions. *Phys. Rev.* **113**:1556–9.

Holton, Gerald. 1977. Fermi's group and the recapture of Italy's place in physics. In Gerald Holton, ed. *The Scientific Imagination.* Cambridge: Cambridge University Press.

Hoyle, Fred. 1967. Concluding remarks. *Proc. Roy. Soc. A* **301**:171.

Hulth, Per Olof, et al. (AMANDA Collaboration). 1997. The AMANDA experiment. In Kari Enqvist, Katri Huitu, and Jukka Maalampi, eds. 1997. *NEUTRINO '96: Proceedings of the 17th international conference on neutrino physics and astrophysics* (Helsinki, Finland, June 13–20, 1996). River Edge, New Jersey: World Scientific, 518–23. [arXiv:astro-ph/9612068]

Huntford, Roland. 1999. *The last place on earth*. New York: Random House, Modern Library.

Hutchison, Kristan. 2001. Causes for the common crud. *The Antarctic Sun*, 18 November.

IceCube-Gen2 Collaboration. 2016. PINGU: A vision for neutrino and particle physics at the South Pole. [arXiv:1607.02671 [hep-ex]]

INR (Institute for Nuclear Research, Russian Academy of Sciences). 1997. *Baksan Neutrino Observatory* (a brochure).

Iwanenko, Dmitrij. 1932. *Comptes Rendus* **195**:439.

Jacobsen, John E. 1996a. A first look at v_μ candidates in AMANDA-B. AMANDA internal memo, 8 July.

Jacobsen, John E. 1996b. Simulating the detection of muons and neutrinos in deep Antarctic ice. Ph.D. thesis, University of Wisconsin–Madison.

Janka, H.-Th., et al. 2007. Theory of core-collapse supernovae. *Phys. Rep.* **442**:38–74.

Jayawardhana, Ray. 2013. *Neutrino Hunters: The thrilling chase for a ghostly particle to unlock the secrets of the universe*. New York: Scientific American/Farrar, Straus and Giroux.

JINR (Joint Institute for Nuclear Research (Dubna)). 1998. "About B. Pontecorvo." http://pontecorvo.jinr.ru/about.html (accessed December 10, 2014).

Jung, Carl Gustav. 1958. *Psyche and symbol: A selection from the writings of C. G. Jung*. Ed. Violet de Laszlo. New York: Doubleday.

Jung, Carl Gustav. 1967. *Symbols of transformation (Collected Works of C.G. Jung, Vol. 5)*. Eds. G. Adler and R. F. C. Hull. Princeton: Princeton University Press.

Jung, Carl, and Wolfgang Pauli. 1952. *Naturerklaerung und Psyche*. Zürich: Rascher Verlag. (English translation. 1954. *The interpretation of nature and the psyche*. Bollingen Series 51. New York: Pantheon.)

Kaempffert, Waldemar. 1931. Cosmic problems engage scientists. *New York Times*, 21 June.

Kampa, Elizabeth M. 1979. Serpents in an astrophysical Eden. In G. A. Wilkins, ed. *Proceedings of the 1978 DUMAND summer workshop, Vol. 3: Oceanographic and ocean engineering studies*. La Jolla: Scripps Institution of Oceanography, 47.

Keuffel, J. W. 1963. Research proposal to the National Science Foundation, Dec. 20.

———. 1965. *In Proc. Ninth Int. Conf. Cosmic Rays*.

Kirshner, Robert P. 2002. *The extravagant universe: Exploding stars, dark energy, and the accelerating cosmos*. Princeton: Princeton University Press.

———. 2004. The full story on neutrino detection. *Science* **303**:1976–7. [doi: 10.1126/science.303.5666.1976b]

Kleinfelder, Stuart A. 1992. A multi-gigahertz analog transient waveform recorder integrated circuit. Master's thesis, University of California, Berkeley.

———. 2003. Gigahertz waveform sampling and digitization circuit design and implementation. *IEEE T. Nucl. Sci.* **50**:955–62.

Kodama, K., et al. (DONUT Collaboration). 2001. Observation of tau neutrino interactions. *Phys. Lett. B* **504**:218–24.

Korff, Serge A. 1985. High altitude observatories for cosmic rays and other purposes. In Yataro Sekido and Harry Elliot, eds. *Early history of cosmic ray studies: Personal reminiscences with old photographs*. Dordrecht, Boston: D. Reidel, 171–9.

Kotzer, P., ed. 1976. *DUMAND 1975: proceedings of DUMAND 1975 Summer Study, Bellingham, WA*. Bellingham: Western Washington State College.

Kronig. R. 1928. *Die Naturwissenschaften* **16**:335.

Kropp, William, Jonas Schultz, and Henry Sobel. 2009. *Frederick Reines 1918–1998: A biographical memoir*. Washington: National Academy of Sciences.

Landau, L. 1932. *Phys. Z. Sowjetunion* **1**:285.

Lande, Kenneth. 2009. The life of Raymond Davis, Jr. and the beginning of neutrino astronomy. *Annu. Rev. Nucl. Part. Sci.* **59**:21–39. [doi: 10.1146/annurev.nucl.010909.083753]

Lang, Kenneth. 2010. Serendipitous astronomy. *Science* **327**:39.

LANL (Los Alamos National Laboratory). 1997. The Reines-Cowan experiments. *Los Alamos Science* 25. http://library.lanl.gov/cgi-bin/getfile?00326606.pdf (accessed December 10, 2010).

Lattes, C. M. G., H. Muirhead, G. P. S Occhialini, and C. F. Powell. 1947. Processes involving charged mesons. *Nature* **159**:694–7. [doi:10.1038/159694a0]

Lattes, C. M. G., G. P. S. Occhialini, and C. F. Powell. 1947. Observations on the tracks of slow mesons in photographic emulsions. *Nature* **160**:453–456.

Learned, John G., ed. 1980. *Proceedings of the 1979 DUMAND Summer workshops at Khabarovsk and Lake Baikal*. Honolulu: Hawaii DUMAND Center, University of Hawaii.

Learned, John G. 1991. Introduction to section 8: Extra-solar neutrinos. In W. Kropp, L. Price, J. Schultz, and H. Sobel, eds. *Neutrinos and other matters: Selected works of Frederick Reines*. Singapore: World Scientific, 309–12.

Learned, John G. 2001. Short biography of John Gregory Learned. http://www.phys.hawaii.edu/~jgl /jgl_biog.html (accessed June 8, 2011).

Learned, John G., and Joseph P. Platt. 2005. Arthur Roberts (obituary). *Phys. Today* **58**:68.

Lederman, Leon. 1993. *The god particle*. Boston: Houghton Mifflin.

Lee, Patrick A. 2014. Seeking out Majorana under the microscope. *Science* **346**:545–66. [doi:10.1126/ science.1260282]

Lee, T. D., and C. N. Yang. 1956. Question of parity conservation in weak interactions. *Phys. Rev.* **104**:254.

———. 1957. Parity nonconservation and a two-component theory of the neutrino. *Phys. Rev.* **105**:1671.

———. 1960. Theoretical discussions on possible high-energy neutrino experiments. *Phys. Rev. Lett.* **4**:307–11.

Lee, T. D., R. Oehme, and C. N. Yang. 1957. Remarks on possible noninvariance under time reversal and charge conjugation. *Phys. Rev.* **106**:340.

Levinson, Mark, director. 2015. *Particle Fever*, DVD. Arlington: PBS.

Lindley, David. 2007. Landmarks: Detecting the elusive neutrino. *Phys. Rev. Focus* **19**:13. https:// physics.aps.org/story/v19/st13 (accessed October 24, 2016).

Lloyd-Evans, J., et al. 1983. Observation of gamma rays > 10^{15}-eV from Cygnus X-3. *Nature* **305**:784–7.

Lowder, Doug, et al. (AMANDA collaboration). 1996. 1995–96 Results for the AMANDA neutrino observatory (talk presented by P. B. Price). In Milla Baldo-Ceolin, ed. 1996. *Proc. of the 7th Intl. workshop on neutrino telescopes, Venice, Italy*. Padua: University of Padua, 383.

Lowder, Doug, Timothy Miller, P. Buford Price, Andrew Westphal, Steven W. Barwick, Francis Halzen, and Robert Morse. 1991. Observation of muons using the polar ice cap as a Cerenkov detector. *Nature* **353**:331–3. [doi:10.1038/353331a0]

Lowder, Doug, and Kurt Woschnagg. 1999. Determination of optical parameters in AMANDA-B Ice. In Spiering 1999, 145–56.

Lynch, John T. 1990. A proposal for an international station in Antarctica. In Mullan et al. 1990, 39–51.

Madison.com. 2006. Koci, Bruce. R., obituary. 19 November. http://host.madison.com/news/opinion /editorial/obituaries/article_dbb41248-f4b0-5195-86b9-83f0049c98a1.html (accessed March 17, 2015).

Majorana, Ettore. 1937. Symmetrical theory of the electron and positron. *Nuovo Cimento* **5**:171–84.

Malakoff, David. 2001. U.S. researchers go for scientific gold mine. *Science* **292**:1979. [doi: 10.1126/ science.292.5524.1979a]

March, Robert, Francis Halzen, and John Learned. 1988. Neutrino detection in clear polar ice. Talk presented at the 1st international workshop on neutrino telescopes, Venice, Italy. (University of Wisconsin preprint MAD/PH/462, Dec. 1988.)

Markov, M. A. 1960. On high energy neutrino physics. In E. C. G. Sudarshan, J. H. Tinlot, and A. C. Melissinos, eds. *Proc. 1960 annual international conference on high-energy physics at Rochester*. New York: Interscience, 578.

Markov, M. A., and Igor M. Zheleznykh. 1961. On high energy neutrino physics in cosmic rays. *Nucl. Phys.* **27**:385.

———. 1986. Large-scale Cherenkov detectors in ocean, atmosphere and ice. *Nucl. Instrum. Meth. A* **248**:242–51.

Mather, John C., and John Boslough. 2008. *The very first light: The true inside story of the scientific journey back to the dawn of the universe*. Revised and updated edition. New York: Basic Books.

Meier, C. A., ed. 2001. *Atom and archetype: The Pauli/Jung Letters, 1932–1958*. Princeton: Princeton University Press.

Meitner, Lise, and Wilhelm Oorthman. 1930. *Z. Für Physik* **60**:143.

Menon, M. G. K., S. Naranan, et al. 1967. Studies of cosmic ray neutrino interactions in the Kolar Gold Field experiment. *Proc. Roy. Soc. A* **301**:137–57

Mervis, Jeffrey. 2001a. Funding backlog at NSF sets off free-for-all. *Science* **293**:586–7. [doi: 10.1126/science.293.5530.586]

———. 2001b. NSF research bounces back; Congress funds new facilities. *Science* **294**:1430–1. [doi: 10.1126/science.294.5546.1430a]

———. 2016. NSF director unveils big ideas. *Science* **352**:755–6. [doi: 10.1126/science.352.6287.755]

Metropolis, Kate. 2004. Understanding luminosity through "barn," a unit that helps physicists count particle events. *Stanford Report,* 21 July. http://news.stanford.edu/news/2004/july21/femtobarn -721.html (accessed October 7, 2014).

Miller, Arthur I. 2010. *137: Jung, Pauli, and the pursuit of a scientific obsession.* New York, London: W. W. Norton & Co.

Morse, Robert, and James Gaidos,. 1990. A South Pole facility to observe very high energy gamma ray sources. In Mullan et al. 1990, 24–34.

Mullan, Dermott J., Martin A. Pomerantz, and Todor Stanev, eds. 1990. *Astrophysics in Antarctica.* New York: American Institute of Physics.

Musser, George. 1999. Seven wonders of modern astronomy. *Scientific American,* December.

Neddermeyer, Seth H., and Carl. D Anderson. 1937. Note on the nature of cosmic-ray particles. *Phys. Rev.* **51**:884.

New Scientist. 1978. Cash shortage attenuates neutrinos. 11 May, 349.

Normile, Dennis. 2015. Japanese neutrino physicists think really big. *Science* **347**:598. [doi:10.1126/science.347.6222.598]

Overbye, Dennis. 1998. Remembering David Schramm, gentle giant of cosmology. *New York Times,* 10 February.

———. 2002. The most seductive equation in science: Beauty equals truth. *New York Times,* 26 March.

Pais, Abraham. 1982. *"Subtle is the Lord . . .": The science and life of Albert Einstein.* New York: Oxford University Press.

———. 1986. *Inward Bound: Of matter and forces in the physical world.* New York: Oxford University Press.

Palinkas, Lawrence A. 1989. Antarctica as a model for the human exploration of Mars. In C. Stoker, ed. *Case For Mars III.* San Diego: Univelt, 215–28.

———. 1990. Psychosocial effects of adjustment in Antarctica: Lessons for long-duration space flight. *J. Spacecraft Rockets* **27**:471–7.

———. 2000. On the ice: Individual and group adaptation in Antarctica. Colloquium presentation, Center for Behavior, Evolution and Culture Speaker Series, University of California, Los Angeles, November 13. http://www.sscnet.ucla.edu/anthro/bec/papers/Palinkas_On_The_Ice.pdf (accessed July 16, 2015).

Palinkas, Lawrence A., H. L. Reed, and N. Do. 1997. Association between the Polar T3 syndrome and the winter-over syndrome in Antarctica. *Antarctic Journal of the United States* **32**: 112–14.

Pauli, Wolfgang. 1925. *Z. Für Physik* **31**:373.

———. 1929. Letter to Bohr, 17 July 1929. Niels Bohr Archives, Copenhagen.

———. 1934. *Septieme conseil de physique Solvay 1933: Noyauz atomiques. Discussions,* 324f. Paris: Gautier-Villars.

———. 1945. Exclusion principle and quantum mechanics. Nobel lecture. http://nobelprize.org /nobel_prizes/physics/laureates/1945/pauli-lecture.pdf (accessed December 21, 2010).

———. 1946. Remarks on the history of the exclusion principle. *Science* **103**:213–15.

———. 1948. Sommerfeld's contributions to quantum theory. In Pauli 1994.

———. 1955. Exclusion principle, Lorentz group and reflexion of space-time and charge. In Pauli, Wolfgang, ed., with the assistance of L. Rosenfeld and V. Weisskopf. 1955. *Niels Bohr and the development of physics: Essays dedicated to Niels Bohr on the occasion of his seventieth birthday.* New York: McGraw-Hill.

———. 1957a. Letter to Victor Weisskopf on 27, Jan. In Pauli 1964.

———. 1957b. Phenomenon and physical reality. In Pauli 1994.

———. 1958a. On the earlier and more recent history of the neutrino. In Pauli 1994.

———. 1958b. The violation of reflection symmetry in the laws of atomic physics. In Pauli 1994.

———. 1964. *Collected scientific papers by Wolfgang Pauli, vol. I.* Weisskopf, V. F., and R. Kronig, eds. New York: John Wiley.

———. 1994. *Writings on physics and philosophy.* Enz, Charles P., and Karl von Meyenn, eds. Berlin: Springer-Verlag.

Peierls, Rudolph E. 1960. Wolfgang Ernst Pauli: 1900–1958. *Biogr. Mems. Fell. R. Soc.* **5**:175–92. [doi: 10.1098/rsbm.1960.0014]

———. 1983. The early days of neutrino physics. *Contemp. Phys.* **24**:221–8.

Perl, Martin L., et al. 1975. Evidence for anomalous lepton production in e⁺-e⁻ annihilation. *Phys. Rev. Lett.* **35**:1489–92.

Perrin. Francis. 1930. *Comptes Rendus* **197**:1625.

Perry, Robert L. 1978. Tuning in on the "ghost beam." *Popular Mechanics,* September.

philosophical society.com. 2012. The most ridiculous particle. http://www.philosophicalsociety.com/Archives/The%20Most%20Ridiculous%20Particle.htm. (accessed August 12, 2015).

Pierre Auger Collaboration. 2011. Frequently asked questions about cosmic rays. http://www.auger.org/cosmic_rays/faq.html (accessed May 24, 2011).

Pilachowski, C. A., and Cameron Pace. 2015. The abundance of fluorine in normal G and K stars of the galactic thin disk. *Astron. J.* **150**:66. [doi: 10.1088/0004-6256/150/3/66] [arXiv:1507.01550 [astro-ph.SR]]

Pontecorvo, Bruno. 1945. Report P.D.-141, Chalk River Laboratory, 21 May.

———. 1946. Inverse β process. Report P.D.-205 of the National Research Council of Canada Division of Atomic Energy, Chalk River, Ontario, 20 November. http://pontecorvo.jinr.ru/work/ibp.html (accessed September 11, 2014).

———. 1947. Nuclear capture of mesons and the meson decay. *Phys. Rev.* **72**:246.

———. 1958. Inverse β processes and nonconservation of lepton charge *J. Exp. Theor. Phys. (USSR)* **33**:549.

———. 1959. Electron and muon neutrinos. *J. Exp. Theor. Phys. (USSR)* **37**:1751. (translation: 1960. *Sov. Phys. JETP* **10**:1236.)

———. 1967. Neutrino experiment and the problem of conservation of leptonic charge. *J. Exp. Theor. Phys. (USSR)* **53**:1717.

———. 1982. The infancy and youth of neutrino physics: Some recollections. In Colloque international sur l'histoire de la physique des particules, *J. Phys.-Paris* **43**, suppl. C8:221–236.

Powell, Anthony, director. 2014. *Antarctica: A year on ice*, DVD. Chicago: Music Box Films.

Price, P. Buford. 1995. Kinetics of conversion of air bubbles to air hydrate crystals in Antarctic ice. *Science* **267**:1802–4. [doi:10.1126/science.267.5205.1802]

Pryor, C., C. E. Roos, and M. S. Webster. 1988. Detecting thermal neutrinos from supernovae with DUMAND. *Astrophys. J.* **329**:335–8.

Rachen, J. P., and P. Mészáros. 1998. In C. A. Meegan, R. D. Preece, and T. M. Koshut, eds. *Fourth Huntsville gamma-ray burst symposium.* (American Institute of Physics Conference Proceedings Vol. 428, 1998), 776–80.

Rankin, P., et al. 1998. Report on the scientific assessment group for experiments in non-accelerator physics (SAGENAP), March and November, 1997. Washington D.C.: United States Department of Energy. http://science.energy.gov/~/media/hep/pdf/files/pdfs/sagena98.pdf (accessed July 27, 2015).

Rasetti, F. 1929. On the Raman effect in diatomic molecules. *Proc. Natl. Acad. Sci. U.S.A.* **15**:234–7; **15**:515–19.

Regener, E. 1931. Spectrum of cosmic rays. *Nature* **127**:233–4. [doi:10.1038/127233b0]

Reines, F. 1960. Neutrino interactions. *Ann. Rev. Nucl. Sci.* **10**:1–26.

———. 1967. High energy neutrinos underground: Status of the Case-Wits-Irvine experiment and future prospects. *Proc. Roy. Soc. A* **301**:125–35.

———. 1979. The early days of experimental neutrino physics. *Science* **203**:11–16.

———. 1982. Neutrinos to 1960—personal recollections. In Colloque International sur l'Histoire de la Physique des Particules, *J. de Phys.* **43**, suppl. C8:237–260.

———. 1992. Venice is a great place for a conference. Talk at the IV international workshop on neutrino telescopes, Venice, 10–13 Mar. http://neutrino.pd.infn.it/Reines-1992.pdf (accessed March 13, 2012).

———. 1995. The neutrino: From poltergeist to particle. Nobel lecture. http://nobelprize.org/nobel_prizes/physics/laureates/1995/reines-lecture.html (accessed January 8, 2011).

———. 1996. The neutrino: From poltergeist to particle. *Rev. Mod. Phys.* **68**:317–27.

Reines, F., and C. L. Cowan, Jr. 1953. Detection of the free neutrino. *Phys. Rev.* **92**:830.

———. 1959. Free antineutrino absorption cross section. I. Measurement of the free antineutrino absorption cross section by protons. *Phys. Rev.* **113**:273.

Reines, F., M. F. Crouch et al. 1965. Evidence for high-energy cosmic-ray neutrino interactions. *Phys. Rev. Lett.* **15**:429–33.

Rhodes, Richard. 1986. *The making of the atomic bomb*. New York: Simon and Schuster.

Richards, Charles. 1992. Confessions of an atom spy. *The Independent* (UK), 2 August.

Roberts, Arthur, Howard Blood, John G. Learned, and Frederick Reines. 1977. DUMAND: The ocean as a neutrino detector. In H. Faissner, H. Reithler, P. Zerwas. eds. *Proceedings of the international neutrino conference, Aachen 1976: Held at Rheinisch-Westfalische Technische Hoshschule Aachen, June 8–12, 1976*. Braunschweig: Vieweg.

Roberts, Arthur. 1992. The birth of high-energy neutrino astronomy: A personal history of the DUMAND project. *Rev. Mod. Phys.* **64**:259–312.

Robinson, Arthur L. 1979. Is a diamond really forever? *Science* **270**:670–3. [doi:10.1126/science.206.4419.670]

Romero, Simon. 2016. Antarctic life: No dogs, few vegetables and 'a little intense' in the winter. *New York Times*, 7 January.

Rossi, Bruno, and D. B. Hall. 1941. Variation of the rate of decay of mesotrons with momentum. *Phys. Rev.* **59**:223–8.

Rossi, Bruno. 1939. The disintegration of mesotrons. *Rev. Mod. Phys.* **11**:296–303.

Rossi, Bruno. 1985. Arcetri, 1928–1932. In Yataro Sekido and Harry Elliot, eds. *Early history of cosmic ray studies: Personal reminiscences with old photographs*. Dordrecht, Boston: D. Reidel, 53–73.

Royal Swedish Academy of Sciences. 2015. The 2015 Nobel Prize in physics–Press release. http://www.nobelprize.org/nobel_prizes/physics/laureates/2015/press.html (accessed September 3, 2016).

Rubbia, Carlo. 1984. Experimental observation of the intermediate vector bosons W^+, W^- and Z^0. Nobel lecture. http://www.nobelprize.org/nobel_prizes/physics/laureates/1984/rubbia-lecture.pdf (accessed Feb 14, 2015).

Sáenz, A. W., H. Überall, F. J. Kelly, D. W. Padgett, and N. Seeman. 1977. Telecommunication with neutrino beams. *Science* **198**:295–7.

Samorski, M., and W. Stamm. 1983. Detection of 2×10^{15}-eV to 2×10^{16}-eV gamma rays from Cygnus X-3. *Astrophys. J.* **268**:L17–L21.

Schecter, J., and L. Schecter. 2002. *Sacred secrets: How Soviet intelligence operations changed American history*. Dulles, Virginia: Brassey's.

Schmeck, Harold M., Jr. 1957. Basic concept in physics is reported upset in tests. *New York Times*, 16 Jan.

Schmitz, David W. 2016. Hunting the sterile neutrino. *Physics* **9**:94. http://physics.aps.org/articles/v9/94 (accessed October 5, 2016).

Schneider, Darryn. 2007. Diaries of trips to the South Pole. http://antarctica.kulgun.net/SouthPole/ (accessed February 15, 2013).

Schorn, Ronald A. 1987a. A Supernova in our backyard. *Sky & Telescope* **73**:382 (April).

———. 1987b. Supernova shines on. *Sky & Telescope* **73**:470 (May).

———. 1987c. Neutrinos from hell. *Sky & Telescope* **73**:477 (May).

Schwartz, M. 1960. Possibility of using high energy neutrinos to study weak interactions. *Phys Rev Lett.* **4**:307.

Schwarz, Robert. 2015. iceman's South Pole page. http://www.antarctic-adventures.de/ (accessed June 3, 2015).

Segrè, Emilio. 1970. *Enrico Fermi, physicist*. Chicago: University of Chicago Press.

Segrè, Gino. 2007. *Faust in Copenhagen: A struggle for the soul of physics*. New York: Viking Penguin.

Seife, Charles. 2004. Wanted: One good cosmic blast to shake the neighborhood. *Science* **303**:164–5. [doi: 10.1126/science.303.5655.164]

Simonsen, Mike. 2009. Albert Jones–the interbiew [sic] (blog post). http://simostronomy.blogspot.com/2009/11/albert-jones-interbiew.html (accessed August 21, 2011).

Sobel, Henry W., and Yoichiro Suzuki. 2008. Yoji Totsuka (1942–2008). *Nature* **454**:954.

Sokalski, I., and C. Spiering, eds. (Baikal Collaboration). 1992. The Baikal neutrino telescope NT-200.

Spiering, Christian, ed. 1999. *Simulation and analysis methods for large neutrino telescopes* (Proceedings of an international workshop, Zeuthen, Germany, July 6–9, 1998). Zeuthen: DESY. (DESY-PROC-1999-01.)

Spiering, Christian. 2012. Towards high-energy neutrino astrophysics: A historical review. *Eur. Phys. J. H* 37:515–65. [doi: 10.1140/epjh/e2012-30014-2]

Stark, Jack. 1995. The Wisconsin Idea: The university's service to the state, in *1995–1996 Wisconsin blue book*. Madison: Wisconsin Legislative Reference Bureau. http://legis.wisconsin.gov/lrb/pubs/feature/wisidea.pdf (accessed July 10, 2012).

Stenger, V. J., ed. 1981. *DUMAND 80: Proceedings.* 2 vol. Honolulu: Hawaii DUMAND Center.

Stokstad, Robert G. 2005. Design and performance of the IceCube electronics. Presentation at 11th workshop on electronics for LHC and future experiments, Heidelberg, 12–16 September. http://icecube.lbl.gov/PDFS/RGS_Heidelberg_paper.pdf (accessed November 15, 2015).

Sutton, Christine. 1992. *Spaceship neutrino.* Cambridge: Cambridge University Press.

Taubes, Gary. 1986. *Nobel dreams: Power, deceit, and the ultimate experiment.* New York: Random House.

Tilav, Serap, et al. (AMANDA Collaboration). 1997. Indirect evidence for long absorption lengths in Antarctic ice. In A. Taroni, ed. 1997. *XXIV ICRC Rome 1995.* (24th International cosmic ray conference: Proceedings) Vol. 1. Bologna: Ed. Compositori, 1011–4. (*Il Nuovo Cim.* **19C** (1997) 623–804).

Time. 1951. SPIES: Worse than murder, 15 April. http://www.time.com/time/magazine/article/0,9171,814669-1,00.html (accessed March 15, 2011).

Totsuka, Yoji. 1999. Evidence for neutrino oscillation observed by Super-Kamiokande. In Spiering 1999, 76–95.

Turchetti, Simone. 2003. Atomic secrets and governmental lies: Nuclear science, politics and security in the Pontecorvo case. *Brit. J. Hist. Sci.* **36**(4):389–415.

———. 2012. *The Pontecorvo Affair: A cold war defection and nuclear physics.* Chicago: Univ. of Chicago Press.

Überall, H., and C. Cowen. 1965. In C. Franzinetti, ed. *Proc. CERN conf. on exptl. neutrino physics:* CERN Report 65–32. Geneva: CERN, 231.

Uhlenbeck, G. E., and S. Goudsmit. 1925. Ersetzung der hypothese von unmechanischen zwang durch eine forderung bezuglich des inneren verhaltens jedes einzelnen elektrons. *Naturwissenschaften* **13**:953–4.

———. 1926. Spinning electrons and the structure of spectra. *Nature* **117**:264–5.

University of Wisconsin-Madison Archives, Oral History Project. 2007. Interview #842. Halzen, Francis.

Vander Velde, John C. 2002. The IMB experiment: A brief historical sketch 1979–1989. http://www-personal.umich.edu/~jcv/imb/imb.html (accessed July 21, 2011).

von Meyenn, Karl, and Engelbert Schucking. 2001. Wolfgang Pauli. *Phys. Today* **54**:43–8. [doi: 10.1063/1.1359709]

Waxman, Eli. 1995. Cosmological gamma-ray bursts and the highest energy cosmic rays. *Phys. Rev. Lett.* **75**:386. [doi: 10.1103/PhysRevLett.75.386]

———. 2014. The beginning of extra-galactic neutrino astronomy. *Physics* **7**:88. http://physics.aps.org/articles/v7/88 (accessed July 10, 2016).

Waxman, Eli, and John Bahcall. 1997. High energy neutrinos from cosmological gamma-ray burst fireballs. *Phys. Rev. Lett.* **78**:2292. [doi: 10.1103/PhysRevLett.78.2292]

Weekes, Trevor C., et al. 1989. Observation of TeV gamma rays from the Crab nebula using the atmospheric Cerenkov imaging technique. *Astrophys. J.* **342**:379.

Weisskopf, Victor F. 1972. In C. Weiner, ed. *Exploring the history of nuclear physics.* New York: American Institute of Physics.

Westphal, Andrew J., P. Buford Price, and Daniel P. Snowden-Ifft. 1990. A measurement of the isotopic composition of iron-group elements in the galactic cosmic rays using balloon-borne track-recording detectors in Antarctica. In Mullan et al. 1990, 184–9.

Wheeler, John A., and Kenneth Ford. 1998. *Geons, black holes & quantum foam: A life in physics.* New York: W. W. Norton.

Wiebusch, Christopher H. V. 1994. Signal processing with JULIA. In Peter C. Bosetti, ed. *Trends in astroparticle-physics.* (Proceedings of a conference held in Aachen, Germany, in 1991). Stuttgart: B.G. Teubner Verlagsgesellschaft, 217.

Wigner, Eugene P. 1960. The unreasonable effectiveness of mathematics in the natural sciences. *Commun. Pur. Appl. Math.* **13**:1–14.

Wilczek, Frank. 2009. Majorana returns. *Nature Phys.* **5**:614–8. [doi:10.1038/nphys1380]

Wilkes, R. Jeffrey, John G. Learned, and Peter W. Gorham. 2003. Deep ocean neutrino detector development: Contributions by the DUMAND project. http://www.phys.hawaii.edu/~dumand/dumacomp .html (accessed June 9, 2011).

Williams, P. K., et al. 1996. Report on the scientific assessment group for experiments in nonaccelerator physics (SAGENAP), February 20–21, 1996. Washington D.C.: United States Department of Energy. http://science.energy.gov/~/media/hep/pdf/files/pdfs/sagenap.pdf (accessed October 3, 2012).

Wilson, Fred L. 1968. Fermi's theory of beta decay. *Am. J. Phys.* **36**:1150–60.

Wischnewski, Ralf, et al. (AMANDA Collaboration). 1997. A System to search for supernova bursts with the AMANDA detector. In A. Taroni, ed. 1997. *XXIV ICRC Rome 1995* Vol. 1. Bologna: Ed. Compositori, 658–61. (*Il Nuovo Cim.* **19C** (1997) 623–804.)

Wolchover, Natalie. 2015. The particle that broke a cosmic speed limit. *Quanta* (online magazine), 14 May. https://www.quantamagazine.org/20150514-the-particle-that-broke-a-cosmic-speed-limit/ (accessed May 18, 2015).

Woosley, S. E., and H.-T. Janka. 2005. The physics of core-collapse supernovae. *Nature Phys.* **1**:147. [arXiv:astro-ph/0601261v1]

Woosley, S. E., and M. M. Phillips. 1988. Supernova 1987A! *Science* **240**:750–9.

Woosley, S. E., and T. Weaver. 1989. The great supernova of 1987. *Sci. Am.* **261**:32 (August).

Wright, John H. 2012. *Blazing Ice: Pioneering the twenty-first century's road to the South Pole.* Dulles, Virginia: Potomac.

Wu, C. S. 1960. The neutrino. In Fierz and Weisskopf 1960, 249–303.

Wu, C. S., E. R. Ambler, W. Hayward, D. D. Hoppes, and R. P. Hudson. 1957. Experimental test of parity conservation in beta decay. *Phys. Rev.* **105**:1413.

Yang, C. N. 1957. The law of parity conservation and other symmetry laws of physics. Nobel lecture. http://nobelprize.org/nobel_prizes/physics/laureates/1957/yang-lecture.pdf (accessed January 22, 2011).

Zas, Enrique, Francis Halzen, and Todor Stanev. 1992. Electromagnetic pulses from high-energy showers: Implications for neutrino detection. *Phys. Rev. D* **45**:362–76.

Zheleznykh, I. M. 1958. On the interaction of high energy neutrinos in cosmic rays with matter. Diploma thesis, Moscow State University, Physical Faculty.

———. 2006. Early years of high-energy neutrino physics in cosmic rays and neutrino astronomy (1957–1962). *Int. J. Mod. Phys. A* **21**:1. [doi: 10.1142/S0217751X06033271]

INDEX